Lecture Notes in Computer Science   10239

Commenced Publication in 1973
Founding and Former Series Editors:
Gerhard Goos, Juris Hartmanis, and Jan van Leeuwen

More information about this series at http://www.springer.com/series/7410

Marc Joye · Abderrahmane Nitaj (Eds.)

Progress in Cryptology - AFRICACRYPT 2017

9th International Conference on Cryptology in Africa
Dakar, Senegal, May 24–26, 2017
Proceedings

 Springer

Editors
Marc Joye
NXP Semiconductors
San Jose, CA
USA

Abderrahmane Nitaj
University of Caen
Caen
France

ISSN 0302-9743   ISSN 1611-3349   (electronic)
Lecture Notes in Computer Science
ISBN 978-3-319-57338-0   ISBN 978-3-319-57339-7   (eBook)
DOI 10.1007/978-3-319-57339-7

Library of Congress Control Number: 2017937579

LNCS Sublibrary: SL4 – Security and Cryptology

Printed on acid-free paper

This Springer imprint is published by Springer Nature
The registered company is Springer International Publishing AG
The registered company address is: Gewerbestrasse 11, 6330 Cham, Switzerland

Preface

The 9th International Conference on the Theory and Application of Cryptographic Techniques in Africa, Africacrypt 2017, took place May 24–26, 2017, in Dakar, Senegal. The conference was organized by Cheikh Anta Diop University, Dakar, Senegal, in cooperation with the International Association for Cryptologic Research (IACR). We heartily thank our general chairs, Mamadou Sangharé, Djiby Sow, and Abdoul Aziz Ciss, as well as the whole Organizing Committee for their efforts in making the conference a success.

The aim of Africacrypt is to provide an international forum for practitioners and researchers from industry, academia, and government from all over the world for a wide-ranging discussion of all forms of cryptography and its applications.

The conference received a total of 40 full papers, out of which 13 papers were selected for publication in these proceedings. Each submission was assigned at least three Program Committee (PC) members. In addition to the PC members, many external reviewers joined the review process in their particular areas of expertise. We were fortunate to have this energetic team of experts, and are deeply grateful to all of them for their hard work, which included a very active discussion phase. The paper submission, review, and discussion processes were effectively and efficiently made possible by the Web-based system developed by Shai Halevi. We thank him for his support and the IACR for hosting the review system. The program was completed with two keynote talks by Johannes Buchmann and Damien Stehlé, and by an invited talk by Luca De Feo. We are very grateful to them for accepting our invitation.

Last but not least, we would like to thank all the authors who submitted papers to this conference, the Organizing Committee members, colleagues, and student helpers for their valuable time and effort, and all the conference attendees who made this event a truly intellectually stimulating one through their active participation.

March 2017

Marc Joye
Abderrahmane Nitaj

Organization

AFRICACRYPT 2017

9th International Conference on Cryptology in Africa, Dakar, Senegal, May 24–26, 2017

Africacrypt is the annual International Conference on the Theory and Applications of Security and Cryptography.

General Chairs

Mamadou Sangharé	Université Cheikh Anta Diop de Dakar, Senegal
Djiby Sow	Université Cheikh Anta Diop de Dakar, Senegal
Abdoul Aziz Ciss	École Polytechnique de Thiès, Senegal

Program Chairs

Marc Joye	NXP Semiconductors, USA
Abderrahmane Nitaj	Université de Caen, France

Program Committee

Riham Altawy	University of Waterloo, Canada
Muhammad R.K. Ariffin	UPM Kuala Lumpur, Malaysia
Abdelhak Azhari	Université de Casablanca, Morocco
Hussain Benazza	Université de Meknes, Morocco
Colin Boyd	NTNU, Norway
Dario Catalano	Università di Catania, Italy
Pierre-Louis Cayrel	Université Saint Etienne, France
Sherman S.M. Chow	CU Hong Kong, SAR China
Nadia El Mrabet	EMSE, France
Pierre-Alain Fouque	Université Rennes I, France
Georg Fuchsbauer	ENS Paris, France
Jens Groth	University College London, UK
Javier Herranz	Universidad Politècnica de Catalunya, Spain
Tetsu Iwata	Nagoya University, Japan
Saqib Kakvi	University of Bristol, UK
Seny Kamara	Brown University, USA
Fabien Laguillaumie	Université de Lyon I, France
Mark Manulis	University of Surrey, UK
Tarik Moataz	Brown University, USA
Ayoub Otmani	Université de Rouen, France

Thomas Peters	UCL, Belgium
Tajje-eddine Rachidi	Al Akhawayn University in Ifrane, Morocco
Vanishree Rao	PARC, USA
Magdy Saeb	Arab Academy for Science, Egypt
Rei Safavi-Naini	University of Calgary, Canada
Kazue Sako	NEC, Japan
Palash Sarkar	Indian Statistical Institute, India
Peter Schwabe	Radboud Universiteit, The Netherlands
Francesco Sica	Nazarbayev University, Kazakhstan
Djiby Sow	Université de Dakar, Senegal
François-Xavier Standaert	UCL, Belgium
Willy Susilo	University of Wollongong, Australia
Christine Swart	University of Cape Town, South Africa
Joseph Tonien	University of Wollongong, Australia
Amr M. Youssef	Concordia University, Canada

External Reviewers

Ali Akhavi
Lejla Batina
Christof Beierle
Olivier Blazy
Andrea Cerulli
Qian Chen
Noureddine Chikouche
Abdoul Aziz Ciss
Michael Clear
Edouard Cuvelier
Gareth Davies
Julien Devigne
Dario Fiore
Ryo Furukawa
Romain Gay
Benoît Gérard
Essam Ghadafi
Aurore Guillevic
Mohammad Hajiabadi
Tsuchida Hikaru
Vincenzo Iovino
Sune K. Jakobsen
Jérémy Jean
Abdel Alim Kamal
Sabyasachi Karati
Ahmed Abdel Khalek

Elena Kirshanova
Stefan Koelbl
François Koeune
Baptiste Lambin
Liran Lerman
Fuchun Lin
Mary Maller
Paz Morillo
Thierry Mefenza Nountu
Kazuma Ohara
Michele Orrù
Romain Poussier
Raghvendra Rohit
Olivier Sanders
Ben Smith
Martin Strand
Isamu Teranishi
Nicolas Thériault
Yosuke Todo
Mohamed Tolba
Christine van Vredendaal
Vesselin Velichkov
Alexandre Wallet
Fredrich Wiemer
Yongjun Zhao

Contents

Cryptographic Schemes

RingRainbow – An Efficient Multivariate Ring Signature Scheme.   3
  Mohamed Saied Emam Mohamed and Albrecht Petzoldt

Pinocchio-Based Adaptive zk-SNARKs and Secure/Correct Adaptive
Function Evaluation. 21
  Meilof Veeningen

Revisiting and Extending the AONT-RS Scheme: A Robust
Computationally Secure Secret Sharing Scheme 40
  Liqun Chen, Thalia M. Laing, and Keith M. Martin

Side-Channel Analysis

Climbing Down the Hierarchy: Hierarchical Classification for Machine
Learning Side-Channel Attacks. 61
  Stjepan Picek, Annelie Heuser, Alan Jovic, and Axel Legay

Multivariate Analysis Exploiting Static Power on Nanoscale CMOS
Circuits for Cryptographic Applications 79
  Milena Djukanovic, Davide Bellizia, Giuseppe Scotti,
  and Alessandro Trifiletti

Differential Bias Attack for Block Cipher Under Randomized Leakage
with Key Enumeration. 95
  Haruhisa Kosuge and Hidema Tanaka

Differential Cryptanalysis

Impossible Differential Cryptanalysis of Reduced-Round SKINNY 117
  Mohamed Tolba, Ahmed Abdelkhalek, and Amr M. Youssef

Impossible Differential Attack on Reduced Round SPARX-64/128 135
  Ahmed Abdelkhalek, Mohamed Tolba, and Amr M. Youssef

Applications

Private Conjunctive Query over Encrypted Data 149
  Tushar Kanti Saha and Takeshi Koshiba

Efficient Oblivious Transfer from Lossy Threshold
Homomorphic Encryption 165
   Isheeta Nargis

Privacy-Friendly Forecasting for the Smart Grid Using Homomorphic
Encryption and the Group Method of Data Handling.................. 184
   Joppe W. Bos, Wouter Castryck, Ilia Iliashenko,
   and Frederik Vercauteren

Number Theory

On Indifferentiable Hashing into the Jacobian of Hyperelliptic Curves
of Genus 2... 205
   Michel Seck, Hortense Boudjou, Nafissatou Diarra,
   and Ahmed Youssef Ould Cheikh Khlil

Cryptanalysis of Some Protocols Using Matrices over Group Rings........ 223
   Mohammad Eftekhari

Author Index 231

Cryptographic Schemes

RingRainbow – An Efficient Multivariate Ring Signature Scheme

Mohamed Saied Emam Mohamed[1]([✉]) and Albrecht Petzoldt[2]

[1] Technische Universität Darmstadt, Darmstadt, Germany
mohamed@cdc.informatik.tu-darmstadt.de
[2] National Institute for Standards and Technology, Gaithersburg, MD, USA
albrecht.petzoldt@nist.gov

Abstract. Multivariate Cryptography is one of the main candidates for creating post-quantum cryptosystems. Especially in the area of digital signatures, there exist many practical and secure multivariate schemes. However, there is a lack of more advanced schemes, such as schemes for oblivious transfer and signature schemes with special properties. While, in the last years, a number of multivariate ring signature schemes have been proposed, all of these have weaknesses in terms of security or efficiency. In this paper we propose a simple and efficient technique to extend arbitrary multivariate signature schemes to ring signature schemes and illustrate it using the example of Rainbow. The resulting scheme provides perfect anonymity for the signer (as member of a group), as well as shorter ring signatures than all previously proposed post-quantum ring signature schemes.

Keywords: Multivariate cryptography · Ring signatures · Rainbow signature scheme

1 Introduction

Cryptographic techniques are an essential tool to guarantee the security of communication in modern society. Today, the security of nearly all of the cryptographic schemes used in practice is based on number theoretic problems such as factoring large integers and solving discrete logarithms. The best known schemes in this area are RSA [22], DSA [13] and ECC. However, schemes like these will become insecure as soon as large enough quantum computers are built. The reason for this is Shor's algorithm [24], which solves number theoretic problems like integer factorization and discrete logarithms in polynomial time on a quantum computer. Therefore, one needs alternatives to those classical public key schemes which are based on hard mathematical problems not affected by quantum computer attacks (so called post-quantum cryptosystems).

Besides lattice, code and hash based cryptosystems, multivariate cryptography is one of the main candidates for this [4]. Multivariate schemes are in general very fast and require only modest computational resources, which makes

© Springer International Publishing AG 2017
M. Joye and A. Nitaj (Eds.): AFRICACRYPT 2017, LNCS 10239, pp. 3–20, 2017.
DOI: 10.1007/978-3-319-57339-7_1

them attractive for the use on low cost devices like smart cards and RFID chips [5,6]. However, while there exist many practical multivariate standard signature schemes such as UOV [14], Rainbow [9] and Gui [21], there is a lack of more advanced multivariate schemes such as schemes for oblivious transfer and signature schemes with special properties.

Ring signature schemes allow a user to sign messages anonymously as a member of a group $\mathcal{R}$. The verifier can check, if the message was indeed signed by a member of the group, but has no means to reveal the concrete identity of the signer. Therefore, ring signature schemes are an important tool to secure the privacy of the users. In the last years, a number of multivariate ring signature schemes have been proposed [19,27,28,31]. However, as we find, all of these schemes share certain weaknesses with regard to efficiency or security.

In this paper, we present a new general technique to extend multivariate signature schemes to ring signature schemes. By doing so, we obtain a much simpler construction for multivariate ring signature schemes, which is therefore much easier to understand and analyze than previous constructions. By applying our technique to Rainbow, we obtain a ring signature scheme whose ring signatures are not longer than standard signatures of many other post-quantum signature (e.g. lattice, hash based) schemes. Furthermore, due to the efficiency of the Rainbow scheme, our scheme is very fast.

The rest of this paper is organized as follows. Section 2 reviews the concept of ring signatures and discusses the basic security notions. In Sect. 3 we give an overview of multivariate cryptography and introduce the Rainbow signature scheme, which is one of the best studied and most efficient multivariate signature schemes. Furthermore, in this section, we consider the existing multivariate ring signature schemes and analyze them with regard to security and performance. Section 4 presents our technique to extend multivariate signature schemes such as Rainbow to ring signature schemes and discusses the security of our construction. In Sect. 5 we give concrete parameter sets for our scheme based on Rainbow, while Sect. 6 presents an alternative construction of multivariate ring signatures reducing key and signature sizes. In Sect. 7 we describe a technique to reduce the public key size further. Section 8 deals with the implementation of our scheme and presents performance results, whereas Sect. 9 compares our construction with other existing ring signature schemes (both from the classical and the post-quantum world). Finally, Sect. 10 concludes the paper.

2   Ring Signatures

Ring signature schemes as proposed by Rivest et al. [23] allow a signer to sign a message anonymously on behalf of a group $\mathcal{R} = \{u_1, \ldots, u_k\}$ of possible signers. The receiver of a signed message can check, if the message was indeed signed by a member of the group, but can not reveal the concrete identity of the signer.

For example, the group of signers could be the set of employees of a company. By verifying the ring signature of a signed document (e.g. a bill), the receiver can ensure that it really was signed by an employee of the given company. By hiding the identity of the actual signer, ring signatures make therefore an important contribution to secure the privacy of the signer.

The concept of ring signatures is closely related to *group signatures*. However, while, in a group signature scheme, there exists a group manager who can, in the case of a controversy, connect a group signature to the actual signer, such a function does not exist in a ring signature scheme. Therefore, a ring signature scheme provides full anonymity to the signers (as members of the group).

Another related notion is that of *threshold ring signatures*. A threshold ring signature allows a verifier to check if, for any given number $s \in \{1, \ldots, k\}$, at least s members of the group $\mathcal{R}$ contributed to a signature. A basic ring signature scheme is therefore a special case of a threshold ring signature scheme with $s = 1$. Threshold ring signature schemes on the basis of multivariate polynomials have been proposed in [19,31]. However, by restricting to the case of ring signatures, we can reduce the key and signature sizes of the scheme drastically.

Formally, we can define a ring signature scheme $\mathcal{RS}$ as follows [3].

Let $\mathcal{R} = \{u_1, u_2, \ldots, u_k\}$ be a group (called ring) of users. A ring signature scheme consists of the three algorithms KeyGen, RingSign and Verify.

- KeyGen(1^λ): The probabilistic algorithm KeyGen takes as input a security parameter λ and outputs a key pair (sk, pk). In a ring signature scheme, this algorithm is performed by every user $u_i \in \mathcal{R}$.
- RingSign(d, sk_i, $\{pk_1, \ldots, pk_k\}$): The (probabilistic) algorithm RingSign takes as input the message d to be signed, the secret key sk_i of one user u_i and a list of the public keys $\{pk_1, \ldots, pk_k\}$ of all users $u_j \in \mathcal{R}$. The algorithm outputs a ring signature σ for the message d on behalf of the ring $\mathcal{R}$.
- Verify($(d, \sigma), \{pk_1, \ldots, pk_k\}$): The deterministic algorithm Verify takes as input a message/signature pair (d, σ) and a list of public keys $\{pk_1, \ldots, pk_k\}$. It outputs **TRUE**, if σ is a valid ring signature for the message d on behalf of the ring $\mathcal{R}$, and **FALSE** otherwise.

We assume that the ring signature scheme $\mathcal{RS}$ is *correct*, i.e.

$$\Pr[\text{Verify}((d, \text{RingSign}(d, sk_i, \{pk_1, \ldots, pk_k\})), \{pk_1, \ldots, pk_k\}) = 1$$

for all $i \in \{1, \ldots, k\}$.

The basic security criteria of a ring signature scheme are anonymity and unforgeability.

- **Anonymity**: The receiver of a signed message should not be able to detect the concrete identity of the signer. More formally, anonymity can be defined using the following security game.

Game[Anonymity]:

1. The algorithm KeyGen is used to generate k key pairs $((sk_1, pk_1), \ldots, (sk_k, pk_k))$. The set of public keys $\{pk_1, \ldots, pk_k\}$ is given to the adversary $\mathcal{A}$.
2. The adversary $\mathcal{A}$ is given access to a signing oracle $\mathcal{OS}(i, d)$, which, on input of an index $i \in \{1, \ldots, k\}$ and a message d returns a valid ring signature σ for the message d on behalf of the ring $\mathcal{R} = \{u_1, \ldots, u_k\}$. Hereby, in order to create the signature σ, the signing oracle $\mathcal{OS}$ uses the secret key sk_i of the user u_i.
3. $\mathcal{A}$ outputs a message $d^\star$ as well as two indices i_0 and $i_1 \in \{1, \ldots, k\}$. He is given a signature $\sigma \leftarrow \texttt{RingSign}(d^\star, sk_{i_b}, \{pk_1, \ldots, pk_k\})$, where b is randomly chosen from $\{0, 1\}$.
4. The adversary $\mathcal{A}$ outputs a bit b'. He wins the game, if and only if $b' = b$ holds.

The ring signature scheme $\mathcal{RS}$ is said to provide anonymity, if the advantage

$$\mathrm{Adv}_{\mathcal{A}} = 2 \cdot \Pr[b' = b] - 1$$

is, for every PPT adversary $\mathcal{A}$, negligible.

- **Unforgeability:** Given a message d, an adversary $\mathcal{A}$ not belonging to the ring $\mathcal{R}$ of legitimate signers is not able to forge a valid ring signature σ for the message d on behalf of the ring $\mathcal{R}$.

  More formally, we can define unforgeability using the following game.

Game[Unforgeability]:

1. The algorithm KeyGen is used to generate k key pairs $((sk_1, pk_1), \ldots, (sk_k, pk_k))$. The set of public keys $\{pk_1, \ldots, pk_k\}$ is given to the adversary $\mathcal{A}$.
2. The adversary $\mathcal{A}$ is given access to a signing oracle $\mathcal{OS}(d)$, which, on the input of a message d, returns a valid ring signature σ for the message d on behalf of the ring $\mathcal{R} = \{u_1, \ldots, u_k\}$.
3. $\mathcal{A}$ is given a challenge message $d^\star$. He wins the game, if he is able to produce a valid ring signature $\sigma^\star$ for $d^\star$ on behalf of the ring $\mathcal{R}$.

The ring signature scheme $\mathcal{RS}$ is said to provide unforgeability, if the success probability

$$\Pr_{\mathcal{A}}[\text{success}] = \Pr[\texttt{Verify}((d^\star, \sigma^\star), \{pk_1, \ldots, pk_k\}) = \textbf{TRUE}]$$

is, for any PPT adversary $\mathcal{A}$, negligible.

3 Multivariate Cryptography

The basic objects of multivariate cryptography are systems of multivariate quadratic polynomials (see Eq. (1)).

$$p^{(1)}(x_1,\ldots,x_n) = \sum_{i=1}^{n}\sum_{j=i}^{n} p_{ij}^{(1)} \cdot x_i x_j + \sum_{i=1}^{n} p_i^{(1)} \cdot x_i + p_0^{(1)}$$

$$p^{(2)}(x_1,\ldots,x_n) = \sum_{i=1}^{n}\sum_{j=i}^{n} p_{ij}^{(2)} \cdot x_i x_j + \sum_{i=1}^{n} p_i^{(2)} \cdot x_i + p_0^{(2)}$$

$$\vdots$$

$$p^{(m)}(x_1,\ldots,x_n) = \sum_{i=1}^{n}\sum_{j=i}^{n} p_{ij}^{(m)} \cdot x_i x_j + \sum_{i=1}^{n} p_i^{(m)} \cdot x_i + p_0^{(m)} \qquad (1)$$

The security of multivariate schemes is based on the

MQ Problem: Given m multivariate quadratic polynomials $p^{(1)}(\mathbf{x}),\ldots,p^{(m)}(\mathbf{x})$ in n variables $x_1,\ldots,x_n$ as shown in Eq. (1), find a vector $\bar{\mathbf{x}} = (\bar{x}_1,\ldots,\bar{x}_n)$ such that $p^{(1)}(\bar{\mathbf{x}}) = \ldots = p^{(m)}(\bar{\mathbf{x}}) = 0$.

The MQ problem (for $m \approx n$) is proven to be NP-hard even for quadratic polynomials over the field GF(2) [12].

To build a public key cryptosystem on the basis of the MQ problem, one starts with an easily invertible quadratic map $\mathcal{F} : \mathbb{F}^n \to \mathbb{F}^m$ (central map). To hide the structure of $\mathcal{F}$ in the public key, one composes it with two invertible affine (or linear) maps $\mathcal{S} : \mathbb{F}^m \to \mathbb{F}^m$ and $\mathcal{T} : \mathbb{F}^n \to \mathbb{F}^n$. The *public key* of the scheme is therefore given by $\mathcal{P} = \mathcal{S} \circ \mathcal{F} \circ \mathcal{T} : \mathbb{F}^n \to \mathbb{F}^m$. The *private key* consists of $\mathcal{S}$, $\mathcal{F}$ and $\mathcal{T}$ and therefore allows to invert the public key.

Note: Due to the above construction, the security of multivariate public key schemes is not only based on the MQ-Problem, but also on the EIP-Problem ("Extended Isomorphism of Polynomials") of finding the composition of $\mathcal{P}$.

In this paper we concentrate on multivariate signature schemes. The standard signature generation and verification process of a multivariate signature scheme works as shown in Fig. 1.

Signature Generation. To generate a signature for a message d, the signer uses a hash function $\mathcal{H} : \{0,1\}^\star \to \mathbb{F}^m$ to compute the hash value $\mathbf{w} = \mathcal{H}(d) \in \mathbb{F}^m$ and computes recursively $\mathbf{x} = \mathcal{S}^{-1}(\mathbf{w}) \in \mathbb{F}^m$, $\mathbf{y} = \mathcal{F}^{-1}(\mathbf{x}) \in \mathbb{F}^n$ and $\mathbf{z} = \mathcal{T}^{-1}(\mathbf{y})$. The signature of the message $\mathbf{w}$ is $\mathbf{z} \in \mathbb{F}^n$. Here, $\mathcal{F}^{-1}(\mathbf{x})$ means finding one (of possibly many) pre-image of $\mathbf{x}$ under the central map $\mathcal{F}$.

Verification. To check, if $\mathbf{z} \in \mathbb{F}^n$ is indeed a valid signature for a message d, one computes $\mathbf{w} = \mathcal{H}(d)$ and $\mathbf{w}' = \mathcal{P}(\mathbf{z}) \in \mathbb{F}^m$. If $\mathbf{w}' = \mathbf{w}$ holds, the signature is accepted, otherwise rejected.

A good overview of existing multivariate schemes can be found in [8].

Signature Generation

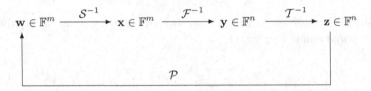

Signature Verification

Fig. 1. General workflow of multivariate signature schemes

3.1   The Rainbow Signature Scheme

The Rainbow signature scheme [9] is one of the most promising and best studied multivariate signature schemes. The scheme can be described as follows:

Let $\mathbb{F} = \mathbb{F}_q$ be a finite field with q elements, $n \in \mathbb{N}$ and $0 < v_1 < v_2 < \ldots < v_\ell < v_{\ell+1} = n$ be a sequence of integers. We set $m = n - v_1$, $O_i = \{v_i + 1, \ldots, v_{i+1}\}$ and $V_i = \{1, \ldots, v_i\}$ $(i = 1, \ldots \ell)$.

Key Generation. The *private key* of the scheme consists of two invertible affine maps $\mathcal{S} : \mathbb{F}^m \to \mathbb{F}^m$ and $\mathcal{T} : \mathbb{F}^n \to \mathbb{F}^n$ and a quadratic map $\mathcal{F}(\mathbf{x}) = (f^{(v_1+1)}(\mathbf{x}), \ldots, f^{(n)}(\mathbf{x})) : \mathbb{F}^n \to \mathbb{F}^m$. The polynomials $f^{(i)}$ $(i = v_1 + 1, \ldots, n)$ are of the form

$$f^{(i)} = \sum_{k,l \in V_j} \alpha_{k,l}^{(i)} \cdot x_k \cdot x_l + \sum_{k \in V_j, l \in O_j} \beta_{k,l}^{(i)} \cdot x_k \cdot x_l + \sum_{k \in V_j \cup O_j} \gamma_k^{(i)} \cdot x_k + \eta^{(i)} \quad (2)$$

with coefficients randomly chosen from $\mathbb{F}$. Here, j is the only integer such that $i \in O_j$. The *public key* is the composed map $\mathcal{P} = \mathcal{S} \circ \mathcal{F} \circ \mathcal{T} : \mathbb{F}^n \to \mathbb{F}^m$.

Signature Generation. To generate a signature for a document d, one uses a hash function $\mathcal{H} : \{0,1\}^* \to \mathbb{F}^m$ to compute the hash value $\mathbf{w} = \mathcal{H}(d) \in \mathbb{F}^m$ and computes recursively $\mathbf{x} = \mathcal{S}^{-1}(\mathbf{w}) \in \mathbb{F}^m$, $\mathbf{y} = \mathcal{F}^{-1}(\mathbf{x}) \in \mathbb{F}^n$ and $\mathbf{z} = \mathcal{T}^{-1}(\mathbf{y})$. Here, $\mathcal{F}^{-1}(\mathbf{x})$ means finding one (of approximately q^{v_1}) pre-image of $\mathbf{x}$ under the central map $\mathcal{F}$. This is done as shown in Algorithm 1.

It might happen that one of the linear systems in step 3 of the algorithm does not have a solution. In this case one has to choose other values for $y_1, \ldots, y_{v_1}$ and start again.

The signature of the document d is $\mathbf{z} \in \mathbb{F}^n$.

Signature Verification. To check, if $\mathbf{z} \in \mathbb{F}^n$ is indeed a valid signature for a document d, one computes $\mathbf{w} = \mathcal{H}(d)$ and $\mathbf{w}' = \mathcal{P}(\mathbf{z}) \in \mathbb{F}^m$. If $\mathbf{w}' = \mathbf{w}$ holds, the signature is accepted, otherwise rejected.

3.2   Multivariate Ring Signature Schemes

In the last years, a number of multivariate ring signature schemes have been proposed [19,27,28,31]. In this section, we give an overview of the main constructions and analyze them with regard to security and performance.

Algorithm 1. Inversion of the Rainbow central map

Input: Rainbow central map $\mathcal{F}$, vector $\mathbf{x} \in \mathbb{F}^m$
Output: vector $\mathbf{y} \in \mathbb{F}^n$ such that $\mathcal{F}(\mathbf{y}) = \mathbf{x}$

1: Choose random values for the variables $y_1, \ldots, y_{v_1}$ and substitute these values into the polynomials $f^{(i)}$ $(i = v_1 + 1, \ldots, n)$.
2: **for** $k = 1$ to ℓ **do**
3:     Perform Gaussian Elimination on the polynomials $f^{(i)}$ $(i \in O_k)$ to get the values of the variables y_i $(i \in O_k)$.
4:     Substitute the values of y_i $(i \in O_k)$ into the polynomials $f^{(i)}$ $(i \in \{v_{k+1} + 1, \ldots, n\})$.
5: **end for**

The Schemes of Petzoldt _et al._ [19] and Zhang _et al._ [31]. These two schemes are threshold ring signature schemes, i.e. they allow the verifier to check if a minimal number s of users contributed to the signature $(1 \leq s \leq k)$. Both of the schemes are based on the multivariate identification scheme of Sakumoto et al. [25], but use different techniques to extend the identification into a signature scheme: In the case of [19] this is the Fiat-Shamir protocol, the authors of [31] use the Γ-transformation. By both techniques it is possible to obtain a threshold ring signature scheme whose security is only based on the MQ Problem of solving a system of multivariate quadratic equations, which makes the schemes provable secure. However, due to the additional functionality of a _threshold_ ring signature scheme, both schemes produce very long signatures. By restricting to a simple ring signature scheme (i.e. $s = 1$), we can reduce the signature length and improve the performance of the scheme drastically.

The Scheme of Wang [27]. The ring signature scheme proposed by Wang in [27] is also based on the multivariate identification scheme of Sakumoto et al. [25]. Each user u_i $(u = 1, \ldots, k)$ chooses a vector $\mathbf{s}_i \in \mathbb{F}^n$ as his private key and a multivariate quadratic system $\mathcal{P}_i : \mathbb{F}^n \rightarrow \mathbb{F}^m$ with $\mathcal{P}_i(\mathbf{s}_i) = 0$ as his public key. In order to generate a ring signature for a message d, the signer produces for each user a transcript of the identification scheme (using the "secret" $\mathbf{0}$ for the non signers). Unfortunately, the verifier has no means to check how many zero vectors were used during the signature generation. Therefore it is possible for an adversary which is no member of the ring and therefore does not know any of the private keys s_i to forge a valid ring signature (using $\mathbf{0}$ for all the secret vectors $\mathbf{s}_i$ $(i = 1, \ldots, k)$). Furthermore, the scheme proposed in [27] contains only one round of the identification scheme, enabling an adversary to forge a signature with probability $\frac{2}{3}$. Therefore, the scheme of [27] does not provide any security at all.

The Scheme of Wang _et al._ [28]. The scheme of Wang et al. is similar to our construction in the sense that it provides a general technique to extend an arbitrary multivariate signature scheme to a ring signature scheme. Therefore, as it is in the case of our construction, the security of the resulting ring signature scheme is based on the security of the underlying multivariate signature scheme. However, in our construction, the ring signatures are generated in a much simpler and faster way. To generate a ring signature with the scheme of [28], one needs k hash function evaluations, $2k + 1$ evaluations of public keys and one signature generation of the underlying signature scheme, while our scheme requires only $k - 1$ evaluations of multivariate systems and one signature generation. During verification, [28] requires $k - 1$ hash function evaluations and $2k - 2$ evaluations of a multivariate quadratic system, while our scheme needs only k evaluations of the public key. Furthermore, these simple signature generation and verification algorithms make our scheme much easier to understand and to analyze and lead to (slightly) shorter ring signatures. Moreover, with regard to security, the paper [28] does not take attacks against underdetermined multivariate systems (see Sects. 4.1 and 5 of this paper) into consideration. Therefore, especially for large sizes of the group $\mathcal{R}$, the authors of [28] overestimate the security of their scheme significantly (or propose too small parameters).

4   Our Ring Signature Scheme

In this section we present our technique to extend arbitrary multivariate signature schemes such as UOV [14], Rainbow [9] and Gui [21] to ring signature schemes. Whereas, in this section, we present our technique in a very general way, we concentrate in the following sections on ring signatures based on the Rainbow signature scheme (see Sect. 3.1), which offers both good performance and short signatures. Furthermore, the key sizes of Rainbow are acceptable and can be further reduced by the technique of [18] (see Sect. 7).

Let $\mathcal{R} = \{u_1, \ldots, u_k\}$ be a ring of users.

Key Generation. Each user u_i generates a key pair $((\mathcal{S}_i, \mathcal{F}_i, \mathcal{T}_i), \mathcal{P}_i)$ of the underlying multivariate signature scheme. The public key $\mathcal{P}$ of the group is the concatenation of all individual public keys, i.e. $\mathcal{P} = \mathcal{P}_1 || \mathcal{P}_2 || \ldots || \mathcal{P}_k$, while each user u_i keeps $\mathcal{S}_i, \mathcal{F}_i$ and $\mathcal{T}_i$ as his private key sk_i.

Signature Generation. In order to sign a message d on behalf of the ring $\mathcal{R}$, a user u_i uses a hash function $\mathcal{H}$ to compute the hash value $\mathbf{w} = \mathcal{H}(d) \in \mathbb{F}^m$ of the message. He then chooses random vectors $\mathbf{z}_1, \ldots, \mathbf{z}_{i-1}, \mathbf{z}_{i+1}, \ldots, \mathbf{z}_k \in \mathbb{F}^n$. He computes

$$\tilde{\mathbf{w}} = \mathbf{w} - \sum_{\substack{j=1 \\ j \neq i}}^{k} \mathcal{P}_j(\mathbf{z}_j) \in \mathbb{F}^m \tag{3}$$

and uses his private key to compute a vector $\mathbf{z}_i \in \mathbb{F}^n$ such that $\mathcal{P}(\mathbf{z}_i) = \tilde{\mathbf{w}}$.

The ring signature for the message d is $(\mathbf{z}_1, \mathbf{z}_2, \ldots, \mathbf{z}_k) \in \mathbb{F}^{k \cdot n}$.

Signature Verification. In order to check if $(\mathbf{z}_1, \mathbf{z}_2, \ldots, \mathbf{z}_k) \in \mathbb{F}^{k \cdot n}$ is indeed a valid ring signature for the message d, the receiver computes the hash value $\mathbf{w} = \mathcal{H}(d) \in \mathbb{F}^m$ of the message d and uses the public keys $\mathcal{P}_1, \ldots, \mathcal{P}_k$ to compute

$$\hat{\mathbf{w}} = \sum_{j=1}^{k} \mathcal{P}_j(\mathbf{z}_j). \tag{4}$$

If $\hat{\mathbf{w}} = \mathbf{w}$ holds, the signature is accepted, otherwise it is rejected.

Remark: In case of an honestly computed ring signature $(\mathbf{z}_1, \mathbf{z}_2, \ldots, \mathbf{z}_k) \in \mathbb{F}^{kn}$ we have

$$\hat{\mathbf{w}} = \sum_{j=1}^{k} \mathcal{P}_j(\mathbf{z}_j) = \sum_{\substack{j=1 \\ j \neq i}}^{k} \mathcal{P}_j(\mathbf{z}_j) + \mathcal{P}_i(\mathbf{z}_i) = \mathbf{w} - \tilde{\mathbf{w}} + \tilde{\mathbf{w}} = \mathbf{w}. \tag{5}$$

Therefore, an honestly generated ring signature is always accepted.

4.1   Security

In this section we analyze the security of our construction. We do not consider the security of the underlying multivariate signature schemes in this paper and refer to the original papers [14,17,21] for a security analysis of the different schemes. Here, we concentrate on our construction of a ring signature scheme. For this, we have to show the anonymity and unforgeability of the resulting scheme.

Anonymity.

Theorem 1. *Our construction provides perfect anonymity for the actual signer as a member of the group, i.e. the final ring signature contains no information, which member of the group generated the signature and even a computationally unrestricted adversary can not reveal the identity of the signer.*

Proof (sketch). We assume that $\mathcal{R} = \{u_1, u_2\}$ and perform **Game[Anonymity]** (see Sect. 2) for this situation. Then we show that, independently of the fact which secret key is used during the generation of the ring signature, the signing oracle $\mathcal{OS}$ outputs each of the q^{n+v_1} possible ring signatures of the message $d^\star$ with probability $\approx q^{-n-v_1}$. For each possible ring signature $\sigma^\star$ of $d^\star$ we therefore have

$$\Pr[\sigma^\star \text{generated using } sk_1] = \Pr[\sigma^\star \text{ generated using } sk_2] \approx 1/2.$$

Therefore, an adversary can only guess whether $\sigma^\star$ was computed with sk_1 or sk_2 and his advantage is exactly 0 (independent from his resources).

Unforgeability. To forge a ring signature with respect to a ring of signers $\mathcal{R} = \{u_1, \ldots, u_k\}$, an attacker has to find a solution $\mathbf{z}_1, \ldots, \mathbf{z}_k$ of the equation

$$\mathcal{P}_1(\mathbf{z}_1) + \mathcal{P}_2(\mathbf{z}_2) + \cdots + \mathcal{P}_k(\mathbf{z}_k) = \mathbf{w}. \tag{6}$$

Basically, there are two possibilities to do this.

1. The adversary could proceed similar to a legitimate user of the ring signature scheme and choose $k - 1$ random vectors $\mathbf{z}_1, \ldots, \mathbf{z}_{k-1} \in \mathbb{F}^n$, compute $\tilde{\mathbf{w}} = \mathbf{w} - \sum_{i=1}^{k-1} \mathcal{P}_i(\mathbf{z}_i)$ and try to find a solution to the system $\mathcal{P}_k(\mathbf{z}_k) = \tilde{\mathbf{w}}$.

2. The adversary could try to solve the system (6) directly as an underdetermined system of multivariate quadratic equations.

Note that the first case is equivalent to breaking an instance of the underlying multivariate signature scheme. We do not consider this case here and refer to the papers [14, 17, 21] for a security analysis of the various schemes, We assume that, if we choose the parameters of our scheme according to the recommendations given in these papers, our scheme is secure against attacks of this kind. Hence, we concentrate in the following on the second case.

Unfortunately, solving (6) directly is not as hard as breaking the underlying scheme, where the attacker has to find a solution $\mathbf{z_k} \in \mathbb{F}^n$ of $\mathcal{P}_k(\mathbf{z_k}) = \tilde{\mathbf{w}}$.

The reason for this is that the system (6) is a highly underdetermined multivariate quadratic system. For systems of this type we have to consider the following two important results.

1. If the number of variables n in an underdetermined multivariate quadratic system $\mathcal{P}$ of m equations is given by $n = \omega \cdot m$, then a solution of the system $\mathcal{P}$ can be found in the same time as finding a solution of a determined system of $m - \lfloor \omega \rfloor + 1$ equations [26].
2. If the number of variables n in the multivariate quadratic system $\mathcal{P}$ exceeds $n \geq \frac{m(m+3)}{2}$, $\mathcal{P}$ can be solved in polynomial time [15].

In our parameter choice (see next section), we have to consider these two results. Therefore, the parameters of our scheme depend not only on the required level of security, but, since the number of variables in the public system $\mathcal{P}$ (6) depends on k, also on the size of the ring $\mathcal{R}$.

5 Parameters

In this section we give concrete parameter proposals for our ring signature scheme. We define our scheme over the field GF(256) and instantiate it on the basis of the Rainbow signature scheme of Sect. 3.1, which offers both good performance and short signatures. The proposed parameter sets are obtained as follows.

1. Direct attacks against the scheme should be infeasible, i.e. the parameters of the scheme have to be chosen in such a way that the two attacks against underdetermined quadratic systems mentioned in the previous section become infeasible.
2. Attacks of the Rainbow type against the single systems $\mathcal{P}(\mathbf{z}_i) = \mathbf{w}_i$ ($i = 1, \ldots k$) must be infeasible. With regard to this, we follow the results of [17].

As we find, for small numbers of k (e.g. $k = 5$), the parameters of our scheme are very similar to the parameters recommended for Rainbow in [17]. For larger values of k, attacks against underdetermined systems play an increasing role.

The resulting parameter sets and key sizes can be found in Table 1.

Table 1. Proposed parameters for our ring signature scheme ($\mathbb{F} = \mathrm{GF}(256)$; Rainbow)

Security level (bit)		5 users	10 users	20 users	50 users
80	Parameters	(16,17,15)	(15,20,18)	(14,26,24)	(13,56,53)
	Public key size (kB)	191	551	2,095	40,588
	Signature size (bit)	1,920	4,240	10,240	48,800
100	Parameters	(25,21,19)	(24,25,22)	(22,31,28)	(20,60,55)
	Public key size (kB)	432	1,206	3,921	52,312
	Signature size (bit)	2,600	5,680	12,960	54,000
128	Parameters	(36,23,20)	(34,26,23)	(32,33,29)	(30,64,58)
	Public key size (kB)	680	1,708	5,522	70,180
	Signature size (bit)	3,160	6,640	15,040	60,800

As the table shows, especially for small values of k, the signature sizes of our scheme are quite small. The size of a ring signature is of range several kbit and therefore not longer than standard signatures of many other post-quantum (e.g. lattice, hash based) signature schemes. However, for larger values of k, key and signature sizes of our scheme increase significantly.

6   Alternative Construction of a Multivariate Ring Signature Scheme

As can be seen from Table 1, the key sizes (especially the size of the public key) increase drastically if the number of users in the ring gets larger. To avoid this, we present in this section an alternative way to construct a ring signature scheme on the basis of multivariate signature schemes such as Rainbow. In particular, we use here instead of component wise addition of the single signatures component wise multiplication. By doing so, we can prevent attacks against highly underdetermined multivariate quadratic systems, since the degree of the corresponding system becomes very large. Our alternative construction can be described as follows.

Key Generation. The key generation of our alternative construction works just as presented in Sect. 4. Each user u_i generates a key pair $((\mathcal{S}_i, \mathcal{F}_i, \mathcal{T}_i), \mathcal{P}_i)$ of the underlying multivariate signature scheme. The public key $\mathcal{P}$ of the group is the set of all individual public keys, i.e. $\mathcal{P} = \{\mathcal{P}_1, \mathcal{P}_2, \ldots, \mathcal{P}_k\}$, while each user u_i keeps $\mathcal{S}_i, \mathcal{F}_i$ and $\mathcal{T}_i$ as his private key sk_i.

Signature Generation. In order to sign a message d on behalf of the ring $\mathcal{R}$, a user u_i uses a hash function $\mathcal{H} : \{0,1\}^\star \to \{0,\ldots,q-1\}^m$ to compute the hash value $\mathbf{w} = \mathcal{H}(d) + 1^m \in \mathbb{F}^m$ of the message, where 1^m is a vector with all entries equal to one. He then chooses random vectors $\mathbf{z}_1, \ldots, \mathbf{z}_{i-1}, \mathbf{z}_{i+1}, \ldots, \mathbf{z}_k \in \mathbb{F}^n$ satisfying

$$(\mathcal{P}_j(\mathbf{z}_j))_s \neq 0, \quad j \in \{1,\ldots,k\} \setminus \{i\}, \ s \in \{1,\ldots,m\}.$$

He computes

$$\tilde{\mathbf{w}} = \mathbf{w} \cdot (\prod_{\substack{j=1 \\ j \neq i}}^{k} \mathcal{P}_j(\mathbf{z}_j))^{-1} \in \mathbb{F}^m \tag{7}$$

and uses his private key to compute a vector $\mathbf{z}_i \in \mathbb{F}^n$ such that $\mathcal{P}_i(\mathbf{z}_i) = \tilde{\mathbf{w}}$.

The ring signature for the message d is $(\mathbf{z}_1, \mathbf{z}_2, \ldots, \mathbf{z}_k) \in \mathbb{F}^{kn}$. Note that in Eq. (7) multiplication and inversion work component wise on the elements of the corresponding vectors.

Remark: The reason of constructing the hash value of the document d in the way shown above is to generate a hash value without zero elements. By doing so we can ensure that all vectors $\mathbf{w}_i$ have the same structure. This guarantees the anonymity of the actual signer.

Signature Verification. In order to check if $(\mathbf{z}_1, \mathbf{z}_2, \ldots, \mathbf{z}_k) \in \mathbb{F}^{kn}$ is indeed a valid ring signature for the message d, the receiver computes the hash value $\mathbf{w} = \mathcal{H}(d) \in \mathbb{F}^m$ of the message d and uses the public keys $\mathcal{P}_1, \ldots, \mathcal{P}_k$ to compute

$$\hat{\mathbf{w}} = \prod_{j=1}^{k} \mathcal{P}_j(\mathbf{z}_j). \tag{8}$$

If $\hat{\mathbf{w}} = \mathbf{w}$ holds, the signature is accepted, otherwise it is rejected. Again note that the multiplication works component wise.

Remark: In the case of an honestly computed ring signature $(\mathbf{z}_1, \mathbf{z}_2, \ldots, \mathbf{z}_k) \in \mathbb{F}^{kn}$ we have

$$\hat{\mathbf{w}} = \prod_{j=1}^{k} \mathcal{P}_j(\mathbf{z}_j) = \prod_{\substack{j=1 \\ j \neq i}}^{k} \mathcal{P}_j(\mathbf{z}_j) \cdot \mathcal{P}_i(\mathbf{z}_i) = \mathbf{w} \cdot (\tilde{\mathbf{w}})^{-1} \cdot \tilde{\mathbf{w}} = \mathbf{w}. \tag{9}$$

Therefore, an honestly generated ring signature is always accepted.

6.1   Unforgeability

While the anonymity of our ring signature scheme can be shown exactly as in Sect. 4.1, we here concentrate on the unforgeability. Similar to Sect. 4.1, an attacker can try to forge a ring signature in two different ways:

1. The adversary could proceed similar to a legitimate user of the ring signature scheme and choose $k - 1$ random vectors $\mathbf{z}_1, \ldots, \mathbf{z}_{k-1} \in \mathbb{F}^n$, compute $\tilde{\mathbf{w}} = \mathbf{w} \cdot (\prod_{j=1}^{k-1} \mathcal{P}_j(\mathbf{z}_j))^{-1}$ and try to find a solution of the system $\mathcal{P}_k(\mathbf{z}_k) = \tilde{\mathbf{w}}$.
2. The adversary could try to solve the system

$$\mathcal{P}_1(\mathbf{z}_1) \cdot \ldots \cdot \mathcal{P}_k(\mathbf{z}_k) = \mathbf{w}$$

directly as an underdetermined system of multivariate equations.

Again, forging a ring signature by the first method is equivalent to breaking an instance of the underlying multivariate scheme, which is, by our assumption, infeasible.

When attacking our scheme in the second way, the attacker is faced with an underdetermined system of multivariate polynomial equations. But, in contrast to Sect. 4.1, this system is no longer quadratic, but the polynomials are, for a ring of k users, of degree $2k$. The methods to solve underdetermined quadratic systems mentioned in Sect. 4.1 do not work in this case[1]. It is therefore infeasible for the attacker to forge a ring signature using this strategy. This means that we do not have to increase the parameters of our scheme when the number of users in the ring gets large. Beyond the significant reduction of key size this also makes it much easier to add additional users to the ring.

Table 2 shows our parameter recommendations and resulting key and signature sizes for our alternative construction of a multivariate ring signature scheme.

Table 2. Proposed parameters for our alternative construction of a multivariate ring signature scheme ($\mathbb{F} = \mathrm{GF}(256)$; Rainbow)

Security level (bit)		5 users	10 users	20 users	50 users
80	Parameters (v_1, o_1, o_2)	(17,13,13)	(17,13,13)	(17,13,13)	(17,13,13)
	Public key size (kB)	125.7	251.4	502.7	1,257
	Signature size (bit)	1,720	3,440	6,880	17,200
100	Parameters (v_1, o_1, o_2)	(26,16,17)	(26,16,17)	(26,16,17)	(26,16,17)
	Public key size (kB)	294.9	589.7	1,179	2,949
	Signature size (bit)	2,6360	4,720	9,440	23,600
128	Parameters (v_1, o_1, o_2)	(36,21,22)	(36,21,22)	(36,21,22)	(36,21,22)
	Public key size (kB)	680.3	1,361	2,721	6,803
	Signature size (bit)	3,160	6,320	12,640	31,600

7 Reduction of Public Key Size

In [18], Petzoldt et al. proposed a technique to reduce the public key size of the UOV and Rainbow signature schemes. In particular, they were able to construct a Rainbow key pair $((\mathcal{S}, \mathcal{F}, \mathcal{T}), \mathcal{P})$, where the coefficient matrix P of the public key has the form (for Rainbow schemes with two layers) (Fig. 2).

---

[1] Of course, the attacker could try to transform the given system of high degree into a quadratic one. However, even if the given system is very sparse, this increases the number of equations and variables in the quadratic system drastically. Furthermore, the ratio between the number of variables and the number of equations gets close to 1.

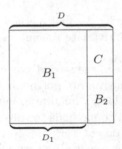

Fig. 2. Structure of the public key $\mathcal{P}$

Here we have $D = \frac{(n+1)\cdot(n+2)}{2}$ and $D_1 = D - \frac{(o_2+1)\cdot(o_2+2)-2}{2}$. The matrices $B_1 \in \mathbb{F}^{m \times D_1}$ and $B_2 \in \mathbb{F}^{o_2 \times (D-D_1)}$ can be arbitrarily set by the user. In particular, B_1 and B_2 can be chosen in a structured way which reduces the public key size of the Rainbow scheme significantly.

Note that, when applying this technique to our ring signature scheme, we can choose the same matrices B_1 and B_2 for all users $u_1, \ldots, u_k$. By doing so, we can reduce the public key size of our scheme by up to 68% (see Table 3).

Table 3. Possible reduction of public key size

Security level (bit)		5 users	10 users	20 users	50 users
80	Parameters	(17,13,13)	(17,13,13)	(17,13,13)	(17,13,13)
	Public key size (standard) (kB)	125.7	251.4	502.7	1,257
	Public key size (reduced) (kB)	47.3	93.8	186.7	465.5
	Reduction (%)	62.4	62.7	62.9	63.0
100	Parameters	(26,16,17)	(26,16,17)	(26,16,17)	(26,16,17)
	Public key size (standard) (kB)	294.9	589.7	1,179	2,949
	Public key size (reduced) (kB)	99.5	197.3	393.0	980.2
	Reduction (%)	66.3	66.5	66.7	66.8
128	Parameters	(36,21,22)	(36,21,22)	(36,21,22)	(36,21,22)
	Public key size (standard) (kB)	680.3	1,361	2,721	6,803
	Public key size (reduced) (kB)	219.4	435.9	868.8	2,168
	Reduction (%)	67.7	68.0	68.1	68.1

Furthermore, when choosing the matrices B_1 and B_2 in a cyclic way, we can speed up the evaluation of the Rainbow public key by up to 60% [20]. Since this step is used both in the signature generation and verification processes of our scheme, both processes can be sped up drastically (see Table 4).

8 Implementation and Efficiency Results

In this section we present our results regarding the efficiency of our construction. To generate a ring signature on behalf of a ring $\mathcal{R} = \{u_1, \ldots, u_k\}$ of k members, a user u_i has to perform

- $k - 1$ evaluations of public systems $\mathcal{P}_i$ and
- 1 Rainbow signature generation.

The verification process of our scheme consists of k evaluations of the Rainbow public keys $\mathcal{P}_1, \ldots, \mathcal{P}_k$.

Since both the evaluation and inversion of Rainbow systems are very efficient, our scheme offers good performance. By using structured public keys (see last section), we can speed up our scheme further (see Table 4).

To study the efficiency of our construction in practice, we created a straightforward C implementation of Rainbow and our ring signature scheme and ran it for the parameter sets proposed in Sect. 5. Table 4 shows the results. In each cell, the first number shows the running time of the signature generation/verification process of the standard scheme, while the second number shows the corresponding timings for the structured scheme.

Table 4. Running times of signature generation and verification

Security level (bit)		5 users	10 users	20 users	50 users
80	Parameters	(16,17,15)	(15,20,18)	(14,26,24)	(13,56,53)
	Sign. generation (ms)	13.31/9.15	28.73/17.31	58.98/37.04	1225/738.0
	Sign. verification (ms)	10.81/5.08	26.23/13.54	50.51/27.42	1200/703.0
100	Parameters	(25,21,19)	(24,25,22)	(22,31,28)	(20,60,55)
	Sign. generation (ms)	16.04/10.30	41.07/23.81	123.80/68.10	1580.32/905.4
	Sign. verification (ms)	12.75/5.57	35.55/ 16.38	115.97/57.35	1547.69/859.0
128	Parameters	(36,23,20)	(34,26,23)	(32,33,29)	(30,64,58)
	Sign. generation (ms)	27.56/17.54	58.97/32.00	175.37/88.94	3177.13/1706
	Sign. verification (ms)	20.03/7.70	50.43/20.45	163.22/72.24	3110.79/1610

9 Discussion

Especially for small values of k, our ring signature scheme is very efficient. In this case, the resulting ring signatures are not larger than standard signatures of other post-quantum signature schemes such as lattice and hash based constructions.

However, when the number of users in the ring $\mathcal{R}$ increases, key sizes and signature sizes of our scheme increase significantly.

Another disadvantage of the first version of our scheme is that it is very difficult to add additional users to the ring $\mathcal{R}$, since this might made it necessary to change the parameters. One therefore has to fix the maximal number $k_{\max}$ of users in the ring $\mathcal{R}$ a priori and choose the parameters of the scheme according to $k_{\max}$. This problem can be solved by switching from addition to componentwise multiplication (see Sect. 6). Table 5 shows a comparison of our scheme with other (post-quantum and classical) ring signature schemes.

As the Table shows, our scheme outperforms the other post-quantum ring signature schemes in terms of signature size. In this sense, it can also compete

Table 5. Comparison of our scheme with other ring signature schemes (80 bit security)

	Scheme	Our	[29]	[16]	[7]	[1]	[2]	[10]
5 users	Pk size (kB)	47.3	1.3	179	147	751	1.0	1.0
	Sign. size (bit)	1,720	26,000	301,546	659,632	6,973,251	7,810	15,820
50 users	Pk size (kB)	465.5	12.5	1,785	7,513	15,020	9.8	9.8
	Sign. size (bit)	17,200	260,000	3,015,462	6,596,320	69,732,510	78,100	158,200
		Mult	Lattice	Mult	Lattice	Code	RSA	DL
				Threshold ring signatures				

with the RSA and DL based constructions of [2,10]. On the other hand, the key sizes of our scheme are much larger than those of the classical and the lattice based construction of [29].

Furthermore, there exist some pairing based constructions of ring signature schemes, which offer a sublinear signature size [11]. For large values of k, the ring signatures of these schemes are significantly smaller than those of the above constructions. However, these schemes can be easily broken by quantum computers. Therefore, our scheme offers the shortest ring signatures of all post-quantum constructions.

10   Conclusion

In this paper we proposed a new multivariate ring signature scheme on the basis of the Rainbow signature scheme [9]. However, we can construct our scheme on the basis of every other multivariate signature scheme such as UOV and HVEv-, too. Our scheme is one of the first multivariate signature schemes with special properties and one of few candidates for post-quantum ring signatures. The scheme is very efficient, especially when the number of users in the ring is small, and produces the shortest ring signatures of all existing post-quantum constructions.

Future work includes the development of other multivariate signature schemes with special properties such as blind and group signatures.

References

1. Aguilar, C., Cayrel, P.L., Gaborit, P., Laguillaumie, F.: A new efficient threshold ring signature scheme based on coding theory. IEEE Trans. Inf. Theory **57**(7), 4833–4842 (2011)
2. Asaar, M.R., Salmasizadeh, M., Susilo, W.: A short identity-based proxy ring signature scheme from RSA. Comput. Stand. Interfaces **38**, 144–151 (2015)
3. Bender, A., Katz, J., Morselli, R.: Ring signatures: stronger definitions and constructions without random oracles. IACR eprint 2005/304
4. Bernstein, D.J., Buchmann, J., Dahmen, E. (eds.): Post Quantum Cryptography. Springer, Heidelberg (2009)

5. Bogdanov, A., Eisenbarth, T., Rupp, A., Wolf, C.: Time-area optimized public-key engines: $\mathcal{MQ}$-cryptosystems as replacement for elliptic curves? In: Oswald, E., Rohatgi, P. (eds.) CHES 2008. LNCS, vol. 5154, pp. 45–61. Springer, Heidelberg (2008). doi:10.1007/978-3-540-85053-3_4

6. Chen, A.I.-T., Chen, M.-S., Chen, T.-R., Cheng, C.-M., Ding, J., Kuo, E.L.-H., Lee, F.Y.-S., Yang, B.-Y.: SSE implementation of multivariate PKCs on modern x86 CPUs. In: Clavier, C., Gaj, K. (eds.) CHES 2009. LNCS, vol. 5747, pp. 33–48. Springer, Heidelberg (2009). doi:10.1007/978-3-642-04138-9_3

7. Cayrel, P.-L., Lindner, R., Rückert, M., Silva, R.: A lattice-based threshold ring signature scheme. In: Abdalla, M., Barreto, P.S.L.M. (eds.) LATINCRYPT 2010. LNCS, vol. 6212, pp. 255–272. Springer, Heidelberg (2010). doi:10.1007/978-3-642-14712-8_16

8. Ding, J., Gower, J.E., Schmidt, D.S.: Multivariate Public Key Cryptosystems. Springer, USA (2006)

9. Ding, J., Schmidt, D.: Rainbow, a new multivariable polynomial signature scheme. In: Ioannidis, J., Keromytis, A., Yung, M. (eds.) ACNS 2005. LNCS, vol. 3531, pp. 164–175. Springer, Heidelberg (2005). doi:10.1007/11496137_12

10. Franklin, M., Zhang, H.: Unique ring signatures: a practical construction. In: Sadeghi, A.-R. (ed.) FC 2013. LNCS, vol. 7859, pp. 162–170. Springer, Heidelberg (2013). doi:10.1007/978-3-642-39884-1_13

11. Fujisaki, E.: Sub-linear size traceable ring signatures without random oracles. In: Kiayias, A. (ed.) CT-RSA 2011. LNCS, vol. 6558, pp. 393–415. Springer, Heidelberg (2011). doi:10.1007/978-3-642-19074-2_25

12. Garey, M.R., Johnson, D.S.: Computers and Intractability: A Guide to the Theory of NP-Completeness. W.H. Freeman and Company, New York (1979)

13. Kravitz, D.: Digital signature algorithm. US patent 5231668, July 1991

14. Kipnis, A., Patarin, J., Goubin, L.: Unbalanced oil and vinegar signature schemes. In: Stern, J. (ed.) EUROCRYPT 1999. LNCS, vol. 1592, pp. 206–222. Springer, Heidelberg (1999). doi:10.1007/3-540-48910-X_15

15. Miura, H., Hashimoto, Y., Takagi, T.: Extended algorithm for solving underdefined multivariate quadratic equations. In: Gaborit, P. (ed.) PQCrypto 2013. LNCS, vol. 7932, pp. 118–135. Springer, Heidelberg (2013). doi:10.1007/978-3-642-38616-9_8

16. Petzoldt, A., Bulygin, S., Buchmann, J.: A multivariate based threshold ring signature scheme. Appl. Algebra Eng. Commun. Comput. **24**(3–4), 255–275 (2012)

17. Petzoldt, A., Bulygin, S., Buchmann, J.: Selecting parameters for the rainbow signature scheme. In: Sendrier, N. (ed.) PQCrypto 2010. LNCS, vol. 6061, pp. 218–240. Springer, Heidelberg (2010). doi:10.1007/978-3-642-12929-2_16

18. Petzoldt, A., Bulygin, S., Buchmann, J.: CyclicRainbow – a multivariate signature scheme with a partially cyclic public key. In: Gong, G., Gupta, K.C. (eds.) INDOCRYPT 2010. LNCS, vol. 6498, pp. 33–48. Springer, Heidelberg (2010). doi:10.1007/978-3-642-17401-8_4

19. Petzoldt, A., Bulygin, S., Buchmann, J.: A multivariate threshold ring signature scheme. AAECC **25**(3–4), 255–275 (2012)

20. Petzoldt, A., Bulygin, S., Buchmann, J.: Fast verification for improved versions of the UOV and rainbow signature schemes. In: Gaborit, P. (ed.) PQCrypto 2013. LNCS, vol. 7932, pp. 188–202. Springer, Heidelberg (2013). doi:10.1007/978-3-642-38616-9_13

21. Petzoldt, A., Chen, M.-S., Yang, B.-Y., Tao, C., Ding, J.: Design principles for HFEv-based multivariate signature schemes. In: Iwata, T., Cheon, J.H. (eds.) ASIACRYPT 2015. LNCS, vol. 9452, pp. 311–334. Springer, Heidelberg (2015). doi:10.1007/978-3-662-48797-6_14

22. Rivest, R.L., Shamir, A., Adleman, L.: A method for obtaining digital signatures and public-key cryptosystems. Commun. ACM **21**(2), 120–126 (1978)
23. Rivest, R.L., Shamir, A., Tauman, Y.: How to leak a secret. In: Boyd, C. (ed.) ASIACRYPT 2001. LNCS, vol. 2248, pp. 552–565. Springer, Heidelberg (2001). doi:10.1007/3-540-45682-1_32
24. Shor, P.: Polynomial-time algorithms for prime factorization and discrete logarithms on a quantum computer. SIAM J. Comput. **26**(5), 1484–1509 (1997)
25. Sakumoto, K., Shirai, T., Hiwatari, H.: Public-key identification schemes based on multivariate quadratic polynomials. In: Rogaway, P. (ed.) CRYPTO 2011. LNCS, vol. 6841, pp. 706–723. Springer, Heidelberg (2011). doi:10.1007/978-3-642-22792-9_40
26. Thomae, E., Wolf, C.: Solving underdetermined systems of multivariate quadratic equations revisited. In: Fischlin, M., Buchmann, J., Manulis, M. (eds.) PKC 2012. LNCS, vol. 7293, pp. 156–171. Springer, Heidelberg (2012). doi:10.1007/978-3-642-30057-8_10
27. Wang, L.L.: A new multivariate-based ring signature scheme. In: Proceeedings of ISCCCA (2013)
28. Wang, S., Ma, R., Zhang, Y., Wang, X.: Ring signature scheme based on multivariate public key cryptosystems. Comput. Math. Appl. **62**, 3973–3979 (2011)
29. Wang, S., Zhao, R.: Lattice-based ring signature scheme under the random oracle model (2014). CoRR abs/1405.3177
30. Yang, B.-Y., Chen, J.-M., Chen, Y.-H.: TTS: high-speed signatures on a low-cost smart card. In: Joye, M., Quisquater, J.-J. (eds.) CHES 2004. LNCS, vol. 3156, pp. 371–385. Springer, Heidelberg (2004). doi:10.1007/978-3-540-28632-5_27
31. Zhang, J., Zhao, Y.: A new multivariate based threshold ring signature scheme. In: Au, M.H., Carminati, B., Kuo, C.-C.J. (eds.) NSS 2014. LNCS, vol. 8792, pp. 526–533. Springer, Cham (2014). doi:10.1007/978-3-319-11698-3_42

Pinocchio-Based Adaptive zk-SNARKs and Secure/Correct Adaptive Function Evaluation

Meilof Veeningen[(✉)]

Philips Research, Eindhoven, The Netherlands
meilof.veeningen@philips.com

Abstract. Pinocchio is a practical zk-SNARK that allows a prover to perform cryptographically verifiable computations with verification effort potentially less than performing the computation itself. A recent proposal showed how to make Pinocchio adaptive (or "hash-and-prove"), i.e., to enable proofs with respect to computation-independent commitments. This enables computations to be chosen after the commitments have been produced, and for data to be shared between different computations in a flexible way. Unfortunately, this proposal is not zero-knowledge. In particular, it cannot be combined with Trinocchio, a system in which Pinocchio is outsourced to three workers that do not learn the inputs thanks to multi-party computation (MPC). In this paper, we show how to make Pinocchio adaptive in a zero-knowledge way; apply this to make Trinocchio work on computation-independent commitments; present tooling to easily program flexible verifiable computations (with or without MPC); and use it to build a prototype in a medical research case study.

1 Introduction

Recent advances in SNARKs (Succinct Non-interactive ARguments of Knowledge) are making it more and more feasible to outsource computations to the cloud while obtaining cryptographic guarantees about the correctness of their outputs. In particular, the Pinocchio system [8,11] achieved for the first time for a practical computation a verification time of a computation proof that was actually faster than performing the computation itself.

In Pinocchio, proofs are verified with respect to plaintext inputs and outputs of the verifier; but in many cases, it is useful to have computation proofs that also refer to committed data, e.g., provided by a third party. Ideally, such proofs should be *adaptive*, i.e., multiple different computations can be performed on the same commitment, that are chosen after the data has been committed to; and *zero-knowledge*, i.e., the commitments and proofs should reveal no information about the committed data. This latter property allows proofs on sensitive data, and it allows extensions like Trinocchio [13] that additionally hide this sensitive data from provers by multi-party computation.

Although several approaches are known from the literature, no really satisfactory practical adaptive zk-SNARK exists. The recent "hash first" proposal [7]

© Springer International Publishing AG 2017
M. Joye and A. Nitaj (Eds.): AFRICACRYPT 2017, LNCS 10239, pp. 21–39, 2017.
DOI: 10.1007/978-3-319-57339-7_2

shows how to make Pinocchio adaptive at low overhead, but is unfortunately not zero-knowledge. On the other hand, Pinocchio's successor Geppetto [3] is zero-knowledge but not adaptive: multiple computations can be performed on the same data but they need to be known before committing. The asymptotically best known SNARKS combining the two properties have $\Theta(n \log n)$ non-cryptographic and $\Theta(n)$ cryptographic work for the prover, a $\Theta(n)$-sized CRS, and constant-time verification (where n is the size of the computation), but with a large practical overhead: [10] because it relies on the impractical subset-sum language; other constructions (e.g., [3,7]) because they rely on including hash evaluation in the computation[1]. Finally, [1] enables Pinocchio proofs on authenticated data with prover complexity as above, but verification time is linear in the number of committed inputs.

In this work, we give a new Pinocchio-based adaptive zk-SNARK that solves the above problems. We match the best asymptotic performance (i.e., $\Theta(n \log n)$ non-cryptographic work and $\Theta(n)$ cryptographic work for the prover; a $\Theta(n)$-size CRS and constant-time verification); but obtain the first practical solution by adding only minor overhead to "plain" Pinocchio (instead of relying on expensive approaches such as subset-sum or bootstrapping).

As additional contributions, we apply our zk-SNARK in the Trinocchio setting, and present tooling to easily perform verifiable computations. Trinocchio [13] achieves privacy-preserving outsourcing to untrusted workers by combining the privacy guarantees of multi-party computation with the correctness guarantees of the Pinocchio zk-SNARK. With our adaptive zk-SNARK, computation can be chosen *after* the inputs were provided and more complex functionalities can be achieved by using the output of one computation as input of another. We also improve the generality of [13] by proving security for *any* suitable MPC protocol and adaptive zk-SNARK. Our tooling consists of a Python frontend and a C++ backend. The frontend allows easy programming of verifiable computations (with libraries for zero testing, oblivious indexing and fixed-point computations), and execution either directly (for normal outsourcing scenarios) or with MPC (for privacy-preserving outsourcing). The backend provides key generation, proving, and verification functionality for both scenarios.

2   Preliminaries

2.1   Algebraic Tools, Notation, and Complexity Assumptions

Our constructions aim to prove correctness of computations over a prime order field $\mathbb{F} = \mathbb{F}_p$. We make use of pairings, i.e., groups $(\mathbb{G}, \mathbb{G}', \mathbb{G}_T)$ of order p and an efficient bilinear map $e : \mathbb{G} \times \mathbb{G}' \to \mathbb{G}_T$, where for any generators $g_1 \in \mathbb{G}, g_2 \in \mathbb{G}'$, $e(g_1, g_2) \neq 1$ and $e(g_1^a, g_2^b) = e(g_1, g_2)^{ab}$. Throughout, we will make use of polynomial evaluations in a secret point $s \in \mathbb{F}$. For $f \in \mathbb{F}[x]$, write $\langle f \rangle_1$ for $f(s) \cdot g_1$ and $\langle f \rangle_2$ for $f(s) \cdot g_2$.

---

[1] In practice, computing the hash is complex itself. It can be avoided with bootstrapping [10], giving slightly worse asymptotics and again a large practical overhead.

Let $(\mathbb{G}, \mathbb{G}', \mathbb{G}_T, e) \leftarrow \mathcal{G}(1^\kappa)$ denote parameter generation in this setting. We require the following assumptions from [4] (that generalise those from [11] to asymmetric pairings):

Definition 1. *The q-power Diffie Hellman (q-PDH) assumption holds for $\mathcal{G}$ if, for any NUPPT adversary $\mathcal{A}$:* $\Pr[(\mathbb{G}, \mathbb{G}', \mathbb{G}_T, e) \leftarrow \mathcal{G}(1^\kappa); \; g \in_R \mathbb{G}^*; g' \in_R \mathbb{G}'^*; \; s \in_R \mathbb{Z}_p^*; y \leftarrow \mathcal{A}(\mathbb{G}, \mathbb{G}', \mathbb{G}_T, e, \{g^{s^i}, g'^{s^i}\}_{i=1,\ldots,q,q+2,\ldots,2q}) : y = g^{s^{q+1}}] \approx_\kappa 0.$

Definition 2. *The q-power knowledge of exponent (q-PKE) assumption holds for $\mathcal{G}$ and a class $\mathcal{Z}$ of auxiliary input generators if, for every NUPPT auxiliary input generator $Z \in \mathcal{Z}$ and any NUPPT adversary $\mathcal{A}$ there exists a NUPPT extractor $\mathcal{E}_\mathcal{A}$ such that:* $\Pr[crs := (\mathbb{G}, \mathbb{G}', \mathbb{G}_T, e) \leftarrow \mathcal{G}(1^\kappa); \; g \in_R \mathbb{G}^*; \; s \in_R \mathbb{Z}_p^*; \; z \leftarrow Z(crs, g, \ldots, g^{s^q}); g' \in_R \mathbb{G}'^*; (c, c'||a_0, \ldots, a_q) \leftarrow (\mathcal{A}||\mathcal{E}_\mathcal{A})(crs, \{g^{s^i}, g'^{s^i}\}_{i=0,\ldots,q}, z) : e(c, g') = e(g, c') \wedge c \neq \prod_{i=0}^q g_1^{a_i s^i}] \approx_\kappa 0.$

Here, $(a||b) \leftarrow (\mathcal{A}||\mathcal{E}_\mathcal{A})(c)$ denotes running both algorithms on the same inputs and random tape, and assigning their results to a respectively b. For certain auxiliary input generators, the q-PKE assumption does not hold, so we have to conjecture that our auxiliary input generators are "benign", cf. [4].

Definition 3. *The q-target group strong Diffie Hellman (q-SDH) assumption holds for $\mathcal{G}$ if, for any NUPPT adversary $\mathcal{A}$,* $\Pr[crs := (\mathbb{G}, \mathbb{G}', \mathbb{G}_T, e) \leftarrow \mathcal{G}(1^\kappa); g \in_R \mathbb{G}^*; g' \in_R \mathbb{G}'^*; s \in_R \mathbb{Z}_p(r, Y) \leftarrow \mathcal{A}(crs, \{g^{s^i}, g'^{s^i}\}_{i=0,\ldots,q}) : r \in \mathbb{Z}_p \setminus \{s\} \wedge Y = e(g, g')^{\frac{1}{s-r}}] \approx_\kappa 0.$

2.2 Adaptive zk-SNARKs in the CRS Model

We now define adaptive zk-SNARKs as in [10] with minor modifications. We first define extractable trapdoor commitment families. This is a straightforward generalisation of an extractable trapdoor commitment scheme [10] that explicitly captures multiple commitment keys generated from the same CRS:

Definition 4. *Let $(G0, Gc, C)$ be a scheme where $(crs, td) \leftarrow G0(1^\kappa)$ outputs a system-wide CRS and a trapdoor; $(ck, ctd) \leftarrow Gc(crs)$ outputs a commitment key and a trapdoor; and $c \leftarrow ck(m; r)$ outputs a commitment with the given key. Such a scheme is called an* extractable trapdoor commitment family *if:*

- *(Computationally binding) For every NUPPT $\mathcal{A}$,* $\Pr[(crs, \cdot) \leftarrow G0(1^\kappa); (ck, \cdot) \leftarrow Gc(crs); (v; r; v'; r') \leftarrow \mathcal{A}(crs; ck) : C_{ck}(v; r) = C_{ck}(v'; r')] \approx 0.$
- *(Perfectly hiding) Letting $(crs, \cdot) \leftarrow G0(1^\kappa); (ck, \cdot) \leftarrow Gc(crs)$, for all v, v', $C_{ck}(v; r)$ and $C_{ck}(v'; r')$ are identically distributed given random r, r'*
- *(Trapdoor) There exists a NUPPT algorithm $\mathcal{T}$ such that if $(crs, td) \leftarrow G0(1^\kappa); (ck, ctd) \leftarrow Gc(crs); (u; t) \leftarrow \mathcal{T}(crs; td; ck; ck); r \leftarrow \mathcal{T}(t; u; v)$, then u is distributed identically to real commitments and $C_{ck}(v; r) = u$.*
- *(Extractable) For every NUPPT committer $\mathcal{A}$, there exists a NUPPT extractor $\mathcal{E}_\mathcal{A}$ such that $\Pr[(crs; \cdot) \leftarrow G0(1^\kappa); (ck; \cdot) \leftarrow Gc(crs); (u||v; r) \leftarrow (\mathcal{A}||\mathcal{E}_\mathcal{A})(crs; ck) : u \in Range(C_{ck}) \wedge u \neq C_{ck}(v; r)] \approx 0.$*

Given relation $\mathcal{R}$ and commitment keys $\mathsf{ck}_1, \ldots, \mathsf{ck}_n$ from the same commitment family, define: $\mathcal{R}_{\mathsf{ck}_1,\ldots,\mathsf{ck}_n} := \{(\boldsymbol{u}; \boldsymbol{v}, \boldsymbol{r}, \boldsymbol{w}) : \boldsymbol{u}_i = \mathsf{C}_{\mathsf{ck}_i}(\boldsymbol{v}_i; \boldsymbol{r}_i) \wedge (\boldsymbol{v}; \boldsymbol{w}) \in \mathcal{R}\}$. Intuitively, an adaptive zk-SNARK is a zk-SNARK for relation $\mathcal{R}_{\mathsf{ck}_1,\ldots,\mathsf{ck}_n}$.

Definition 5. *An adaptive zk-SNARK for extractable trapdoor commitment family* $(\mathsf{G0}, \mathsf{Gc}, \mathsf{C})$ *and relation* $\mathcal{R}$ *is a scheme* $(\mathsf{G}, \mathsf{P}, \mathsf{V})$ *where[2]:*

- $(\mathsf{crsp}; \mathsf{crsv}; \mathsf{tdp}) \leftarrow \mathsf{G}(\mathsf{crs}; \{\mathsf{ck}_i\})$, *given a CRS and commitment keys, outputs evaluation and verification keys, and a trapdoor;*
- $\pi \leftarrow \mathsf{P}(\mathsf{crs}; \{\mathsf{ck}_i\}; \mathsf{crsp}; \boldsymbol{u}; \boldsymbol{v}; \boldsymbol{r}; \boldsymbol{w})$, *given a CRS; commitment keys; an evaluation key; commitments; openings; and a witness, outputs a proof;*
- $0/1 \leftarrow \mathsf{V}(\mathsf{crs}; \{\mathsf{ck}_i\}; \mathsf{crsv}; \boldsymbol{u}; \pi)$, *given a CRS; commitment keys; a verification key; commitments; and a proof, verifies the proof,*

satisfying the following properties (let $\mathsf{setup} := (\mathsf{crs}; \mathsf{td}) \leftarrow \mathsf{G0}(1^\kappa); \forall i : (\mathsf{ck}_i; \mathsf{ctd}_i) \leftarrow \mathsf{Gc}(\mathsf{crs}); (\mathsf{crsp}; \mathsf{crsv}; \mathsf{tdp}) \leftarrow \mathsf{G}(\mathsf{crs}; \{\mathsf{ck}_i\}))$:

- Perfect completeness *("proofs normally verify")* $\Pr[\mathsf{setup}; (\boldsymbol{u}; \boldsymbol{v}; \boldsymbol{r}; \boldsymbol{w}) \leftarrow \mathcal{R}_{\{\mathsf{ck}_i\}} : \mathsf{V}(\mathsf{crs}; \{\mathsf{ck}_i\}; \mathsf{crsv}; \boldsymbol{u}; \mathsf{P}(\mathsf{crs}; \{\mathsf{ck}_i\}; \mathsf{crsp}; \boldsymbol{u}; \boldsymbol{v}; \boldsymbol{r}; \boldsymbol{w})) = 1] = 1$.
- Argument of knowledge *("the commitment openings and a valid witness can be extracted from an adversary producing a proof")*: *for every NUPPT* $\mathcal{A}$ *there exists NUPPT extractor* $\mathcal{E}_\mathcal{A}$ *such that, for every auxiliary information* $\mathsf{aux} \in \{0, 1\}^{poly(\kappa)}$: $\Pr[\mathsf{setup}; (\boldsymbol{u}; \pi || \boldsymbol{v}; \boldsymbol{r}; \boldsymbol{w}) \leftarrow (\mathcal{A} || \mathcal{E}_\mathcal{A})(\mathsf{crs}; \{\mathsf{ck}_i\}; \mathsf{crsp}; \mathsf{aux} || \ldots; \mathsf{td}; \mathsf{ctd}_1; \ldots; \mathsf{ctd}_n; \mathsf{tdp}) : (\boldsymbol{u}; \boldsymbol{v}; \boldsymbol{r}; \boldsymbol{w}) \notin \mathcal{R}_{\{\mathsf{ck}_i\}} \wedge \mathsf{V}(\mathsf{crs}; \{\mathsf{ck}_i\}; \mathsf{crsv}; \boldsymbol{u}; \pi) = 1] \approx_\kappa 0$. *Here,* $(\mathcal{A} || \mathcal{E}_\mathcal{A})(\cdot || \ldots; \cdot')$ *is parallel execution with extra input* $\cdot'$ *for* $\mathcal{E}_\mathcal{A}$.
- Perfectly composable zero knowledge *("proofs can be simulated using the commitments and trapdoor")*: *there exists a PPT simulator* $\mathcal{S}$ *such that, for all stateful NUPPT adversaries* $\mathcal{A}$, $\Pr[\mathsf{setup}; (\boldsymbol{u}; \boldsymbol{v}; \boldsymbol{r}; \boldsymbol{w}) \leftarrow \mathcal{A}(\mathsf{crs}, \{\mathsf{ck}_i\}, \mathsf{crsp}); \pi \leftarrow \mathsf{P}(\mathsf{crs}, \{\mathsf{ck}_i\}, \mathsf{crsp}; \boldsymbol{u}; \boldsymbol{v}; \boldsymbol{r}; \boldsymbol{w}) : (\boldsymbol{u}, \boldsymbol{v}, \boldsymbol{r}, \boldsymbol{w}) \in \mathcal{R}_{\{\mathsf{ck}_i\}} \wedge \mathcal{A}(\pi) = 1] = \Pr[\mathsf{setup}; (\boldsymbol{u}; \boldsymbol{v}; \boldsymbol{r}; \boldsymbol{w}) \leftarrow \mathcal{A}(\mathsf{crs}, \{\mathsf{ck}_i\}, \mathsf{crsp}); \pi \leftarrow \mathcal{S}(\mathsf{crs}, \{\mathsf{ck}_i\}, \mathsf{crsp}; \boldsymbol{u}; \mathsf{td}, \{\mathsf{ctd}_i\}, \mathsf{tdp}) : (\boldsymbol{u}, \boldsymbol{v}, \boldsymbol{r}, \boldsymbol{w}) \in \mathcal{R}_{\{\mathsf{ck}_i\}} \wedge \mathcal{A}(\pi) = 1]$.

We base our definitions on [10] because it is closest to what we want to achieve. Unlike in [3], we do not guarantee security when relation $\mathcal{R}$ is chosen adaptively based on the commitment keys; this is left as future work.

2.3   The Pinocchio zk-SNARK Construction from [11]

QAPs. Pinocchio models computations as *quadratic arithmetic programs* (QAPs) [8]. A QAP over a field $\mathbb{F}$ is a triple $(\mathbf{V}, \mathbf{W}, \mathbf{Y}) \in (\mathbb{F}^{d \times k})^3$, where d is called the *degree* of the QAP and k is called the *size*. A vector $\boldsymbol{x} \in \mathbb{F}^k$ is said to be a

---

[2] We differ from [10] in three minor ways: (1) we generalise from commitment schemes to families because we need this in Adaptive Trinocchio; (2) we allow witnesses that are not committed to separately, giving a slight efficiency improvement; (3) the extractor has access to the trapdoor, as needed when using Pinocchio [8].

solution to the QAP if $(\mathbf{V} \cdot \boldsymbol{x}) \times (\mathbf{W} \cdot \boldsymbol{x}) = \mathbf{Y} \cdot \boldsymbol{x}$, where $\times$ denotes the pairwise product and $\cdot$ denotes normal matrix-vector multiplication. A QAP Q is said to *compute* function $f : \mathbb{F}^i \rightarrow \mathbb{F}^j$ if $\boldsymbol{b} = f(\boldsymbol{a})$ if and only if there exists a *witness* $\boldsymbol{w}$ such that $(\boldsymbol{a}; \boldsymbol{b}; \boldsymbol{w})$ is a solution to Q. For example, consider the QAP

$$\mathbf{V} = \begin{pmatrix} 1\,1\,0\,0 \\ 1\,1\,0\,0 \end{pmatrix}, \quad \mathbf{W} = \begin{pmatrix} 1\,1\,0\,0 \\ 0\,0\,0\,1 \end{pmatrix}, \quad \mathbf{Y} = \begin{pmatrix} 0\,0\,0\,1 \\ 0\,0\,1\,0 \end{pmatrix}.$$

Intuitively, the first row of this QAP represents equation $(x_1 + x_2) \cdot (x_1 + x_2) = x_4$ in variables (x_1, x_2, x_3, x_4) whereas the second row represents equation $(x_1 + x_2) \cdot x_4 = x_3$. Note that $x_3 = (x_1 + x_2)^3$ if and only if there exists x_4 satisfying the two equations, so this QAP computes function $f : (x_1, x_2) \mapsto x_3$.[3]

Fixing d distinct, public $\omega_1, \ldots, \omega_d \in \mathbb{F}$, then a QAP can equivalently be described by a collection of interpolating polynomials in these points. Namely, let $\{v_i(x)\}$ be the unique polynomials of degree $< d$ such that $v_i(\omega_j) = \mathbf{V}_{j,i}$, and similarly for $\{w_i(x)\}$, $\{y_i(x)\}$. Then $\{v_i(x), w_i(x), y_i(x)\}$ is an equivalent description of the QAP. Defining $t(x) = (x - \omega_1) \cdot \ldots \cdot (x - \omega_d) \in \mathbb{F}[x]$, note that $\boldsymbol{x}_1, \ldots, \boldsymbol{x}_n$ is a solution to Q if and only if, for all j, $(\sum_i \boldsymbol{x}_i \cdot v_i(\omega_j)) \cdot (\sum_i \boldsymbol{x}_i \cdot w_i(\omega_j)) = (\sum_i \boldsymbol{x}_i \cdot y_i(\omega_j))$, or equivalently, if $t(x)$ divides $p(x) := (\sum_i \boldsymbol{x}_i \cdot v_i(x)) \cdot (\sum_i \boldsymbol{x}_i \cdot w_i(x)) - (\sum_i \boldsymbol{x}_i \cdot y_i(x)) \in \mathbb{F}[x]$.

Security Guarantees. Pinocchio is a zk-SNARK, which is essentially the same as an adaptive zk-SNARK except proving and verifying are with respect to plaintext values instead of commitments. In Pinocchio, relation $\mathcal{R}$ is that, for given $\boldsymbol{v}$, there exists witness $\boldsymbol{w}$ such that $(\boldsymbol{v}; \boldsymbol{w})$ is a solution to a given QAP Q. Pinocchio replies on a pairing secure under the $(4d + 4)$-PDH, d-PKE and $(8d + 8)$-SDH assumptions discussed above, where d is the degree of the QAP.

Construction. Fix random, secret $s, \alpha_v, \alpha_w, \alpha_y, \beta, r_v, r_w, r_y(x) := r_v r_w$. The central idea of Pinocchio is to prove satisfaction of all QAP equations using evaluations of the interpolating polynomials in a secret point. Namely, the prover computes quotient polynomial $h = p/t$ and basically provides evaluations "in the exponent" of h, $\sum_i \boldsymbol{x}_i \cdot v_i$, $\sum_i \boldsymbol{x}_i \cdot w_i$, $\sum_i \boldsymbol{x}_i \cdot y_i$ in the point s that is unknown to him, that can then be verified using the pairing. Precisely, the prover algorithm, given solution $\boldsymbol{x} = (\boldsymbol{v}; \boldsymbol{w})$ to the QAP, generates random $\delta_v, \delta_w, \delta_y$; computes coefficients $\boldsymbol{h}$ of the polynomial $(\sum_i \boldsymbol{x}_i \cdot v_i(x) + \delta_v t(x)) \cdot (\sum_i \boldsymbol{x}_i \cdot w_i(x) + \delta_w t(x)) - (\sum_i \boldsymbol{x}_i \cdot y_i(x) + \delta_y t(x))/t(x)$ (with δ terms added to make the proof zero-knowledge), and outputs (all $\sum_i$ over witness indices $|\boldsymbol{v}| + 1, \ldots, |\boldsymbol{x}|$; recall that for polynomial f, $\langle f \rangle_1 := f(s) \cdot g_1$ and $\langle f \rangle_2 := f(s) \cdot g_2$):

---

[3] In Pinocchio, the linear terms corresponding to $\mathbf{V}$, $\mathbf{W}$, $\mathbf{Y}$ can also contain constant values. This is achieved by assigning special meaning to a "constant" wire with value 1. We do not formalise this separately, instead leaving it up to the user to include a special variable and an equation $x_i \cdot x_i = x_i$ that forces this variable to be one.

$$\langle V \rangle_1 = \sum_i \boldsymbol{x}_i \langle r_v v_i \rangle_1 + \delta_v \langle r_v t \rangle_1, \langle \alpha_v V \rangle_2 = \sum_i \boldsymbol{x}_i \langle \alpha_v r_v v_i \rangle_2 + \delta_v \langle \alpha_v r_v t \rangle_2,$$
$$\langle W \rangle_2 = \sum_i \boldsymbol{x}_i \langle r_w w_i \rangle_2 + \delta_w \langle r_w t \rangle_2, \langle \alpha_w W \rangle_1 = \sum_i \boldsymbol{x}_i \langle \alpha_w r_w w_i \rangle_1 + \delta_w \langle \alpha_w r_w t \rangle_1,$$
$$\langle Y \rangle_1 = \sum_i \boldsymbol{x}_i \langle r_y y_i \rangle_1 + \delta_y \langle r_y t \rangle_1, \langle \alpha_y Y \rangle_2 = \sum_i \boldsymbol{x}_i \langle \alpha_y r_y y_i \rangle_2 + \delta_y \langle \alpha_y r_y t \rangle_2.$$
$$\langle Z \rangle_1 = \sum_i \boldsymbol{x}_i \langle r_v \beta v_i + r_w \beta w_i + r_y \beta y_i \rangle_1 + \delta_v \langle r_v \beta t \rangle_1 + \delta_w \langle r_w \beta t \rangle_1 + \delta_y \langle r_y \beta t \rangle_1,$$
$$\langle H \rangle_1 = \sum_{j=0}^d \boldsymbol{h}_j \langle x^j \rangle_1.$$

The evaluation key consists of all $\langle \cdot \rangle_1$, $\langle \cdot \rangle_2$ items used in the formulas above.[4]

The verification algorithm, given statement $\boldsymbol{v}$, extends $\langle V \rangle_1, \langle W \rangle_1, \langle Y \rangle_1$ to include also the input/output wires ($\sum_i$ over I/O wire indices $1, \ldots, |\boldsymbol{v}|$): $\langle V^+ \rangle_1 = \langle V \rangle_1 + \sum_i \boldsymbol{x}_i \langle r_v v_i \rangle_1$, $\langle W^+ \rangle_2 = \langle W \rangle_2 + \sum_i \boldsymbol{x}_i \langle r_w w_i \rangle_2$, $\langle Y^+ \rangle_1 = \langle Y \rangle_1 + \sum_i \boldsymbol{x}_i \langle r_y y_i \rangle_1$. Then, it checks (the verification key are the needed $\langle \cdot \rangle_1$, $\langle \cdot \rangle_2$ items):

$$e(\langle V \rangle_1, \langle \alpha_v \rangle_2) = e(\langle 1 \rangle_1, \langle \alpha_v V \rangle_2); \tag{V}$$
$$e(\langle \alpha_w \rangle_1, \langle W \rangle_2) = e(\langle \alpha_w W \rangle_1, \langle 1 \rangle_2); \tag{W}$$
$$e(\langle Y \rangle_1, \langle \alpha_y \rangle_2) = e(\langle 1 \rangle_1, \langle \alpha_y Y \rangle_2); \tag{Y}$$
$$e(\langle V \rangle_1 + \langle Y \rangle_1, \langle \beta \rangle_2) \cdot e(\langle \beta \rangle_1, \langle W \rangle_2) = e(\langle Z \rangle_1, \langle 1 \rangle_2); \tag{Z}$$
$$e(\langle V^+ \rangle_1, \langle W^+ \rangle_2) \cdot e(\langle Y^+ \rangle_1, \langle 1 \rangle_2)^{-1} = e(\langle H \rangle_1, \langle r_y t \rangle_2). \tag{H}$$

At a high level, checks (**V**), (**W**), (**Y**) guarantee that the proof is a proof of knowledge of the witness $\boldsymbol{w}$; check (**Z**) guarantees that the same witness $\boldsymbol{w}$ was used for $\langle V \rangle_1, \langle W \rangle_2, \langle Y \rangle_1$; and check (**Z**) guarantees that indeed, $p(x) = h(x) \cdot t(x)$ holds, which implies a solution to the QAP.

3   Adaptive zk-SNARKs Based on Pinocchio

This section presents the central contribution of this paper: an adaptive zk-SNARK based on Pinocchio. We obtain our Pinocchio-based adaptive zk-SNARK by generalising the role of the $\langle Z \rangle_1$ element of the Pinocchio proof. Recall that in Pinocchio, proof elements $\langle V \rangle_1$, $\langle W \rangle_1$, and $\langle Y \rangle_1$ are essentially weighted sums $\sum_j \boldsymbol{x}_j \langle v_j \rangle_1$, $\sum_j \boldsymbol{x}_j \langle w_j \rangle_2$, $\sum_j \boldsymbol{x}_j \langle y_j \rangle_1$ over elements $\langle v_j \rangle_1$, $\langle w_j \rangle_2$, $\langle y_j \rangle_1$ from the CRS, with the weights given by the witness part of the QAP's solution vector $\boldsymbol{x}$. The $\langle Z \rangle_1$ element ensures that these weighted sums consistently use the same witness. This is done by forcing the prover to come up essentially with $\beta \cdot (\langle V \rangle_1 + \langle W \rangle_2 + \langle Y \rangle_1)$ given only elements $\langle \beta \cdot (v_j + w_j + y_j) \rangle_1$ in which v_j, w_j, and y_j occur together. The essential idea is of our construction is to use the $\langle Z \rangle_1$ element also to ensure consistency to external commitments.

In more detail, in earlier works [3,13], it was noted that the Pinocchio $\langle V \rangle_1, \langle W \rangle_2, \langle Y \rangle_1$ elements can be divided into multiple "blocks" ($\langle V_i \rangle_1, \langle W_i \rangle_2, \langle Y_i \rangle_1, \langle Z_i \rangle_1$). Each block contains the values of a number of variables of the

---

[4] We use $\langle \alpha_v V \rangle_2$ etc. instead of $\langle \alpha_v V \rangle_1$ from [13], so that we can rely on the asymmetric q-PKE assumption from [4] (which [13] did not spell out).

Extractable Trapdoor Commitment Scheme Family $(\mathsf{G0}^1, \mathsf{Gc}^1, \mathsf{C}^1)$:

- $\mathsf{G0}^1$: Fix $\mathbb{G}_1, \mathbb{G}_2, \mathbb{G}_3$ and random s. Return $\mathsf{crs} = (\{\langle x^i\rangle_1, \langle x^i\rangle_2\}_{i=0,\dots,d}, \mathsf{td} = s$.
- Gc^1: Pick random α. Return $\mathsf{ck} = (\langle 1\rangle_1, \langle\alpha\rangle_2, \langle x\rangle_1, \langle\alpha x\rangle_2, \dots, \langle x^d\rangle_1, \langle\alpha x^d\rangle_2)$
- C^1: Return $(r\langle 1\rangle_1 + v_1\langle x\rangle_1 + v_2\langle x^2\rangle_1 + \dots, r\langle\alpha\rangle_2 + v_1\langle\alpha x\rangle_2 + v_2\langle\alpha x^2\rangle_2 + \dots)$

Key generation G^1: Fix a QAP of degree at most d, and let $v_j(x), w_j(x), y_j(x)$ be as in Pinocchio. Fix random, secret $\alpha_v, \alpha_w, \alpha_y, \beta, r_v, r_w$. Let $r_y = r_v r_w$. Let $z_j(x) = x^j + r_v v_j + r_w w_j + r_y y_j$ if $j \le W$ and $z_j(x) = r_v v_j + r_w w_j + r_y y_j$ otherwise. Evaluation key $(i = 1, \dots, n, \ j = 1, \dots, d)$:

$\langle x^j\rangle_1, \langle r_v v_j\rangle_1, \langle r_v t\rangle_1, \langle\alpha_v r_v v_j\rangle_2, \langle\alpha_v r_v t\rangle_2 \langle r_w w_j\rangle_1, \langle r_w t\rangle_1, \langle\alpha_w r_w w_j\rangle_1, \langle\alpha_w r_w t\rangle_1 \langle r_y y_j\rangle_1,$
$\langle r_y t\rangle_1, \langle\alpha_y r_y y_j\rangle_2, \langle\alpha_y r_y t\rangle_2 \langle\beta_i z_{(i-1)d+j}\rangle_1, \langle\beta_i z_{nd+j}\rangle_1, \langle\beta_i\rangle_1, \langle\beta_i r_v t\rangle_1, \langle\beta_i r_w t\rangle_1, \langle\beta_i r_y t\rangle_1$

Verification key $(i = 1, \dots, n)$: $(\langle\alpha_v\rangle_2, \langle\alpha_w\rangle_1, \langle\alpha_y\rangle_2, \langle\beta_i\rangle_2, \langle\beta_i\rangle_1, \langle r_y t\rangle_2)$.

Proof generation P^1: Let $u_i = \mathsf{C}^1_{\mathsf{ck}_i}(v_i; r_i)$, and let w be the witness such that $(v_1, \dots, v_n; w)$ is a solution to the QAP. Generate random $\delta_{v,i}, \delta_{w,i}, \delta_{y,i}$. Compute h as the coefficients of polynomial $((\sum_j x_j \cdot v_j(x) + \delta_v \cdot t(x)) \cdot (\sum_j x_j \cdot w_j(x) + \delta_w \cdot t(x)) - (\sum_j x_j \cdot y_j(x) + \delta_y \cdot t(x)))/t(x)$. Return $(i = 1, \dots, n; \ [\cdot]$ means only if $i = 1)$:

$$\langle V_i\rangle_1 = \sum_{j=1}^d v_{i,j}\langle r_v v_{(i-1)d+j}\rangle_1 \left[+ \sum_{j=1}^N w_j\langle r_v v_{nd+j}\rangle_1\right] + \delta_{v,i}\langle r_v t\rangle_1, \langle\alpha_v V_i\rangle_2 = \dots$$

$$\langle W_i\rangle_1 = \sum_{j=1}^d v_{i,j}\langle r_w w_{(i-1)d+j}\rangle_1 \left[+ \sum_{j=1}^N w_j\langle r_w w_{nd+j}\rangle_1\right] + \delta_{w,i}\langle r_w t\rangle_1, \langle\alpha_w W_i\rangle_1 = .$$

$$\langle Y_i\rangle_1 = \sum_{j=1}^d v_{i,j}\langle r_y y_{(i-1)d+j}\rangle_1 \left[+ \sum_{j=1}^N w_j\langle r_y y_{nd+j}\rangle_1\right] + \delta_{y,i}\langle r_y t\rangle_1, \langle\alpha_y Y_i\rangle_2 = \dots$$

$$\langle Z_i\rangle_1 = \sum_{j=1}^d v_{i,j}\langle\beta_i z_{(i-1)d+j}\rangle_1 \left[+ \sum_{j=1}^N w_j\langle\beta_i z_{nd+j}\rangle_1\right] + r_i\langle\beta_i\rangle_1 + \delta_{v,i}\langle\beta_i r_v t\rangle_1$$

$$\langle H\rangle_1 = \sum_j h_j\langle x^j\rangle_1. \qquad\qquad + \delta_{w,i}\langle\beta_i r_w t\rangle_1 + \delta_{y,i}\langle\beta_i r_y t\rangle_1$$

Proof verification V^1: Letting $\mathsf{ck}_i = (\dots, \langle\alpha_i\rangle_2)$, $u_i = (\langle C_i\rangle_1, \langle\alpha_i C_i\rangle_2)$, check that:

$$e(\langle C_i\rangle_1, \langle\alpha_i\rangle_2) = e(\langle 1\rangle_1, \langle\alpha_i C_i\rangle_2); \quad e(\langle V_i\rangle_1, \langle\alpha_v\rangle_2) = e(\langle\alpha_v V_i\rangle_1, \langle 1\rangle_2); \qquad \textbf{(C,V)}$$
$$e(\langle\alpha_w\rangle_1, \langle W_i\rangle_2) = e(\langle 1\rangle_1, \langle\alpha_w W_i\rangle_2); \quad e(\langle Y_i\rangle_1, \langle\alpha_y\rangle_2) = e(\langle 1\rangle_1, \langle\alpha_y Y_i\rangle_2); \qquad \textbf{(W,Y)}$$
$$e(\langle V_i\rangle_1 + \langle Y_i\rangle_1 + \langle C_i\rangle_1, \langle\beta_i\rangle_2) \cdot e(\langle\beta_i\rangle_1, \langle W_i\rangle_2) = e(\langle Z_i\rangle_1, \langle 1\rangle_2); \qquad \textbf{(Z)}$$
$$e(\langle V\rangle_1, \langle W\rangle_2) \cdot e(\langle Y\rangle_1, \langle 1\rangle_2)^{-1} = e(\langle H\rangle_1, \langle r_y t\rangle_2). \qquad\qquad \textbf{(H)}$$

(where $\langle V\rangle_1 = \langle V_1\rangle_1 + \dots + \langle V_n\rangle_1$, $\langle W\rangle_2 = \langle W_1\rangle_2 + \dots + \langle W_n\rangle_2$, $\langle Y\rangle_1 = \langle Y_1\rangle_1 + \dots)$

Fig. 1. Pinocchio-Based Adaptive zk-SNARK $(\mathsf{G}^1, \mathsf{P}^1, \mathsf{V}^1)$

QAP solution, which is enforced by providing $\langle z_j\rangle_1 = \langle\beta_i \cdot (v_j + w_j + y_j)\rangle_1$ elements only for the indices j of those variables. Our core idea is use external commitments of the form $\sum_k v_k \cdot \langle x^k\rangle_1$ (that can be re-used across Pinocchio computations) and link the kth component of this commitment to the jth variable of the block using a modified $\langle z_j\rangle_1 = \langle\beta_i \cdot (x^k + v_j + w_j + y_j)\rangle_1$. We use one block per external commitment that the proof refers to. The witness (which is not committed to externally) is included in the first block, with the normal

Pinocchio element $\langle z_j \rangle_1 = \langle \beta_1 \cdot (v_j + w_j + y_j) \rangle_1$ just checking internal consistency as usual. The verification procedure changes slightly: $\langle V \rangle_1$ is no longer extended to $\langle V^+ \rangle_1$ to include public I/O (which we do not have); instead, the $(\mathbf{Z})$ check ensures consistency with the corresponding commitment, for which there is a new correctness check $(\mathbf{C})$.

The precise construction is shown in Fig. 1. This construction contains details on how to add randomness to make the proof zero-knowledge; and it shows how additional $\langle \alpha \cdot \rangle_1$ elements are added to obtain an extractable trapdoor commitment family $(\mathsf{G0}^1, \mathsf{Gc}^1, \mathsf{C}^1)$. In [14], we show that:

Theorem 1. *Under the* $(4d+3)$-*PDH, d-PKE, and* $(8d+6)$-*SDH assumptions (with d the maximal QAP degree),* $(\mathsf{G0}^1, \mathsf{Gc}^1, \mathsf{C}^1)$ *is an extractable trapdoor commitment scheme family, and* $(\mathsf{G}^1, \mathsf{P}^1, \mathsf{V}^1)$ *is an adaptive zk-SNARK.*

4   Smaller Proofs and Comparison to Literature

We now present two optimization that decrease the size of the above zk-SNARK, and compare the concrete efficiency of our three proposals to two related proposals from the literature. Note that, in the above construction, seven Pinocchio proof elements $\langle V \rangle_1, \langle \alpha_v V \rangle_2, \langle W \rangle_2, \langle \alpha_w W \rangle_1, \langle Y \rangle_1, \langle \alpha_y Y \rangle_2, \langle Z \rangle_1$ are repeated for each input commitment. We present two different (but, unfortunately, mutually incompatible) ways in which this can be avoided.

In our first optimization, inspired by a similar proposal to reduce verification work in Pinocchio ([3], later corrected by [12]), we decrease proof size and verification time at the expense of needing a larger-degree QAP. Namely, suppose that all variables in a given commitment occur only in the right-hand side of QAP equations. In this case, $v_j(x) = w_j(x) = 0$ for all j, so proof elements $\langle V_i \rangle_1, \langle \alpha_v V_i \rangle_2, \langle W_i \rangle_2, \langle \alpha_w W_i \rangle_1, \langle Y_i \rangle_1, \langle \alpha_y Y_i \rangle_2$ contain only randomness and, setting $\delta_{v,j} = \delta_{w,j} = 0$, can be omitted. As a consequence, the marginal costs per commitment used decrease from 7 to 3; the $(\mathbf{V})$ and $(\mathbf{W})$ verification steps can be skipped and the $(\mathbf{Z})$ step simplified. To guarantee that a committed variable a only occurs in the right-hand of equations, we can introducing a witness b and equation $0 \cdot 0 = a - b$, slightly increasing the overall QAP size and degree. (This cannot be done for the first commitment since $\langle V_1 \rangle_1, \ldots$ also contain the witness, which occur in the left-hand side of equations as well.)

Our second proposal is a modified zk-SNARK that also reduces the marginal cost per commitment from 7 to 3, but gives more efficient verification when using many commitments. The core idea is to first concatenate all commitments $\mathbf{u}_1, \ldots, \mathbf{u}_n$ into one "intermediate commitment $\mathbf{u}'$, and then use our original zk-SNARK with respect to $\mathbf{u}'$. More precisely, we build intermediate commitment $\mathbf{u}'_1$ with the first ℓ_1 values of $\mathbf{u}_1$; $\mathbf{u}'_2$ with ℓ_1 zeros followed by the first ℓ_2 values of $\mathbf{u}_2$; etcetera. Then, $\mathbf{u}' = \sum_i \mathbf{u}'_i$ is a commitment to the first $\ell_1, \ldots, \ell_n$ values of the respective commitments $\mathbf{u}_1, \ldots, \mathbf{u}_n$. To avoid ambiguity between normal and intermediate commitments, to normal commitments we add a random factor r_c, i.e. $(r \langle r_c \rangle_1 + \sum_i \mathbf{v}_i \langle r_c x^i \rangle_1, r \langle \alpha r_c \rangle_2 + \ldots)$ and intermediate commitments are as above[5]. Proving correspondence between normal and intermediate commitments is done similarly to the (**Z**) check above: we generate random β_i and give $\langle \beta_i' \cdot (r_c x^j + x^{\ell_1 + \cdots + \ell_{i-1} + j}) \rangle_1$ to the prover, who needs to produce proof element $\langle Z_i' \rangle_1$ such that $\langle Z_i' \rangle_1 = \beta_i' \cdot (\langle C_i \rangle_1 + \langle C_i' \rangle_1)$, which he can only do if $\langle C_i' \rangle_1$ is formed correctly. Details and the security proof appear in [14].

Table 1. Comparison between Pinocchio-based SNARKs (n: number of commitments; d is QAP degree; $d' \leq d$ is QAP degree with optimization; $D \geq d$ is fixed QAP degree)

Construction	Comm. size	Proof size	Prover computation		Verif comp
			Non-crypt. op	Crypt. op	
Geppetto	3 gr. el	8 gr. el	$\Theta(D \log D)$	$\Theta(D)$	$4n + 12$ pair.
Hash First+Pinocchio	2 gr. el	9n+1 gr. el	$\Theta(d \log d)$	$\Theta(d)$	$13n + 3$ pair.
Hash First+Pinocchio[a]	2 gr. el	5n+5 gr. el	$\Theta(d' \log d')$	$\Theta(d')$	$8n + 8$ pair.
Our zk-SNARK I	2 gr. el	7n+1 gr. el	$\Theta(d \log d)$	$\Theta(d)$	$11n + 3$ pair.
Our zk-SNARK I[a]	2 gr. el	3n+5 gr. el	$\Theta(d' \log d')$	$\Theta(d')$	$7n + 7$ pair.
Our zk-SNARK II	2 gr. el	3n+8 gr. el	$\Theta(d \log d)$	$\Theta(d)$	$6n + 12$ pair

[a] First optimization from Sect. 4 applied.

In Table 1, we provide a detailed comparison of our zk-SNARKs with two similar constructions: the Geppetto protocol due to [3] (which is also zero-knowledge but not adaptive); and the "hash first" approach applied to Pinocchio [7] (which is adaptive but not zero-knowledge). Geppetto is Protocol 2 from [3]. We assume QAP witnesses of $\mathcal{O}(d)$. In Geppetto, a fixed set of QAPs of degree d_i are combined into one large "MultiQAP" of degree D slightly larger than $\max d_i$. As a consequence, if both small and large computations need to be applied on the same data, then the small computations take over the much worse performance of the large computations. For Hash First+Pinocchio, we took the extractable scheme $\mathsf{XP}_{\mathcal{E}}$ since the Geppetto and our construction are extractable as well. To make it work on multiple commitments (which is described for neither Hash First nor Pinocchio), we assume natural generalisations of Hash First and of Pinocchio along the lines of [3,13]. Our first optimization can be applied to this construction; we mark the result with a star and write $d' \geq d$ for the increased degree due to the use of this optimization. Finally, we show our zk-SNARK without and with the first optimization; and our second zk-SNARK construction (to which the optimization does not apply).

In conclusion, Geppetto is the most efficient construction, but apart from not being adaptive, it also requires all computations to be fixed and of the same size, making it inefficient for small computations when they are combined with large ones. Our construction outperforms Hash First+Pinocchio, essentially adding zero knowledge for free; which variant is best depends on n and $d' - d$. Note that Hash First allows using the same commitment in different zk-SNARK schemes;

---

[5] Hence this construction can only handle inputs of combined size at most d.

our scheme only allows this for zk-SNARKs based on the kind of polynomial commitments used in Pinocchio.

5   Secure/Correct Adaptive Function Evaluation

In this section, we sketch how our zk-SNARK can be used to perform "adaptive function evaluation": privacy-preserving verifiable computation on committed data. We consider a setting in which multiple mutually distrusting *data owners* want to allow privacy-preserving outsourced computations on their joint data. A *client* asks a computation to be performed on this data by a set of *workers*. The input data is sensitive, so the workers should not learn what data they are computing on (assuming up to a maximum number of workers are passively corrupted). On the other hand, the client wants to be guaranteed the computation result is correct, for instance, with respect to a commitment to the data published by the data owner (making no assumption on which data owners and/or workers are actively corrupted). The difference in assumptions for the privacy and correctness guarantees is motivated by settings where data owners together choose the computation infrastructure (so they feel active corruption is unlikely) but need to convince an external client (e.g. a medical reviewer) of correctness. We work in the CRS model, where a trusted party (who is otherwise not involved in the system) performs one-time key generation.

In [14], we provide a precise security model that captures the above security guarantees by ideal functionalities. We define two ideal functionalities. The first guarantees privacy *and* correctness (*secure adaptive function evaluation*), and is realised by our construction if at most a threshold of workers are passively corrupted (but all other parties can be actively corrupted). The second guarantees only correctness (*correct adaptive function evaluation*), and is realised by our construction regardless of corruptions.

5.1   Our Construction

We now present our general construction based on multi-party computation and *any* adaptive zk-SNARK (as we will see later, our adaptive zk-SNARK gives a particularly efficient instantiation). At a high level, to achieve secure adaptive function evaluation, the workers compute the function using multi-party computation (MPC), guaranteeing privacy and correctness under certain conditions. However, when these conditions are not met, we still want to achieve correct adaptive function evaluation, i.e., we still want to ensure a correct computation result. To achieve this, the workers also produce, using MPC, a zk-SNARK proof of correctness of the result.

We require a MPC protocol in the outsourcing setting, i.e., with separate inputters (in our case, the data owners and the client), recipients (the client) and workers. The protocol needs to be reactive, so that the data owners can provide their input before knowing the function to be computed[6]; and secure even if any number of data owners the client are actively corrupted. Security of the MPC protocol will generally depend on how many workers are corrupted; our construction will realise secure adaptive function evaluation (as opposed to just correct adaptive function evaluation) exactly when the underlying MPC protocol is secure. (As we show below, MPC protocols based on (t, n)-Shamir secret sharing (e.g., [5]) between $n = 2t + 1$ workers satisfy these requirements.)

Protocol Adaptive Trinocchio

(Data provider has $a_i \in \mathbb{F}^d$; client has $a_c \in \mathbb{F}^{d'}$, function $f : (\mathbb{F}^d)^n \times \mathbb{F}^{d'} \to \mathbb{F}^{d-d'}$.)

1. The trusted party generates a system-wide CRS crs of the trapdoor commitment family, and commitment keys $ck_1, \ldots, ck_n, ck_c$ for the data owner and client. This material is distributed to all parties.
2. Each data owner computes commitment $c_i = C_{ck_i}(a_i, r_i)$ to its input $a_i \in \mathbb{F}^d$ using randomness r_i, and publishes it on a bulletin board.
3. The data owners, workers, and client use the MPC protocol to do the following:
   - Each data owner provides input a_i and randomness r_i
   - For each i, compute $c'_i = C_{ck_i}(\llbracket a_i \rrbracket, \llbracket r_i \rrbracket)$; if $c_i \neq c'_i$ then abort
4. The client provides function f to the trusted party. The trusted party determines a QAP Q computing f and a function f' solving Q, and performs key generation of the adaptive zk-SNARK (where one commitment combines the client's input and output). The client gets verification key crsv; the workers get Q, f', and the corresponding evaluation key crsp.
5. The data owners, workers, and client continue with the MPC from step 3:
   - Client: provide input a_c
   - Compute $(\llbracket b \rrbracket; \llbracket w \rrbracket) \leftarrow f'(\llbracket a_1 \rrbracket; \ldots; \llbracket a_n \rrbracket; \llbracket a_c \rrbracket)$
   - Compute $\llbracket c_c \rrbracket \leftarrow C_{ck_c}(\llbracket a_c \rrbracket, \llbracket b \rrbracket; \llbracket r_c \rrbracket)$ for random r_c
   - Compute $\llbracket \pi \rrbracket \leftarrow P(crs, \{ck_i\}, \ldots, ck_n, ck_c, crsp; c_1, \ldots, c_n, \llbracket c_c \rrbracket; \llbracket a_1 \rrbracket; \ldots; \llbracket a_n \rrbracket; \llbracket a_c \rrbracket, \llbracket b \rrbracket; \llbracket r_1 \rrbracket, \ldots, \llbracket r_n \rrbracket, \llbracket r_c \rrbracket; \llbracket w \rrbracket)$
   - Open outputs $\llbracket b \rrbracket, \llbracket r_c \rrbracket, \llbracket c_c \rrbracket, \llbracket \pi \rrbracket$ to the client
6. The client checks whether $V(crs, ck_1, \ldots, ck_n, ck_c, crsv; c_1, \ldots, c_n, c_c; \pi) = 1$ and $c_c = C_{ck_c}(a_c, b; r_c)$ and if so, returns computation result b.

Fig. 2. The adaptive Trinocchio protocol

Our protocol is shown in Fig. 2. It uses an MPC protocol with the above properties, a trapdoor commitment family, and an adaptive zk-SNARK, instantiated for the function to be computed. The protocol relies on a trusted party that generates the key material of the zk-SNARK, but is otherwise not involved in the computation. Each data owner has an input $a_i \in \mathbb{F}^d$ and the client has an

---

[6] Using non-reactive MPC requires is also possible, but then steps 3 and 4 of the protocol need to be swapped. As a consequence, data owners can abort based on the client's choice of function, leading to a weaker form of correct function evaluation.

input $a_c \in \mathbb{F}^{d'}$ and a function $f : (\mathbb{F}^d)^n \times \mathbb{F}^{d'} \to \mathbb{F}^{d-d'}$ that it wants to compute on the combined data. Internal variables of the MPC protocol are denoted $[\![\cdot]\!]$.

In step 1, the trusted party sets up the trapdoor commitment family, generating separate keys for data providers and the client. (This prevents parties from copying each other's input.) In step 2, each data provider publishes a commitment to his input. In step 3, each data providers inputs its data and the randomness used for the commitment to the MPC protocol. The workers re-compute the commitments based on this opening and abort in case of a mismatch. (This prevents calling P on mismatching inputs in which case it may not be zero-knowledge.) In step 4, the client chooses the function f to be computed, based on which the trusted party performs key generation. (By doing this after the data owners' inputs, we prevent a selective failure attack from their side.) In step 5, the computation is performed. Using MPC, the client's output and witness are computed; a commitment to the client's I/O is produced, and a zk-SNARK proof of correctness with respect to the commitments of the data owners and client is built.[7] The client learns the output, randomness for its commitment, the commitment itself, and the proof. In step 6, the client re-computes the commitment and verifies the proof; in case of success, it accepts the output.

By sharing commitments between proofs, it is possible to generate key material for a number of small building blocks once, and later flexibly combine them into larger computations without requiring new key material. In particular, as we show in the case study, this enables computations on arbitrary-length data using the same key material (which was impossible in Trinocchio). It is also easy to support multiple clients or multiple commitments per data owner.

In [14], we show that indeed, the above construction achieves the formal definitions of secure adaptive function evaluation (under the same conditions of the corruptions of workers as the underlying MPC protocol) and correct adaptive function evaluation (regardless of corruptions).

5.2   Efficient Instantiation Using Secret Sharing and Our zk-SNARK

We now show that our zk-SNARKs and MPC based on Shamir secret sharing give a particularly efficient instantiation of the above framework. The idea is the same as for Trinocchio [13]: our zk-SNARK is essentially an arithmetic circuit of multiplicative depth 1, so given a solution to the QAP, the prover algorithm can be performed under MPC without any communication between the workers.

In more detail, we perform MPC based on Shamir secret sharing between the m workers (e.g., [5]). This guarantees privacy as long as at most t workers are passively corrupted, where $m = 2t + 1$. Inputs are provided by the inputters as an additive sharing between all workers: this way actively corrupted inputters cannot provide an inconsistent sharing. The workers Shamir-share and sum up the additive shares to obtain a Shamir sharing of the input. Outputs are provided

---

[7] Equivalently, the workers can open c_c and π and send them to the client in the plain.

to recipients either as Shamir shares or as freshly randomised additive shares: the latter allows producing our zk-SNARK proof without any communication.

Either of our zk-SNARK constructions can be used; we provide details for the first one. Below, write $[\![\cdot]\!]$ for Shamir sharing and $[\cdot]$ for additive sharing. (Note that Shamir sharings can be converted locally to additive sharings at no cost.) In step 3 of the protocol, to open c_i', the parties apply $\mathsf{C}_{\mathsf{ck}_i}$ on their additive shares of the input and randomness, add a random additive sharing of zero (which can be generated non-interactively using pseudo-random zero sharing), and reveal the result. In step 5, $[\![b]\!]$; $[\![w]\!]$ are computed as Shamir secret shares. Next, $[c_c]$ is computed as an additive sharing by applying $\mathsf{C}_{\mathsf{ck}_c}$ on additive shares and adding a random sharing of zero. Next, P^1 is applied by performing the following steps:

- Generate $\delta_{v,i}, \delta_{w,i}, \delta_{y,i}$ by pseudo-random secret sharing.
- Compute $[h] = ((\sum_j [\![x_j]\!] \cdot v_j(x) + [\![\delta_v]\!] \cdot t(x)) \cdot (\sum_j [\![x_j]\!] \cdot w_j(x) + [\![\delta_w]\!] \cdot t(x)) - (\sum_j [\![x_j]\!] \cdot y_j(x) + [\![\delta_y]\!] \cdot t(x)))/t(x)$. Essentially this is done by performing the computation straight on Shamir secret shares; because there is only layer of multiplications of shares, this directly gives an additive sharing of the result. Smart use of FFTs gives time complexity $\mathcal{O}(d \cdot \log d)$ [2,13].
- All proof elements are now linear combinations of secret-shared data; compute them by taking linear combinations of the (Shamir or additive) shares and adding a random sharing of zero.

What remains is how to compute the solution of the QAP using multi-party computation. Namely, in addition to computing the function result $[\![b]\!]$, the MPC also needs to compute witness $[\![w]\!]$ to the QAP. Actually, if the function to be computed is described as an arithmetic circuit, this is very easy. Namely, in this case, the witness for the natural QAP for the function is exactly the vector of results of all intermediate multiplications; and these results are already available as Shamir secret shares as a by-product of performing the MPC. Hence, in this case, computing $[\![w]\!]$ in addition to $[\![b]\!]$ incurs no overhead.

If a custom MPC protocol for a particular subtask is used, then it is necessary to devise specific QAP equations and an MPC protocol to compute their witness. As an example, consider the MPC operation $[\![b]\!] \leftarrow [\![a \neq 0]\!]$, i.e., b is assigned 1 if $a \neq 0$ and 0 if $a = 0$. For computing $[\![b]\!]$, a fairly complex protocol is needed, cf. [5]. However, proving that b is correct using a QAP is simple [11]: introduce witnesses $c := (a + (1 - b))^{-1}$, $d := 1$ and equations:

$$a \cdot c = b \qquad a \cdot (d - b) = 0 \qquad d \cdot d = d.$$

Indeed, if $a = 0$ then the first equation implies that $b = 0$; if $a \neq 0$ then the second and third equations imply that $b = 1$. In both cases, the given value for c and $d = 1$ make all three equations hold. In our case study, we show similarly how, for complex MPC protocols for fixed-point arithmetic, simple QAPs proving correctness exist with easily computable witnesses.

6 Prototype and Distributed Medical Research Case

In this section, we present a proof-of-concept implementation of our second zk-SNARK construction and our Adaptive Trinocchio protocol. Computations can be specified at a high level using a Python frontend; executed either locally or in a privacy-preserving way using multi-party computation; and then automatically proven and verified to be correct by a C++ backend. We show how two different computations can be performed on the same committed data coming from multiple data owners (with key material independent from input length, and optionally in a privacy-preserving way): aggregate survival statistics on two patient populations, and the "logrank test": a common statistical test whether there is a statistically significant difference survival rate between the populations.

6.1 Prototype of Our zk-SNARK and Adaptive Trinocchio

Our prototype, available at https://github.com/meilof/geppetri, is built on top of VIFF, a Python implementation of MPC based on Shamir secret sharing. In VIFF, computations on secret shares are specified as normal computations by means of operator overloading, e.g., assigning `c=a*b` induces a MPC multiplication protocol. We add a new runtime to VIFF that also allows computations to be performed locally without MPC.

To support computation proofs, we developed the `viffvc` library that provides a new data type: `VcShare`, a wrapper around a secret share. Each `VcShare` represents a linear combination of QAP variables. Addition and multiplication by constants of `VcShares` is performed locally by manipulating the linear combination. Constants v are represented as $v \cdot one$, where witness one satisfies $one \cdot one = one$ so $one = 1$. When two `VcShares` $\lambda_1 x_1 + \ldots$ and $\mu_1 x_1 + \ldots$ are multiplied, a local or MPC multiplication operation is performed on the underlying data, and the result is a new `VcShare` x_k wrapping the result as a new QAP variable. QAP equation $(\lambda_1 x_1 + \ldots) \cdot (\mu_1 x_1 + \ldots) = 1 \cdot x_k$ is written to a file, and the multiplication result x_k or its secret share, when known, is written to another file. Apart from multiplication, some additional operations are supported. For the $[\![b]\!] \leftarrow [\![a \neq 0]\!]$ operation discussed in Sect. 5.2, the implementation computes $[\![b]\!]$ and $[\![c]\!] = ([\![a]\!] + (1 - [\![b]\!]))^{-1}$, and writes these secret shares/values and the equations from Sect. 5.2 to the respective files. We also support secret indexing (e.g., [5]), and fixed-point computations as discussed below.

Computations are performed by this custom VIFF-based system together with an implementation of our zk-SNARK. A first tool, `qapgen`, generates the CRS for our trapdoor commitment scheme. A second tool, `qapinput`, builds a commitment to a given input; and computes secret shares of these inputs that are used for MPC computations. Then, our Python implementation is used to compute the function, either locally or using multi-party computation. At the end of this execution, there is one file with the QAP equations, and one file with values/shares for each QAP variable. Our `qapgenf` tool uses the first file to perform key generation of the QAP (this is done only once and for next executions, previous key material is re-used). Our `qapprove` tool uses the second file to generate the zk-SNARK proof (shares) to be received by the client. Finally, a `qapver` tool verifies the proof based on the committed inputs and outputs.

Algorithm 1. Anonymized survival data computation

Require: $[\![d_1]\!], [\![n_1]\!], [\![d_2]\!], [\![n_2]\!]$: block of survival data points for two populations
Ensure: $([\![d'_1]\!], [\![n'_1]\!], [\![d'_2]\!], [\![n'_2]\!])$ aggregated survival data for the block
 1: **function** SUMM($[\![d_{i,1}]\!], [\![d_{i,2}]\!], [\![n_{i,1}]\!], [\![n_{i,2}]\!]$)
 2:     **return** $(\sum_i [\![d_{1,i}]\!], [\![n_{1,1}]\!], \sum_i [\![d_{2,i}]\!], [\![n_{2,1}]\!])$

6.2 Application to Medical Survival Analysis

We have applied our prototype to (adaptively) perform computations on survival data about two patient populations. In medical research, survival data about a population is a set of tuples (n_j, d_j), where n_j is the number of patients still in the study just before time j and d_j is the number of deaths at time j. We assume both populations are distributed among multiple hospitals, that each commit to their contributions $(d_{j,1}, n_{j,1}, d_{j,2}, n_{j,2})$ to the two populations at each time.

Aggregate Survival Data. Our first computation is to compute an aggregate version of the survival data, where each block $\{d_{j,1}, n_{j,1}, d_{j,2}, n_{j,2}\}_{j=1}^{25}$ of 25 time points is summarised as $(\sum_j d_{j,1}, n_{1,1}, \sum_j d_{j,2}, n_{1,2})$. The function SUMM computing this summary is shown in Algorithm 1. Function SUMM translates into a QAP on 26 commitments: as input, for each time point j, a commitment $\sum_i c_{i,j}$ to the combined survival data ($[\![d_{i,1}]\!], [\![n_{i,1}]\!], [\![d_{i,2}]\!], [\![n_{i,2}]\!]$) from the different hospitals i at that time (using the fact that commitments are homomorphic); as output, a commitment to ($[\![d'_1]\!], [\![n'_1]\!], [\![d'_2]\!], [\![n'_2]\!]$).

Logrank Test. Our second computation is the so-called "Mantel-Haenzel logrank test", a statistical test to decide whether there is a significant difference in survival rate between the two populations (as implemented, e.g., in R's `survdiff` function). Given the survival data from two populations, define:

$$E_{j,1} = \frac{(d_{j,1} + d_{j,2}) \cdot n_{j,1}}{n_{j,1} + n_{j,2}}; \quad V_j = \frac{n_{j,1} n_{j,2} (d_{j,1} + d_{j,2})(n_{j,1} + n_{j,2} - d_{j,1} - d_{j,2})}{(n_{j,1} + n_{j,2})^2 \cdot (n_{j,1} + n_{j,2} - 1)};$$

$$X = \frac{\sum_j E_{j,1} - \sum_j d_{j,1}}{\sum_j V_j}.$$

The null hypothesis for the logrank test, i.e., the hypothesis that the two curves represent the same underlying "survival function", corresponds to $X \sim \chi_1^2$. This null hypothesis is rejected (i.e., the curves are different) if $1 - \mathrm{cdf}(X) > \alpha$, where cdf is the cumulative density function of the χ_1^2 distribution and, e.g., $\alpha = 0.05$. We use MPC to compute X, and then apply the cdf in the clear.

Our implementation consists of two different functions: a function BLOCK (Algorithm 2) that computes $(E_{j,1}, V_j, d_{j,1})$ given the survival data at point j; and a function FIN that, given $\sum E_{j,1}$, $\sum V_j$, and $\sum d_{j,1}$ computes X (Algorithm 3). As above, function BLOCK is applied to commitment $\sum_i c_{i,j}$ to the combined survival data from different hospitals at a particular time, giving output commitment c'_j. Function FIN is applied to commitment $\sum_j c'_j$ to

---

Algorithm 2. Logrank computation for each'time step

Require: $[\![\mathbf{d}_{i,1}]\!], [\![\mathbf{d}_{i,2}]\!], [\![\mathbf{n}_{i,1}]\!], [\![\mathbf{n}_{i,2}]\!]$ survival data at time point i
Ensure: $([\![e_i]\!]^f, [\![v_i]\!]^f, [\![d_i]\!])$ contributions to $\sum_j E_{j,1}, \sum_j V_j, \sum_j d_{j,1}$ for test statistic
1: **function** BLOCK($[\![\mathbf{d}_{i,1}]\!], [\![\mathbf{d}_{i,2}]\!], [\![\mathbf{n}_{i,1}]\!], [\![\mathbf{n}_{i,2}]\!]$)
2:     $[\![ac]\!] \leftarrow [\![\mathbf{d}_{i,1}]\!] + [\![\mathbf{d}_{i,2}]\!]$
3:     $[\![bd]\!] \leftarrow [\![\mathbf{n}_{i,1}]\!] + [\![\mathbf{n}_{i,2}]\!]$
4:     $[\![frc]\!]^f \leftarrow [\![ac]\!]/[\![bd]\!]$
5:     $[\![e_i]\!]^f \leftarrow [\![frc]\!]^f \cdot [\![\mathbf{n}_{i,1}]\!]$
6:     $[\![vn]\!] \leftarrow [\![\mathbf{n}_{i,1}]\!] \cdot [\![\mathbf{n}_{i,2}]\!] \cdot [\![ac]\!] \cdot ([\![bd]\!] - [\![ac]\!])$
7:     $[\![vd]\!] \leftarrow [\![bd]\!] \cdot [\![bd]\!] \cdot ([\![bd]\!] - 1)$
8:     $[\![v_i]\!]^f \leftarrow [\![vn]\!]/[\![vd]\!]$
9:     **return** $([\![e_i]\!]^f, [\![v_i]\!]^f, [\![d_i]\!])$

Algorithm 3. Logrank final computation

Require: $[\![es]\!], [\![vs]\!], [\![ds]\!]$: summed-up values required to compute X
Ensure: $[\![chi]\!]^f$ test statistic comparing two curves; supposedly $chi \sim \chi_1^2$
1: **function** FIN($[\![es]\!], [\![vs]\!], [\![ds]\!]$)
2:     $[\![ds]\!]^f \leftarrow [\![ds]\!] \ll \text{PRECISION}$
3:     $[\![dmi]\!]^f \leftarrow [\![ds]\!]^f - [\![vs]\!]^f$
4:     $[\![chi0]\!]^f \leftarrow [\![dmi]\!]^f/[\![vs]\!]^f$
5:     $[\![chi]\!]^f \leftarrow [\![chi0]\!]^f \cdot [\![dmi]\!]^f$
6:     **return** $[\![chi]\!]^f$

$(\sum E_{j,1}, \sum V_j, \sum d_{j,1})$, again using the fact that commitments are homomorphic; outputting a commitment to X that is output to the client.

Algorithms 2 and 3 use fixed-point numbers $[\![x]\!]^f$, representing value $x \cdot 2^{-k}$ where we use precision $k = 20$. We use the fixed-point multiplication $[\![c]\!]^f \leftarrow [\![a]\!]^f \cdot [\![b]\!]^f$ and division $[\![c]\!]^f \leftarrow [\![a]\!]/[\![b]\!]$, $[\![c]\!]^f \leftarrow [\![a]\!]^f/[\![b]\!]^f$ protocols due to [5]. To prove that $[\![c]\!]^f \leftarrow [\![a]\!]^f \cdot [\![b]\!]^f$ is correct, note that we need to show that $2^k c - a \cdot b \in [-2^k, 2^k]$, or equivalently, that $\alpha := 2^k c - a \cdot b + 2^k \geq 0$ and $\beta := 2^k - (2^k c - a \cdot b) \geq 0$. We prove this by computing, using MPC, bit decompositions [5] $\alpha = \alpha_0 + \alpha_1 \cdot 2 + \ldots + \alpha_k \cdot 2^k$ and $\beta = \beta_0 + \beta_1 \cdot 2 + \ldots + \beta_k \cdot 2^k$ (indeed, α and β are $\leq k + 1$ bits long); these α_i, β_i are the witnesses to QAP equations:

$$\forall i : \alpha_i \cdot (1 - \alpha_i) = 0 \quad c - a \cdot b + 2^k = \alpha_0 + \alpha_1 \cdot 2 + \ldots + \alpha_k \cdot 2^k$$

$$\forall i : \beta_i \cdot (1 - \beta_i) = 0 \quad \beta = 2^k - (c - a \cdot b) = \beta_0 + \beta_1 \cdot 2 + \ldots + \beta_k \cdot 2^k.$$

Similarly, note that $[\![c]\!]^f \leftarrow [\![a]\!]^f/[\![b]\!]^f$ is correct if and only if $2^k a - b \cdot c \in [-b, b]$, i.e., $\gamma := b + 2^k a - b \cdot c \geq 0$ and $\delta := b - (2^k a - b \cdot c) \geq 0$. If b has bitlength at most K (i.e., the represented number has absolute value $\leq 2^{K-k}$), then γ and δ have at most $K + 1$ bits. As above, we prove correctness by determining $(K+1)$-length bit decompositions of γ and δ and proving them correct. Proving correctness of $[\![c]\!]^f \leftarrow [\![a]\!]/[\![b]\!]$ is analogous.

Performance. Table 2 shows the performance of our proof-of-concept implementation for computing aggregate survival data and the logrank test (on a modern

Table 2. Performance: computation/proving/verification; with/without MPC

Aggregate	Computation (function): 0.0 s (w/o MPC)/0.1 s (w/MPC)		
	Computation (function+witness): 0.0 s (w/o MPC)/0.1 s (w/MPC)		
	BS=1	BS=25	BS=175
	QAP degree: 3	QAP degree: 3	QAP degree: 57
	Prover: 0.3 s/0.4 s	Prover: 0.1 s/0.1 s	Prover: 0.0 s/0.0 s
	Verifier: 1.2 s/1.5 s	Verifier: 0.2 s/0.2 s	Verifier: 0.0 s/0.0 s
Logrank	Computation (function): 0.2 s (w/o MPC) / 190.5 s (w/MPC)		
	Computation (function+witness): 0.6 s (w/o MPC)/235.2 s (w/MPC)		
	BS=1	BS=25	BS=175
	QAP deg (block): 173	QAP deg (block): 4304	QAP deg (block): 30104
	QAP deg (fin): 85	QAP deg (fin): 85	QAP deg (fin): 85
	Prover: 13.9 s/78.5 s	Prover: 16.2 s/81.0 s	Prover: 9.8 s/73.5 s
	Verifier: 3.9 s/4.9 s	Verifier: 0.2 s/0.3 s	Verifier: 0.0 s/0.0 s

laptop). As input, we used the "btrial" data set included in R's "kmsurv" package (on which we indeed reproduce R's `survdiff` result) of 175 data points. Apart from having one data point per commitment, we also experiment with having a "block size" of 25 or 175 data points. For the logrank test, we use one QAP per block; larger blocks mean less work for the verifier (since there are fewer proofs) but, in theory, more work for the prover (since the proving algorithm is superlinear in the QAP size). For aggregation, we use one QAP per 25 data points or per commitment, whichever is more.

We time the performance of running the computation, producing the proof, and verifying it, with or without MPC. As expected, MPC induces a large overhead for the computation, especially for the logrank test (due to the many fixed-point computations). MPC also incurs an overhead for proving: this is because of the many exponentiations with $|\mathbb{F}|$-sized secret shares rather than small witnesses. Note that proving is faster than computing with MPC: the underlying operations are slower [13], but the QAP proof is in effect on a verification circuit that is smaller than the circuit of the computation itself [6]. Proving is faster for block size 175 than block size 25, which is unexpected; this may be because our FFT subroutine rounds up QAP degrees to the nearest power of two, which is favourable in the 175-sized case but not in the 25-sized case. As expected, verification is faster for larger block sizes. (The overhead of MPC here is due to recombing the proof shares into one overall proof.)

7 Conclusion

In this work, we have given the first practical Pinocchio-based adaptive zk-SNARK; applied it in the privacy-preserving outsourcing setting; and presented a proof-of-concept implementation. We mention a few promising directions for follow-ups. Concerning our construction for making Pinocchio adaptive, it would be interesting to see if it can be applied to make recent, even more efficient zk-SNARKS (e.g., [9] in the generic group model) adaptive as well. Moreover, apart from providing a non-adaptive zk-SNARK, Geppetto also introduces the interesting idea of proof bootstrapping, where the verification procedure of the zk-SNARK itself can be performed by means of a verifiable computation, so multiple related proofs can be verified in constant time. Applying this technique in our setting should combine our flexibility with their constant-time verification.

Concerning our privacy-preserving outsourcing framework, it is interesting to see if, apart from secret sharing plus our SNARK, other appealing instantiations are possible. Also, the combination of MPC and verifiable computation raises the challenge to construct efficient QAPs for specific operations and build efficient MPC protocols for computing their witnesses. We have presented zero testing and fixed-point computations as examples, but the same idea is applicable to many other operations as well. More generally, extending our zk-SNARK prototype with more basic operations, and improving its user-friendliness, would help bring the techniques closer to practice.

Acknowledgements. This work is part of projects that have received funding from the European Union's Horizon 2020 research and innovation programme under grant agreement No. 643964 (SUPERCLOUD) and No. 731583 (SODA).

References

1. Backes, M., Barbosa, M., Fiore, D., Reischuk, R.M.: ADSNARK: nearly practical and privacy-preserving proofs on authenticated data. In: Proceedings S&P (2015)
2. Ben-Sasson, E., Chiesa, A., Genkin, D., Tromer, E., Virza, M.: SNARKs for C: verifying program executions succinctly and in zero knowledge. In: Canetti, R., Garay, J.A. (eds.) CRYPTO 2013. LNCS, vol. 8043, pp. 90–108. Springer, Heidelberg (2013). doi:10.1007/978-3-642-40084-1_6
3. Costello, C., Fournet, C., Howell, J., Kohlweiss, M., Kreuter, B., Naehrig, M., Parno, B., Zahur, S.: Geppetto: versatile verifiable computation. In: Proceedings S&P, pp. 253–270 (2015)
4. Danezis, G., Fournet, C., Groth, J., Kohlweiss, M.: Square span programs with applications to succinct NIZK arguments. In: Sarkar, P., Iwata, T. (eds.) ASIACRYPT 2014. LNCS, vol. 8873, pp. 532–550. Springer, Heidelberg (2014). doi:10.1007/978-3-662-45611-8_28
5. de Hoogh, S.: Design of large scale applications of secure multiparty computation: secure linear programming. Ph.D. thesis, Eindhoven University of Technology (2012)
6. de Hoogh, S., Schoenmakers, B., Veeningen, M.: Certificate validation in secure computation and its use in verifiable linear programming. In: Pointcheval, D., Nitaj, A., Rachidi, T. (eds.) AFRICACRYPT 2016. LNCS, vol. 9646, pp. 265–284. Springer, Cham (2016). doi:10.1007/978-3-319-31517-1_14
7. Fiore, D., Fournet, C., Ghosh, E., Kohlweiss, M., Ohrimenko, O., Parno, B.: Hash first, argue later: adaptive verifiable computations on outsourced data. In: Proceedings CCS (2016)

8. Gennaro, R., Gentry, C., Parno, B., Raykova, M.: Quadratic span programs and succinct NIZKs without PCPs. In: Johansson, T., Nguyen, P.Q. (eds.) EURO-CRYPT 2013. LNCS, vol. 7881, pp. 626–645. Springer, Heidelberg (2013). doi:10. 1007/978-3-642-38348-9_37

9. Groth, J.: On the size of pairing-based non-interactive arguments. In: Fischlin, M., Coron, J.-S. (eds.) EUROCRYPT 2016. LNCS, vol. 9666, pp. 305–326. Springer, Heidelberg (2016). doi:10.1007/978-3-662-49896-5_11

10. Lipmaa, H.: Prover-efficient commit-and-prove zero-knowledge SNARKs. In: Pointcheval, D., Nitaj, A., Rachidi, T. (eds.) AFRICACRYPT 2016. LNCS, vol. 9646, pp. 185–206. Springer, Cham (2016). doi:10.1007/978-3-319-31517-1_10

11. Parno, B., Howell, J., Gentry, C., Raykova, M.: Pinocchio: nearly practical verifiable computation. In: Proceedings S&P, pp. 238–252 (2013)

12. Parno, B.: A note on the unsoundness of vntinyram's snark. Cryptology ePrint Archive, Report 2015/437 (2015). http://eprint.iacr.org/

13. Schoenmakers, B., Veeningen, M., Vreede, N.: Trinocchio: privacy-preserving outsourcing by distributed verifiable computation. In: Manulis, M., Sadeghi, A.-R., Schneider, S. (eds.) ACNS 2016. LNCS, vol. 9696, pp. 346–366. Springer, Cham (2016). doi:10.1007/978-3-319-39555-5_19

14. Veeningen, M.: Pinocchio-based adaptive zk-SNARKS and secure/correct adaptive function evaluation. Cryptology ePrint Archive, Report 2017/013 (2017). http://eprint.iacr.org/2017/013

Revisiting and Extending the AONT-RS Scheme: A Robust Computationally Secure Secret Sharing Scheme

Liqun Chen[1], Thalia M. Laing[2](✉), and Keith M. Martin[2]

[1] University of Surrey, Guildford, UK
liqun.chen@surrey.ac.uk
[2] Information Security Group, Royal Holloway,
University of London, Egham, UK
{thalia.laing,keith.martin}@rhul.ac.uk

Abstract. In 2010, Resch and Plank proposed a computationally secure secret sharing scheme, called AONT-RS. We present a generalisation of their scheme and discuss two ways in which information is leaked if used to distribute small ciphertexts. We discuss how to prevent such leakage and provide a proof of computational privacy in the random oracle model. Next, we extend the scheme to be robust and prove the robust AONT-RS achieves computational privacy in the random oracle model and computational recoverability under standard assumptions. Finally, we compare the security, share size and complexity of the AONT-RS scheme with Krawczyk's SSMS scheme.

1 Introduction

A (t, n)−threshold secret sharing scheme describes how to distribute data amongst n servers such that t are required to collaborate in order to reconstruct the data. Any fewer than t servers learn nothing about the data. Secret sharing is useful in distributed storage systems where data is stored across multiple servers. A user wishing to access the data must access a threshold number of servers and combine the retrieved shares. This enables greater availability, adds redundancy and offers security without the reliance on cryptographic keys.

In 2010, Resch and Plank proposed a dispersed storage system, called AONT-RS [14], which blends an all-or-nothing transform (AONT) [15] with Reed-Solomon (RS) coding [13]. The result is a computationally secure (t, n)−threshold secret sharing scheme. The AONT-RS is a feature in the object storage system sold by Cleversafe, a company recently acquired by IBM [7], who renamed the product to IBM Cloud Object Storage. In 2016 the system was rated the overall leader in the Gartner Critical Capabilities for Object Storage Report [4].

We present a generalised version of the AONT-RS that enables users the flexibility to choose a block cipher mode of operation and an Information Dispersal Algorithm (IDA). We specify (previously undefined) security properties

© Springer International Publishing AG 2017
M. Joye and A. Nitaj (Eds.): AFRICACRYPT 2017, LNCS 10239, pp. 40–57, 2017.
DOI: 10.1007/978-3-319-57339-7_3

the building blocks of our generalised AONT-RS must satisfy and discuss information leakage and prevention.

Resch and Plank claim the AONT-RS has integrity because they use a canary, which enables an authorised user to confirm whether or not the correct data has been recovered. However, the submission of an incorrect share will prevent the correct data from being recovered. Although the user knows the recovered data is wrong, they are unable to recover the correct data. To address this, we remove the canary and extend the scheme to be robust by using commitment schemes, as in [16]. This ensures that, even if a bounded number of servers submitted false shares, the original data will be uniquely recovered.

Resch and Plank claim the AONT-RS achieves computational security but no thorough security analysis is provided. We prove both the generalised and robust AONT-RS achieve computational privacy in the random oracle (RO) model, then prove the robust AONT-RS achieves computational recoverability under standard assumptions.

We then compare the generalised AONT-RS with Krawczyk's secret sharing made short scheme (SSMS) [9]. This comparison is applicable to the robust AONT-RS with a robust extension of SSMS by Bellare and Rogaway, called HK2 [16]. In our comparison, we consider the security, the share size and the number of bitwise XORs required to distribute and recover data via both schemes.

Related Work. Shamir and Blakely independently introduced secret sharing schemes in 1979 [2,17]. Shamir's scheme is ideal and perfectly secure.

In 1994, Krawczyk published a paper focusing on computationally secure schemes (CSS) in the non-robust setting [9] and proposed his SSMS scheme. Krawczky also proposed goals for a robust CSS scheme, along with a candidate solution. Previously, the CSS goal had been mentioned by Karnin et al. [8]. Prior to Krawczyk's work, robustness had only be studied in the information-theoretic setting in [11,18]. Krawczyk's motivation was to achieve shares smaller than were possible in perfectly secure secret sharing schemes [8].

Krawczyk's work was revisited in 2007 by Bellare and Rogaway [16], in which they proposed formal definitions for a CSS and proved Krawczyk's robust scheme to be secure in the RO model. They then proposed a refined version of Krawczyk's scheme (called HK2), which achieves the robust CSS goals under standard assumptions. Since Bellare and Rogaway's work, based on our best knowledge, there have been no new solutions for robust CSS schemes until Resch and Plank's AONT-RS scheme in 2011 [14], which is studied in detail here.

Contributions. Our contribution can be summarised as follows:

- We present a generalised version of the AONT-RS and highlight the (previously undefined) security properties each element of the scheme must have.
- We discuss and illustrate two examples of information leakage in both the original and our AONT-RS scheme and discuss how to prevent this.
- We prove the AONT-RS achieves computational privacy in the RO model.

- We extend AONT-RS to be robust and provide a proof of computational privacy in the RO model and recoverability under standard assumptions.
- We compare the generalised AONT-RS scheme with Krawczyk's SSMS. Our comparison is applicable to the robust AONT-RS and HK2.

Organisation. In Sect. 2 we present notation and definitions. In Sect. 3 we present a generalised version of the AONT-RS and discuss information leakage. We then prove the generalised AONT-RS achieves computational privacy in the RO model. In Sect. 4 we extend the AONT-RS to be robust and prove computational privacy in the RO model and recoverability under standard assumptions. In Sect. 5 we introduce Krawczyk's SSMS and compare it with the generalised AONT-RS. We conclude in Sect. 6.

2   Preliminaries

In this section we introduce the definitions and notation used throughout.

2.1   Secret Sharing Schemes

Definition 1. *Let $n, t \in \mathbb{N}$ with $2 \leq t \leq n$ and let $\mathcal{P} = \{P_1, \ldots, P_n\}$ be a set of n players. A $(t, n)-$secret sharing scheme Π consists of two algorithms: Share and Recover. Share is probabilistic and takes as input a secret s chosen from a secret space $\mathcal{S}$ and outputs an $n-$vector $\boldsymbol{S}$. Player P_i receives the share $\boldsymbol{S}[i]$. Recover is deterministic and takes as input shares and outputs some $s' \in \{\mathcal{S} \cup \perp\}$. The secret s should be recoverable by any set of at least t players, and private, meaning any fewer than t players (called unauthorised sets) are unable to recover s.*

A $(t, n)-$secret sharing scheme can either have *perfect* or *computational* security. These security models can be defined by two security games [16] as in Fig. 1: one defining privacy and the other recoverability. Note that if an algorithm A is deterministic, we write $x \leftarrow A(\cdot)$. If the algorithm is probabilistic, then $x \xleftarrow{\$} A(\cdot)$ means to choose x according to the distribution induced by A.

In the privacy game *Priv*, for parameters t and n, the challenger chooses a bit b. The adversary chooses secrets $s_0, s_1 \in \mathcal{S}$ and sends them to the challenger, who inputs s_b to *Share*, which outputs $\boldsymbol{S}$. The adversary then makes up to $t - 1$ $Corrupt(i)$ queries for $1 \leq i \leq n$ and receives the share $\boldsymbol{S}[i]$. The adversary then outputs a guess b' for b and wins if $b' = b$.

Let A be an adversary playing the *Priv* game against a secret sharing scheme Π. Call A a *privacy adversary*. Let $\Pr[Priv^A]$ denote the probability A outputs the correct guess $b' = b$ during finalise and define the *advantage* of A as

$$\mathbf{Adv}_{\Pi}^{Priv}(A) = 2 \cdot \Pr[Priv^A] - 1. \tag{1}$$

The recoverability game models an adversary's ability to prevent the recovery of s by deleting or altering shares. The set T denotes the players the adversary

GAME *Priv*

- PROCEDURE *Initialise*(t, n)
  $b \overset{\$}{\leftarrow} \{0, 1\}; \quad j = 1$
- PROCEDURE *Deal*(s_0, s_1)
  If $s_0, s_1 \notin S$
    Return $\perp$
  Else $S \overset{\$}{\leftarrow} Share(s_b)$
- PROCEDURE *Corrupt*(i)
  If $j \leq t - 1$
    Return $S[i]; \quad j = j + 1$
  Else halt.
- PROCEDURE *Finalise*(b')
  Return $b' = b$

GAME *Rec*

- PROCEDURE *Initialise*(t, n)
  $T \leftarrow \emptyset; \quad j = 1$
- PROCEDURE *Deal*(s)
  If $s \notin S$
    Return $\perp$
  Else $S \overset{\$}{\leftarrow} Share(s)$
- PROCEDURE *Corrupt*(i)
  If $j \leq (n - t)$
    Return $S[i]$
    $j = j + 1; \quad T \leftarrow T \cup \{i\}$
  Else halt.
- PROCEDURE *Finalise*(S_T)
  Return $s \neq s' \leftarrow Recover(S_T \cup S_{\overline{T}})$

Fig. 1. Privacy and recoverability games for a $(t, n)-$secret sharing scheme.

corrupts. Let $T = \emptyset$. The adversary chooses and submits a secret $s \in S$ to the challenger, who inputs it to *Share*. The adversary then makes up to $n - t$ queries of the form *Corrupt*(i) for $1 \leq i \leq n$ and receives the share $S[i]$ in return. Each i is noted in T. To finalise, the adversary outputs a partially complete $n-$vector S_T, consisting of at most t altered (and the rest deleted) shares queried during the corrupt procedure. This vector is completed by the challenger filling the remaining t elements with valid shares noted in the vector $S_{\overline{T}}$. The vector $S_T \cup S_{\overline{T}}$ is then submitted to *Recover*. The adversary wins if $s' \neq s$.

Call adversary A playing the game *Rec* a *recoverability adversary*. Let $\Pr[Rec^A]$ denote the probability s is not correctly recovered. Define the advantage of A as

$$\mathbf{Adv}_{\Pi}^{Rec}(A) = \Pr[Rec^A]. \tag{2}$$

Definition 2. *A perfectly secure $(t, n)-$threshold scheme (PSS) is a $(t, n)-$threshold scheme in which a privacy adversary and a recoverability adversary restricted to only deleting shares both have an advantage of zero.*

The size of the share $S[i]$ must be at least the size of the secret [1]; if this bound is met the scheme is *ideal*. This bound can be particularly problematic if s is large or the storage available to each player is small. For the application of AONT-RS, we will focus on distributing large amounts of data. Relaxing the security to be computational can achieve smaller shares.

Definition 3. *A computationally secure $(t, n)-$scheme (CSS) is a $(t, n)-$threshold secret sharing scheme in which a privacy adversary has a negligible advantage and a recoverability adversary restricted to only deleting (and not corrupting) shares has an advantage of zero.*

CSSs are less secure than PSSs but are able to achieve smaller shares. In general, CSSs are sufficient for most applications [9].

In PSS and CSS schemes, the recoverability adversary is limited to only deleting, and not corrupting, shares. A *robust* scheme ensures the recovery of the secret when the recoverability adversary is allowed to both corrupt and delete a (bounded) number of shares.

Definition 4. *A robust, computationally secure (t, n)−secret sharing scheme is a (t, n)−secret sharing scheme in which a privacy adversary and a recoverability adversary both have a negligible advantage at winning their respective games.*

Ramp schemes further relax the security to achieve even smaller shares and are defined information theoretically. Let $\mathcal{S}$ denote the discrete random variable corresponding to the choice of s and let $\mathbf{A}$ denote the discrete random variable corresponding to the set of shares given to the players in the set $A \subseteq \mathcal{P}$.

Definition 5. *A $(t_0, t_1; n)$−ramp scheme distributes a secret s such that any set of at least t_1 players can recover s and a set of t_0 or fewer players reveals no information about the secret. A $(t_0, t_1; n)$−ramp scheme is said to be* linear *if, for any set of players $A \subseteq \mathcal{P}$ such that $|A| = r$, where $t_0 \leq r \leq t_1$,*

$$H(\mathcal{S}|A) = \frac{t_1 - r}{t_1 - t_0} H(\mathcal{S}).\tag{3}$$

Note that in a $(t_0, t_1; n)$−linear ramp scheme, for every player after the initial t_0 players have contributed shares, a fixed amount of information is learnt about s. This continues in a linear fashion until t_1 players have contributed and s is learnt completely. Observe that a (t, n)−PSS is a $(t - 1, t; n)$−ramp scheme.

2.2   Symmetric Key Encryption

Let $\mathcal{E} = (\mathcal{M}, \mathcal{K}, \mathcal{C}, KenGen, Enc, Dec)$ be a symmetric key encryption scheme with message space $\mathcal{M}$, keyspace $\mathcal{K}$, ciphertext space $\mathcal{C}$ and key generation, encryption and decryption algorithms $KeyGen, Enc$ and Dec.

GAME *Ind*

- PROCEDURE *Initialise*
  $k \xleftarrow{\$} KeyGen\{0,1\}^\lambda; \quad b \xleftarrow{\$} \{0,1\}$
- PROCEDURE *Finalise*(b')
  Return $b' = b$

- PROCEDURE *Deal*(M_0, M_1)
  If $M_0 = M_1$, $|M_0| \neq |M_1|$, or $M_0, M_1 \notin \mathcal{M}$
      Return $\perp$
  Else, $C \xleftarrow{\$} Enc_k(M_b)$
  Return C

Fig. 2. Game defining indistinguishability in an encryption scheme $\mathcal{E}$

Figure 2 defines the notion of indistinguishability in $\mathcal{E}$ [16]. Call adversary A playing *Ind* an *indistinguishability adversary*. Let $\Pr[Ind^A]$ denote the probability A outputs the correct guess b'. Define the advantage of A as

$$\mathbf{Adv}_{\mathcal{E}}^{Ind}(A) = 2 \cdot \Pr[Ind^A] - 1 \tag{4}$$

and say $\mathcal{E}$ has indistinguishability if the advantage is negligible.

In *Ind*, A can repeat the deal procedure multiple times. We can limit A to call the procedure only once; this A is called an *ind-1 adversary*.

2.3   Commitment Schemes

Let CS be a *commitment scheme* with parameter generation algorithm *ParGen*, commitment algorithm Ct and verification algorithm Vf. Let M denote the message to be committed to, H be a committal and R a decommittal. A commitment scheme should satisfy the hiding and binding properties, defined in two security games in Fig. 3 and described in [16]. Intuitively, a commitment scheme allows a sender to commit to a message M and reveal it at a later date.

Call adversary A playing *Hide* against CS a *hiding adversary*. Let $\Pr[Hide^A]$ be the probability A correctly guesses $b' = b$. The advantage of A is

$$\mathbf{Adv}_{CS}^{Hide}(A) = 2 \cdot \Pr[Hide^A] - 1. \tag{5}$$

Say CS is $\epsilon(\cdot)-hiding$ if $\mathbf{Adv}_{CS}^{Hide}(A) \leq \epsilon(q)$ for any adversary that makes at most q queries during the deal procedure.

Call adversary A playing *Bind* a *binding adversary*. The advantage of A is

$$\mathbf{Adv}_{CS}^{bind}(A) = \Pr[Bind^A]. \tag{6}$$

GAME *Hide*

- PROCEDURE *Initialise*
  $\pi \xleftarrow{\$} ParGen; \quad b \xleftarrow{\$} \{0,1\}$
- PROCEDURE *Deal*(M_0, M_1)
  If $M_0, M_1 \notin \mathcal{M}$
    Return $\perp$
  Else $(H, R) \xleftarrow{\$} Ct(\pi, M_b)$
  Return H
- PROCEDURE *Finalise*(b')
  Return $b' = b$

GAME *Bind*

- PROCEDURE *Initialise*
  $\pi \xleftarrow{\$} ParGen$
- PROCEDURE *Commit*(M_0)
  If $M_0 \notin \mathcal{M}$
    Return $\perp$.
  Else $(H, R_0) \xleftarrow{\$} Ct(\pi, M_0)$
  Return (H, R_0)
- PROCEDURE *Finalise*(M_1, R_1)
  If $M_1 \notin \mathcal{M}$, return $\perp$.
  Return $M_0 \neq M_1$ and
    $Vf(H, M_0, R_0) = Vf(H, M_1, R_1) = 1$

Fig. 3. Games defining the hiding and binding security properties of CS.

2.4   Error Correcting Codes

An error correcting code (ECC) E is a method of encoding data with some redundant information to ensure the original data can be recovered, even if a number of errors occur during either data transmission or storage [10].

A code E of *length* n over a finite alphabet F is a subset of F^n. Elements of E are called *codewords*. The *size* of E is $|E| = m$. The *minimum distance* d is the minimum Hamming distance between any two distinct codewords.

Let E be a *linear* code, meaning that for all $u, w \in E$, we have $u + w \in E$, where addition is modulo q with $|F| = q$. If $u_1, \ldots, u_t$ is a basis for E, then E has *dimension* t. There are q^t possible codewords and we call E a $[n, t, d]-code$.

One important ECC is a *maximum distance separable* (MDS) code [10], which is a linear code that meets the Singleton bound: $d = n - t + 1$. For any MDS code, recovery of a codeword is possible from any t of the n symbols. Denote such a code as (t, n)−ECC. A Reed Solomon (RS) code [13] is an MDS code. A code where message string appears in the codeword is called *systematic*.

Let $U \xleftarrow{\$} Share^{ECC}(u)$ denote the distribution of a word u via a (t, n)−ECC, resulting in a codeword represented by an n−vector U. The word u is recoverable from any t elements of U via the deterministic algorithm $Recover^{ECC}$.

2.5   Information Dispersal Algorithms

Information dispersal was first introduced by Rabin [12].

Definition 6. *Let* $t, n \in \mathbb{N}, t \leq n$. *A* (t, n)−*information dispersal algorithm (IDA) with message space* $\mathcal{M}$ *consists of two algorithms* $Share^{IDA}$ *and* $Recover^{IDA}$. $Share^{IDA}$ *takes as input a message* $M \in \mathcal{M}$ *and outputs an* n−*vector* S. $Recover^{IDA}$ *takes as input elements of the vector* S. *If at least* t *elements are submitted correctly to* $Recover^{IDA}$, *the algorithm outputs the message* M.

A (t, n)−IDA shares M between n players such that any t players can recover M. This is equivalent to the recoverability property of a (t, n)−secret sharing scheme. A (t, n)−secret sharing scheme satisfies the conditions of an IDA but additionally guarantees privacy, which IDAs do not. IDAs are able to achieve smaller share sizes by taking advantage of the lack of privacy.

Resch and Plank's IDA. In the AONT-RS scheme, a systematic IDA, which is a variant of an RS code [14], is specified. We refer to this as the *systematic RS IDA* with algorithms $Share^{RS-IDA}$ and $Recover^{RS-IDA}$.

Let $F = GF(2^\omega)$ be a Galois field of characteristic 2. $Share^{RS-IDA}$ is a probabilistic algorithm that takes as input M and parses M into t words, treating this as a t−vector, $M \in F^t$. This vector is multiplied on the left by a public $n \times t$ binary matrix G, where multiplication of elements $b \in \{0, 1\}$ and $d \in F$ is defined as: $\{0, 1\} \times F \to F$, where $0 \times d = 0 \in F$ and $1 \times d = d \in F$. The matrix G is constructed such that the first t rows form the $t \times t$ identity matrix and any t of the n rows are linearly independent; the last $n - t$ rows can be generated in any manner as long as this condition is satisfied. The resulting n vector is the codeword vector $G \cdot M = V \in F^n$. Each player receives the share $V[i]$.

In order to recover M, t shares are submitted to $Recover^{RS-IDA}$ and a t−vector V' is created from these shares. A $t \times t$ matrix G' is formed, consisting of the t rows of G corresponding to the shares pooled. This matrix is inverted and multiplied by V' to return $(G')^{-1} \cdot V' = M$, from which M can be constructed.

It is known that an RS code, which is a $[n, t, n - t + 1]$−code, is equivalent to a $(0, t; n)$−linear ramp scheme [5]. Thus the systematic RS IDA used by Resch and Plank is equivalent to a $(0, t; n)$−linear ramp scheme, as in Definition 5.

3   The AONT-RS

In this section, we consider the Resch and Plank's AONT-RS scheme [14]. We present a generalised version, then discuss information leakage. Finally, we prove the AONT-RS scheme achieves computational privacy in the RO model.

3.1   Generalising the AONT-RS

Resch and Plank propose a CSS in [14], which they call AONT-RS. It combines an All or Nothing Transform (AONT) with an RS code. An AONT is an encryption mode that allows the data to be learnt only if all of it is known [3].

Their scheme assumes the existence of a symmetric key encryption scheme $\mathcal{E}$ operating on blocks of plaintext in CBC mode, a cryptographic hash function H and the systematic RS IDA, as in Sect. 2.5. They assume the digest of the hash function is of equal length to the key k use in $\mathcal{E}$. They do not define what security properties $\mathcal{E}$ must have. They also use a canary, which is a known, fixed value concatenated with the plaintext. When a message is recovered, the user can compare the recovered value with the known canary to verify correctness.

We observe that $\mathcal{E}$ requires ind-1 security and must be probabilistic, but need not operate in CBC mode. We allow flexibility of the IDA, as long as it is equivalent to a $(0, t; n)$−linear ramp scheme (which the systematic RS IDA is). We remove the concept of the canary from the definition, noting that if an incorrect message were recovered, a canary would not help recover the correct M. Preventing M from being incorrectly recovered is discussed in Sect. 4.

Let Π denote our generalised AONT-RS scheme, with algorithms defined in Fig. 4. From now on, AONT-RS will refer to algorithms in Fig. 4.

PROCEDURE $Share^{AONT}(M)$

1. $k \xleftarrow{\$} KeyGen(\{0,1\}^\lambda)$
   $C \xleftarrow{\$} Enc_k(M)$
2. $h = H(C); \quad c_d = h \oplus k$
3. $V \leftarrow Share^{IDA}(C||c_d)$
4. Return V

PROCEDURE $Recover^{AONT}(M)$

1. $V \leftarrow Recover^{IDA}(V[0], \ldots, V[n-1])$
2. $C||c_d \leftarrow V$
3. $h = H(C); \quad k = h \oplus c_d$
4. $M \leftarrow Dec_k(C)$
5. Return M

Fig. 4. The dispersal and recovery algorithms defining the AONT-RS scheme.

On input $M \in \mathcal{M}$, $Share^{AONT}$ generates a key k of length λ, encrypts M under k, then computes the hash of the ciphertext $h = H(C)$. The digest h is then XORed with k to give a value c_d, which we call the *difference value*. The difference value and C are concatenated and dispersed via an IDA.

To recover M, at least t players must pool their shares into a vector V. Using the algorithm $Recover^{IDA}$, $C||c_d$ is recovered. The digest $h = H(C)$ is calculated and XORed with c_d to recover k, which is used to decrypt C and return M.

If $\mathcal{E}$ were not probabilistic, an adversary may recognise shares of known ciphertexts and be able to predict C which, if c_d is known, could leak information about k. The scheme also requires ind-1 security; general indistinguishability is not required as each time a new M is shared, a new encryption key k is generated. So each key is only used to encrypt one message.

3.2 Information Leakage

Resch and Plank claim their system is secure because $t-1$ players are unable to recover all of V, due to the security of the IDA. Without all of V, players are unable to learn both C (required to compute $h = H(C)$) and c_d and so learn nothing about k and M. Learning either C or c_d in isolation does not help the adversary. In order to recover M, knowledge of both C and c_d are needed.

It is necessary that an unauthorised set learn at most: some or all of c_d and some (but not all) of C, or none of c_d and all of C. We show that, when C is a short ciphertext (in relation to the security parameter λ and the threshold value t), the adversary may be able to learn enough to leak information about k.

Learning C Completely and c_d Partially. Consider the following example. Let $C, k \in \{0,1\}^{128}$. Let there be $n = 5$ players $P_1, \ldots, P_5$ and let $t = 4$. The string $C \| c_d$ would be parsed into four words to make the t–vector M, where each fragment is 64 bits. Let c_0 and c_1 be the two elements that comprise C and let $c_{d,0}$ and $c_{d,1}$ be the two halves of c_d, each 64 bits. The vector M is then multiplied on the left by the generator matrix $G \in \{0,1\}^{(5 \times 4)}$, which gives

$$
G \cdot M = \begin{pmatrix} 1 & 0 & 0 & 0 \\ 0 & 1 & 0 & 0 \\ 0 & 0 & 1 & 0 \\ 0 & 0 & 0 & 1 \\ G_{4,0} & G_{4,1} & G_{4,2} & G_{4,3} \end{pmatrix} \begin{pmatrix} C_0 \\ C_1 \\ c_{d,0} \\ c_{d,1} \end{pmatrix} = \begin{pmatrix} C_0 \\ C_1 \\ c_{d,0} \\ c_{d,1} \\ x \end{pmatrix},
$$

where $G_{i,0} \xleftarrow{\$} \{0,1\}$, for $i = \{0, \ldots, 3\}$ are chosen such that any 4 rows of G are linearly independent and $x = G_{4,0} \cdot C_0 + G_{4,1} \cdot C_1 + G_{4,2} \cdot c_{d,0} + G_{4,3} \cdot c_{d,1}$.

Players P_1, P_2 and P_3 are an unauthorised set, yet they could learn all of C and $c_{d,0}$. They could then compute $H(C) = h$ and XOR the first half of h with $c_{d,0}$ to recover the first half of k. This reduces the security from 128 to 64 bits.

This attack can be prevented if c_d is contained entirely in one share. So if $c_d \in \{0,1\}^{\lambda}$, then C should be such that $C \in \{0,1\}^{\omega}$, where $\omega \geq (t-1)\lambda$.

Learning C Partially and c_d Completely. An alternative version of this attack utilises the fact that the hash function H is deterministic.

Consider the following example. Assume an attacker knows all of c_d and all but one bit of C. They can construct two possibilities for C (C_0 when the unknown bit is 0, and C_1 when it is 1) and compute the corresponding hashes, $h_0 = H(C_0)$ and $h_1 = H(C_1)$. They can then compute two key candidates

Procedure $Initialise$ G_0, G_1, G_2	Procedure $Initialise$ G_3, G_4, G_5
$k \xleftarrow{\$} \{0,1\}^\lambda;\quad b \xleftarrow{\$} \{0,1\};\quad k' = k$	$k, k' \xleftarrow{\$} \{0,1\}^\lambda;\quad b \xleftarrow{\$} \{0,1\}$

Procedure $Deal(x_0, x_1)$ G_0, G_1, G_4, G_5	Procedure $Deal(x_0, x_1)$ G_2, G_3
$C \leftarrow Enc_k(x_b);\quad H(C) = h;$	$C \leftarrow Enc_k(x_b);\quad \boldsymbol{C} \leftarrow Share^{IDA}(C\|0)$
$h \oplus k' = c_d$	For $i \leftarrow 1$ to n do
$\boldsymbol{V} \leftarrow Share^{IDA}(C\|c_d)$	$\quad (\boldsymbol{H}[i], \boldsymbol{R}[i]) \xleftarrow{\$} Ct(\boldsymbol{C}[i])$
For $i \leftarrow 1$ to n do	$\quad \boldsymbol{S}[i] \leftarrow Share^{ECC}(\boldsymbol{H}[i])$
$\quad (\boldsymbol{H}[i], \boldsymbol{R}[i]) \xleftarrow{\$} Ct(\boldsymbol{V}[i])$	$H(C) = h;\quad h \oplus k' = c_d$
$\quad \boldsymbol{S}_i \xleftarrow{\$} Share^{ECC}(\boldsymbol{H}[i])$	$\boldsymbol{V} \leftarrow Share^{IDA}(C\|c_d)$

Procedure $Corrupt(i)$ G_0, G_5	Procedure $Corrupt(i)$ G_1, G_2, G_3, G_4
$\boldsymbol{X}[i] \leftarrow \boldsymbol{R}[i]\boldsymbol{V}[i]\boldsymbol{S}_1[i]...\boldsymbol{S}_n[i]$	$\boldsymbol{R}[i] \xleftarrow{\$} DCt(\boldsymbol{H}[i], \boldsymbol{V}[i])$
Return $\boldsymbol{X}[i]$	$\boldsymbol{X}[i] \leftarrow \boldsymbol{R}[i]\boldsymbol{V}[i]\boldsymbol{S}_1[i]...\boldsymbol{S}_n[i]$
	Return $\boldsymbol{X}[i]$

Procedure $Finalise(b')$ $G_0 - G_5$	
Return $(b' = b)$	

Fig. 5. Games G_0 and G_5 are used to prove Theorem 1, the privacy of AONT-RS. All games $G_0 - G_5$ are used to prove Theorem 2, the privacy of the RAONT-RS.

$k_0 = c_d \oplus h_0$ and $k_1 = c_d \oplus h_1$ and decrypt the ciphertexts C_0 and C_1 with the corresponding candidate keys to reveal two plaintext messages M_0 and M_1. From these, the adversary can guess which plaintext message is likely to be the true message and has thus learnt k. In general, if the adversary knows c_d and all but j bits of C, if $j < \lambda$ this attack is quicker than brute force. We must ensure an adversary is unable to learn at least λ bits of C if c_d is known. This is true if each $\boldsymbol{M}[i] \in \{0,1\}^\lambda$, meaning that $C \in \{0,1\}^\omega, \omega \geq (t-1)\lambda$.

So both attacks can be prevented if $k \in \{0,1\}^\lambda$ and $C \in \{0,1\}^\omega$, where $\omega \geq (t-1)\lambda$. If C is too small, C should be padded with some random string. This condition on the size of C is a necessary, but not sufficient, condition for the AONT-RS scheme to be secure. To guarantee the security, we must make additional assumptions on H, $\mathcal{CS}$ and $\mathcal{E}$, as discussed next.

3.3 Proving the Privacy of AONT-RS

We now prove the AONT-RS achieves computational privacy in the RO model.

Theorem 1 (Privacy of the non-robust AONT-RS). *Let A be a privacy adversary against the AONT-RS scheme Π (as in Fig. 4) and let the internal hash function H be indistinguishable from a RO. Let the ciphertext be $C \in \{0,1\}^\omega$, where $\omega \geq (t-1)\lambda$. Then there is an ind-1 adversary B attacking the indistinguishability of $\mathcal{E}$ such that*

$$\mathbf{Adv}_\Pi^{Priv}(A) \leq \mathbf{Adv}_\mathcal{E}^{Ind}(B), \tag{7}$$

where B makes only one query during the deal procedure of Game Ind and the running time of B is that of A plus one execution of $Share^{AONT}$.

Proof. The proof relies on games G_0 and G_5, as in Fig. 5. The figure shows multiple procedures, indicating next to each in which games it is included. For example G_0 is defined by the procedures on the left hand side of the figure. The advantage of the AONT-RS privacy adversary A can be defined as

$$\mathbf{Adv}_\Pi^{Priv}(A) = 2 \cdot \Pr[G_0^A] - 1. \tag{8}$$

Game G_5 differs from G_0 only because the key k used to encrypt M is different to the value k' used to compute $c_d = h \oplus k'$. We claim that $\Pr[G_0^A] = \Pr[G_5^A]$, as the hash function H is indistinguishable from a RO. Due to the IDA used and the restriction that $C = \{0,1\}^\omega$, where $\omega \geq \lambda(t-1)$, the adversary is always missing either at least λ bits of C or all of c_d. Thus the adversary can learn either h or c_d. If A learns h, $h = c'_d \oplus k'$. If A learns c_d, then $c_d = h' \oplus k'$, for some $c'_d \neq c_d$ and $h' \neq h$. Thus $\Pr[G_0^A] = \Pr[G_5^A]$ holds true.

We construct an adversary B attacking the privacy of $\mathcal{E}$ such that

$$2 \cdot \Pr[G_5^A] - 1 \leq \mathbf{Adv}_\mathcal{E}^{Ind}(B). \tag{9}$$

Adversary B picks $k' \xleftarrow{\$} \{0,1\}^\lambda$ and runs A. A submits $Deal(x_0, x_1)$ to B, who queries x_0, x_1 to its challenger and receives $C \xleftarrow{\$} Enc_k(x_b)$, where k is the key generated by the challenger. Now B executes the rest of the *Deal* procedure of G_5 using k'; so B computes $H(C) = h$, $h \oplus k' = c_d$ and $V \leftarrow Share^{IDA}(C||c_d)$. When A submits $Corrupt(i)$, B responds with $V[i]$. When A outputs a bit b', B passes this onto their challenger. The advantage of B is $2 \cdot \Pr[b' = b] - 1$.

By combining (8), (9) and $\Pr[G_0^A] = \Pr[G_5^A]$, we see that, as required,

$$\mathbf{Adv}_\Pi^{Priv}(A) \leq \mathbf{Adv}_\mathcal{E}^{Ind}(B).$$

$\square$

4   Extending AONT-RS to be Robust

In [16], Bellare and Rogaway extend Krawczyk's SSMS [9] to be robust by using commitment schemes. Their technique can be applied to the AONT-RS to make it robust. We will call the resulting scheme the RAONT-RS.

Let $\mathcal{E}$ be an ind-1 secure encryption scheme. Assume the existence of a $(t,n)-$ECC, an IDA equivalent to a $(0,t;n)-$linear ramp scheme, an $\epsilon(\cdot)-$hiding commitment scheme $\mathcal{CS}$ and a hash function H that is indistinguishable from a RO. Let Π_R denote the RAONT-RS scheme, with algorithms as in Fig. 6. Let the ciphertext be $C \in \{0,1\}^\omega$, where $\omega \geq (t-1)\lambda$.

Intuitively, the scheme is the same as the AONT-RS. However, in addition to being given the share $V[i]$, each player is given a decommittal $R[i]$ computed on $V[i]$ and fragments of committals $H[i]$ computed on all shares $V[i]$ distributed via a $(t,n)-$ECC. Let the n-vector S_i be the output after the committal $H[i]$ is dispersed via a $(t,n)-$ECC. Let $S_i[j]$ be the jth element of S_i. These values are used to verify each player's share. Let $\Diamond$ denote an empty share.

PROCEDURE $Share^{RAONT}(M)$

1. $k \xleftarrow{\$} \{0,1\}^{\lambda}; \quad C \xleftarrow{\$} Enc_k(M)$
2. $h = H(C); \quad c_d = h \oplus k$
3. $V \leftarrow Share^{IDA}(C\|c_d)$
4. For $i \leftarrow 1$ to n do
   $\quad (H[i], R[i]) \xleftarrow{\$} Ct(V[i])$
   $\quad S_i \xleftarrow{\$} Share^{ECC}(H[i])$
5. For $i \leftarrow 1$ to n do
   $\quad X[i] \leftarrow R[i]V[i]S_1[i]\ldots S_n[i]$
6. Return X

PROCEDURE $Recover^{RAONT}(V)$

1. For $i \leftarrow 0$ to $n-1$ do
   $\quad R[i]V[i]S_1[i]\ldots S_n[i] \leftarrow X[i]$
2. For $i \leftarrow 0$ to $n-1$ do
   $\quad H[i] \leftarrow Recover^{ECC}(S_i, j)$
3. For $i \leftarrow 0$ to $n-1$ do
   $\quad$ If $X[i] \neq \Diamond$ and
   $\quad Vf(H[i], V[i], R[i]) = 0$
   $\quad$ then $V[i] \leftarrow \Diamond$
4. $C\|c_d \leftarrow Recover^{IDA}(V)$
5. $h = H(C); \quad k = h \oplus c_d$
6. $M \leftarrow Dec_k(C)$
7. Return M

Fig. 6. Algorithms defining robust AONT-RS (RAONT-RS).

Unlike a canary, commitment schemes allow recovery of M even if false shares are submitted. Furthermore, the commitment scheme highlights which servers are corrupted and thus take any necessary action. However, it is noted that commitment schemes requires more computation than the use of a canary. In practise, both techniques could be combined; the canary could first be verified and, only if the canary is incorrect, will the shares be individually verified.

4.1 Proof of Privacy

The RAONT-RS scheme Π_R can be proven to achieve computational privacy by adapting the proof of privacy for the HK2 scheme by Bellare and Rogaway [16].

Theorem 2 (Privacy of RAONT-RS). *Let A be a privacy adversary against RAONT-RS Π_R and let H be indistinguishable from a RO. Let the ciphertext be $C \in \{0,1\}^{\omega}$, where $\omega \geq (t-1)\lambda$. Assume Ct is $\epsilon(\cdot)$−hiding (as in Sect. 2.3), then there is an ind-1 adversary B attacking the indistinguishability of $\mathcal{E}$ such that*

$$\mathbf{Adv}_{\Pi_R}^{Priv}(A) \leq \mathbf{Adv}_{\mathcal{E}}^{Ind}(B) \cdot 4\epsilon(n), \tag{10}$$

where B makes only one query during the deal procedure of Game Ind and the running time of B is that of A plus one execution of $Share^{RAONT}$.

Proof. The proof relies on games G_0-G_5, as in Fig. 5. The procedure $Corrupt$ of games G_1-G_4 refers to a probabilistic algorithm DCt that works as follows. On input message M and committal H, it lets $\Omega(M, H)$ denote the set of all coins ω such that Ct, on input M and coins ω, returns a pair whose first component is H. If $\Omega(M, H) = \emptyset$, then DCt returns $\perp$. Else it picks ω at random from $\Omega(M, H)$, runs Ct on input M and coins ω to get a pair (H, R) and returns R. This algorithm is not necessarily efficiently implementable.

The advantage of the RAONT-RS privacy adversary A can be defined as

$$\mathbf{Adv}_{\Pi_R}^{Priv}(A) = 2 \cdot \Pr[G_0^A] - 1. \tag{11}$$

Game G_1 differs from G_0 only in the *Corrupt* procedure, which resamples $R[i]$ using DCt. Clearly,

$$\Pr[G_0^A] = \Pr[G_1^A] = \Pr[G_2^A] + (\Pr[G_1^A] - \Pr[G_2^A]). \tag{12}$$

We construct an adversary D_1 attacking the hiding property of $\mathcal{CS}$ such that

$$\Pr[G_1^A] - \Pr[G_2^A] = \mathbf{Adv}_{\mathcal{CS}}^{Hide}(D_1). \tag{13}$$

Adversary D_1 acts as the challenger to A and wishes to use A's advantage to gain an advantage against the hiding property of $\mathcal{CS}$. Adversary D_1 picks $b \xleftarrow{\$} \{0,1\}$ and runs A. When A submits x_0, x_1 to D_1, D_1 generates $k \xleftarrow{\$} \{0,1\}^\lambda$ and calculates $C \xleftarrow{\$} Enc_k(x_b)$. D_1 then computes $H(C) = h$ and $h \oplus k = c_d$, then calculates both $V \leftarrow Share^{IDA}(C||c_d)$ and $C \leftarrow Share^{IDA}(C||0)$. For i, $1 \le i \le n$, D_1 queries $C[i], V[i]$ (for $V[i] \ne C[i]$) to its challenger. Let $H[i]$ denote the commitment value returned. Let $S_i \leftarrow Share^{ECC}(H[i])$. When A makes a *Corrupt*(i) query to D_1, D_1 computes its reply according to the case of the *Corrupt* procedure of games G_1, G_2; that is, D_1 generates a decommittal value $R[i]$ for $V[i]$ and the given $H[i]$ and passes $X[i] \leftarrow R[i]V[i]S_1[i]\ldots S_n[i]$ to A. When A halts the corruption procedure and finalises with output b', if $b' = b$, adversary D_1 passes 1 to its challenger, guessing the commitment value $H[i]$ was computed on $V[i]$, rather than $C[i]$. Otherwise, D_1 submits 0.

Next, we have that

$$\Pr[G_2^A] = \Pr[G_3^A] + \left(\Pr[G_2^A] - \Pr[G_3^A]\right), \tag{14}$$

where G_3 differs from G_2 only in the initialise procedure which XORs the digest h not with the encryption key k, but with a string k'. We claim that $\Pr[G_2^A] = \Pr[G_3^A]$. because the hash function is indistinguishable from a RO. After A has corrupted at most t shares, they learn at most either

- no information about c_d and all of C, and so can learn $h = H(C)$. In which case $h = k \oplus c_d = k' \oplus c'_d$, where $c'_d \ne c_d$ is some unknown string. Or
- all of c_d, but missing at least λ bits of C. Then $c_d = k' \oplus h'$ where $h' \ne h$ is unknown to A.

In either case, the adversary learns either h or c_d and no information about k. Thus the known value is the XOR of two unknown strings: changing one of these strings does not affect the chances of A winning, thus $\Pr[G_2^A] = \Pr[G_3^A]$.

Next, we have

$$\Pr[G_3^A] = \Pr[G_4^A] + \left(\Pr[G_3^A] - \Pr[G_4^A]\right). \tag{15}$$

Construct adversary D_2, also attacking the hiding property of $\mathcal{CS}$, such that

$$\Pr[G_3^A] - \Pr[G_4^A] = \mathbf{Adv}_{\mathcal{CS}}^{Hide}(D_2). \tag{16}$$

PROCEDURE $Deal(x)$

$\ell \xleftarrow{\$} [1,n]$; $k \xleftarrow{\$} \{0,1\}^\lambda$; $C \xleftarrow{\$} Enc_k(x)$
$H(c) = h$; $h \oplus k = c_d$; $V \leftarrow Share^{IDA}(C \| c_d)$
For $i \leftarrow 1$ to n do
  If $i = \ell$, then $(\boldsymbol{H}[\ell], \boldsymbol{R}[\ell]) \xleftarrow{\$} Commit(\boldsymbol{V}[i])$
  Else $(\boldsymbol{H}[i], \boldsymbol{R}[i]) \xleftarrow{\$} Ct(\boldsymbol{V}[i])$
  $\boldsymbol{S}_i \xleftarrow{\$} Share^{ECC}(\boldsymbol{H}[i])$
For $i \leftarrow 1$ to n do
  $\boldsymbol{X}[i] \leftarrow \boldsymbol{R}[i]\boldsymbol{V}[i]\boldsymbol{S}_i[i]\dots\boldsymbol{S}_n[i]$

PROCEDURE $Corrupt(i)$
Return $\boldsymbol{X}[i]$

PROCEDURE $Finalise(\boldsymbol{x'},j)$
For $i \leftarrow 1$ to n do
  $\boldsymbol{R'}[i]\boldsymbol{V'}[i]\boldsymbol{S'_1}[i]\dots\boldsymbol{S'_n}[i] \leftarrow \boldsymbol{X'}[i]$
Return $(\boldsymbol{V'}[\ell], \boldsymbol{R'}[\ell])$

Fig. 7. Procedures used by adversary B to respond to A for Theorem 3.

The construction of D_2 is similar to D_1, but D_2 generates $k, k' \xleftarrow{\$} \{0,1\}^\lambda$, encrypts x_b under k as before and now calculates $c_d = h \oplus k'$.

Game G_5 and G_4 differ only during *Corrupt*. Clearly $\Pr[G_4^A] = \Pr[G_5^A]$.

Let B be an ind-1 adversary attacking $\mathcal{E}$, as in the proof of Theorem 1. The advantage of B is as in (9).

Now, let D be the hiding-adversary that flips a fair coin and, if it lands head, runs D_1, otherwise D_2. Clearly,

$$\mathbf{Adv}_{CS}^{Hide}(D) = \frac{1}{2}\left(\mathbf{Adv}_{CS}^{Hide}(D_1) + \mathbf{Adv}_{CS}^{Hide}(D_2)\right). \tag{17}$$

Since Ct is assumed to be $\epsilon(\cdot)$−hiding and D makes at most n queries, we have that $\mathbf{Adv}_{CS}^{Hide}(D) \leq \epsilon(n)$. Combining this and (13), (16), (17) gives us

$$\left(\Pr[G_1^A] - \Pr[G_2^A]\right) + \left(\Pr[G_3^A] - \Pr[G_4^A]\right) \leq 2\epsilon(n).$$

By using $\Pr[G_2^A] = \Pr[G_3^A]$ and $\Pr[G_4^A] = \Pr[G_5^A]$ and substituting in the advantages of adversaries A and B, we can simplify and rearrange to give

$$\mathbf{Adv}_{\Pi_R}^{Priv}(A) \leq \mathbf{Adv}_{\mathcal{E}}^{Ind}(B) \cdot 4\epsilon(n),$$

thus completing the proof. □

4.2 Proof of Robustness

The RAONT-RS scheme can be proven to be robust by adapting the proof by Bellare and Rogaway [16].

Theorem 3 (Robustness of RAONT-RS). *Let A be a recoverability adversary against the RAONT-RS scheme Π_R. Then there is an adversary B attacking the binding property of the commitment scheme $\mathcal{CS}$ such that*

$$\mathbf{Adv}_{\Pi_R}^{Rec}(A) \leq n \cdot \mathbf{Adv}_{CS}^{Bind}(B), \tag{18}$$

where the running time of B is that of A plus overhead consisting of an execution of the $Share^{RAONT}$ and $Recover^{RAONT}$ algorithms of Π_R.

Proof. Let A be a recoverability adversary against Π_R. During *Deal*, A submits x to B. Let $k, C, h, c_d, \boldsymbol{V}, \boldsymbol{H}, \boldsymbol{S}_1, \ldots, \boldsymbol{S}_n, \boldsymbol{X} \xleftarrow{\$}$ denote the quantities chosen by $Share^{RAONT}(x)$. Let A corrupt at most $t-1$ shares. Let $(\boldsymbol{X}_T)$ denote the output of A. Let $k', C', h', c'_d, \boldsymbol{V}', \boldsymbol{H}', \boldsymbol{S}'_1, \ldots, \boldsymbol{S}'_n, \boldsymbol{X}'$ denote the quantities recovered from $Recover^{RAONT}$ with input $\boldsymbol{X}'_T \cup \boldsymbol{X}'_{\overline{T}}$. Consider the following events:

E_1: $\exists\, \ell \in [n]$ such that $\boldsymbol{H}[\ell] \neq \boldsymbol{H}'[\ell]$
E_2: $\exists\, \ell \in T$ such that $\boldsymbol{V}[i] \in \{\Diamond, \boldsymbol{V}[i]\}$
E_3: $c_d \neq c'_d$
E_4: $C \neq C'$

If $C' = C$ and $c'_d = c_d$, then the recovered secret x' equals x. This is because $h' = H(C') = H(C) = h$ and so $c'_d \oplus h' = c_d \oplus h = k$. Therefore

$$\mathbf{Adv}_{\Pi_R}^{Rec}(A) \leq \Pr[E_3 \cup E_4] \tag{19}$$

$$\leq \Pr[E_1 \cup E_2 \cup E_3 \cup E_4] \tag{20}$$

$$= \Pr[E_1] + \Pr[\overline{E_1} \cap E_2] + \Pr[\overline{E_2} \cap E_3] + \Pr[\overline{E_2} \cap E_4]. \tag{21}$$

We bound each addend in turn. Let $E_{1,\ell}$ be the event that $\boldsymbol{H}'[\ell] = \boldsymbol{H}[\ell]$. Let T be the set of indexes of the shares corrupted by A. If $i \notin T$, then the submission of $\boldsymbol{X}'[i]$ and the other uncorrupted shares returns $\boldsymbol{X}[i]$. Hence $\boldsymbol{S}'_\ell[i] = \boldsymbol{S}_\ell[i]$. Note that $\boldsymbol{S}_\ell$ is an output of $Share^{ECC}(\boldsymbol{H}[\ell])$. Lemma 10 in [16] discusses perfect recoverability and, when applied to ECCs, $Recover^{ECC}(\boldsymbol{S}_\ell) = \boldsymbol{H}[\ell]$, meaning that $\boldsymbol{H}'[\ell] = \boldsymbol{H}[\ell]$. So $\Pr[E_{1,\ell}] = 0$. By the union bound

$$\Pr[E_1] \leq \sum_{t=1}^{n} \Pr[E_{1,\ell}] = 0. \tag{22}$$

Now we construct adversary B such that

$$\Pr[\overline{E_1} \cup E_2] \leq n \cdot \mathbf{Adv}_{CS}^{Bind}(B). \tag{23}$$

Adversary B runs A, responding to its *Deal* and *Corrupt* calls via the procedures in Fig. 7, where Ct is the committal algorithm of CS run by B and *Commit* is a procedure of the *Bind* game that B plays with its challenger. When A halts with output $(\boldsymbol{X})$, B runs the finalise procedure.

Next, we claim both $\Pr[\overline{E_2} \cap E_3] = 0$ and $\Pr[\overline{E_2} \cap E_4] = 0$. As, if $\boldsymbol{V}'[i] = \boldsymbol{V}[i]$ for all i, then $C = C'$ and $c'_d = c_d$. So now

$$\mathbf{Adv}_{\Pi_R}^{Rec}(A) = \Pr[E_1] + \Pr[\overline{E_1} \cap E_2] + \Pr[\overline{E_2} \cap E_3] + \Pr[\overline{E_2} \cap E_4] \tag{24}$$

$$\leq n \cdot \mathbf{Adv}_{CS}^{Bind}(B), \tag{25}$$

thus completing the proof.   □

5   Comparing RAONT-RS and HK2

We briefly introduce Krawczyk's SSMS scheme [9] and a robust extension, called HK2 [16]. We then compare AONT-RS with SSMS. This comparison can also be applied to the AONT-RS and HK2.

5.1   The SSMS and HK2 Scheme

Krawczyk's SSMS is a CSS [9]. It assumes an ind-1 secure encryption system, an IDA and a $(t, n)-$PSS. Intuitively, SSMS takes as input a message M, generates a key k and calculates $C \xleftarrow{\$} Enc_k(M)$. The ciphertext C is then shared amongst the n participants via an IDA whilst k is shared via a $(t, n)-$PSS. A player's share is one element of C and one of k. Krawczyk then extended his scheme to be robust in the RO model by using hash functions [9], which was proven to be secure in the RO model in [16].

HK2 is a robust extension of SSMS [16] using commitment schemes. HK2 relies on the same assumptions as SSMS, but additionally assumes a $(t, n)-$ECC and an $\epsilon()-$hiding commitment scheme $\mathcal{CS}$. Our extension of AONT-RS to RAONT-RS used similar techniques as in the extension of SSMS to HK2; the commitment scheme is added to SSMS and each player stores their share, along with a decommittal and multiple fragments of committals. For a more detailed description, the reader is directed to [16].

We chose to use commitment schemes to extend AONT-RS to be robust (as was done in HK2) to achieve recoverability under standard assumptions. Instead, hash functions could be used, as in [9], to achieve recoverability in the RO model.

5.2   Comparison

In [14], Resch and Plank only briefly compare AONT-RS to Krawczyk's SSMS [9]. They then conduct a performance comparison of the AONT-RS with Rabin's IDA [12] and Shamir's PSS [17]. However, Rabin's IDA has no security requirements, and would not be used to distribute data if there were any privacy concerns, and Shamir's PSS achieves perfect security, which would not be used to share large data due to the bounds on the share sizes. Krawczyk's SSMS achieves computational security, which is a similar to AONT-RS. Thus we compare SSMS with AONT-RS. Similarly, RAONT-RS can be compared with HK2.

We compare the security, share size and efficiency of AONT-RS with SSMS. We exclude the contribution to the complexity made by $\mathcal{E}$, as this is equal in both schemes. The comparison is applicable to RAONT-RS and HK2, if we exclude the contribution to the share size and efficiency from $\mathcal{CS}$ (which is equal in both).

For the comparison, we will assume both AONT-RS and SSMS use the systematic RS IDA and that SSMS uses an ideal PSS. Such PSSs include Shamir's PSS [17], or Chen et al. [6]. Let $k \in \{0, 1\}^\lambda$.

Assume M is to be distributed and $\mathcal{E}$ is length preserving, so $|C| = |M|$. Let $C \in \{0, 1\}^\omega$, and fix ω such that $\omega \geq \lambda(t - 1)$. This is assumed to prevent attacks described in Sect. 3.2 against the AONT-RS, and to ensure all schemes are distributing a message of equal length, thereby allowing for a fair comparison. It is noted that if $\omega < \lambda(t - 1)$, the AONT-RS will need to pad the message to lengthen C, whereas SSMS can distribute C as is. However, as mentioned previously, CSS schemes are often used when M is large, thus it is reasonable to assume that $\omega \geq \lambda(t - 1)$. To illustrate, we highlight an example presented in [14]: they distribute a 4KB block of data using a 128 bit key amongst 16 servers such that any 10 can recover the data. So $n = 16, t = 10, \lambda = 128$ and $C \in \{0, 1\}^{32000}$ with $\omega = 32000 >> \lambda(t - 1) = 1152$.

Security. AONT-RS achieves computational privacy, assuming $\mathcal{E}$ is ind-1 secure, H is indistinguishable from a RO and the IDA is equivalent to a $(0, t; n)$−linear ramp scheme. SSMS also assumes $\mathcal{E}$ is ind-1 secure and requires a (t, n)−PSS and an IDA (with no privacy requirements).

As SSMS is secure under standard assumptions, whereas AONT-RS is only secure in the RO model, SSMS is considered to be more secure.

Share Size. The share given to each player from the AONT-RS is $\lceil \frac{\omega + \lambda}{t} \rceil$ bits. For SSMS, each share is $\lceil \frac{\omega}{t} \rceil + \lambda$ bits.

The AONT-RS achieves smaller share sizes than SSMS when $t \geq 2$ (which is true in general). The ratio between the share sizes is larger when t is bigger. The main contribution to the share size is from C, meaning the ratio between the share sizes will be small if ω is large and large if ω is small (meaning ω is close to $\lambda(t - 1)$).

Efficiency of *Share*. *Share*^{AONT} requires one hash computation and $\mathcal{O}(\lambda(n + 1) + n\omega)$ bitwise XORs (if multiplication is implemented via a look-up table).

In SSMS distribution of k via either [17] or [6] requires $\mathcal{O}(tn\lambda)$ bitwise XORs. The distribution of C via the IDA requires $\mathcal{O}(n\omega)$ XORs. Thus distribution of SSMS requires $\mathcal{O}(\lambda(tn) + n\omega)$ bitwise XORs.

AONT-RS requires fewer XORs than SSMS. For larger values of t, SSMS requires more XORs, whereas the complexity of AONT-RS is independent of t.

Efficiency of *Recover*. Assume t players pool their shares. *Recover*^{AONT} requires one hash function computation and $\mathcal{O}(t(\omega + \lambda) + \lambda)$ bitwise XORs.

SSMS requires $t(t - 1) \lceil \frac{\omega}{t} \rceil$ bitwise XORs to recover C and either $\mathcal{O}(tn\lambda)$ (if [6] is the chosen PSS), or $\mathcal{O}(t \log^2 t\lambda)$ (for Shamir's PSS [17]) XORs to recover k. The total efficiency is the sum of the recovery of C and k.

Generally, the AONT-RS requires fewer bitwise XORs and is dependent only on t. Recovery of M using SSMS is dependent on the efficiency of the PSS used.

6   Conclusion

We generalised the AONT-RS and showed information is leaked when ciphertexts are shorter than $\lambda(t - 1)$. We proved the AONT-RS scheme has computational privacy in the RO model. We extended the scheme to be robust and proved it achieves computational privacy in the RO model and recoverability under standard assumptions. Finally, we compared AONT-RS with SSMS, which is a comparison that can be used to compare RAONT-RS with HK2. We showed the (R)AONT-RS schemes achieve weaker security than SSMS/HK2 because their proofs are in the RO model, whereas SSMS/HK2 are provable under standard assumptions. However, by compromising security, (R)AONT-RS achieves smaller shares and more efficient dispersal and recovery.

References

1. Beimel, A.: Secret-Sharing schemes: A survey. In: Chee, Y.M., Guo, Z., Ling, S., Shao, F., Tang, Y., Wang, H., Xing, C. (eds.) IWCC 2011. LNCS, vol. 6639, pp. 11–46. Springer, Heidelberg (2011). doi:10.1007/978-3-642-20901-7_2
2. Blakley, G.R.: Safeguarding cryptographic keys. In: Proceeding of the National Computer Conference 1979, vol. 48, pp. 313–317 (1979)
3. Boyko, V.: On the security properties of OAEP as an all-or-nothing transform. In: Wiener, M. (ed.) CRYPTO 1999. LNCS, vol. 1666, pp. 503–518. Springer, Heidelberg (1999). doi:10.1007/3-540-48405-1_32
4. Chandrasekara, A., Bala, R., Landers, G.: Critical capabilities for object storage - Gartner. Technical report (March 2016). https://www.gartner.com/doc/3269531/critical-capabilities-object-storage (Accessed March 2017)
5. Chen, H., Cramer, R.: Algebraic geometric secret sharing schemes and secure multiparty computations over small fields. In: Dwork, C. (ed.) CRYPTO 2006. LNCS, vol. 4117, pp. 521–536. Springer, Heidelberg (2006). doi:10.1007/11818175_31
6. Chen, L., Laing, T.M., Martin, K.M.: Efficient, XOR-based, ideal (t, n)-threshold schemes. In: Foresti, S., Persiano, G. (eds.) CANS 2016. LNCS, vol. 10052, pp. 467–483. Springer, Cham (2016). doi:10.1007/978-3-319-48965-0_28
7. IBM. IBM Cloud Object Storage (2016). https://www.cleversafe.com/platform/why-ibm-cloud-object-storage, Accessed 04 Sept 2016
8. Karnin, E.D., Greene, J.W., Hellman, M.E.: On secret sharing systems. IEEE Trans. Inf. Theory **29**(1), 35–41 (1983)
9. Krawczyk, H.: Secret sharing made short. In: Stinson, D.R. (ed.) CRYPTO 1993. LNCS, vol. 773, pp. 136–146. Springer, Heidelberg (1994). doi:10.1007/3-540-48329-2_12
10. MacWilliams, F.J., Sloane, N.J.A.: The Theory of Error Correcting Codes. Elsevier, New York (1977)
11. McEliece, R.J., Sarwate, D.V.: On sharing secrets and reed-solomon codes. Commun. ACM **24**(9), 583–584 (1981)
12. Rabin, M.O.: Efficient dispersal of information for security, load balancing, and fault tolerance. J. ACM (JACM) **36**(2), 335–348 (1989)
13. Reed, I.S., Solomon, G.: Polynomial codes over certain finite fields. J. Soc. Ind. Appl. Math. **8**(2), 300–304 (1960)
14. Resch, J.K., Plank, J.S.: AONT-RS: blending security and performance in dispersed storage systems. In: FAST-2011: 9th Usenix Conference on File and Storage Technologies, pp. 191–202, February 2011
15. Rivest, R.L.: All-or-nothing encryption and the package transform. In: Biham, E. (ed.) FSE 1997. LNCS, vol. 1267, pp. 210–218. Springer, Heidelberg (1997). doi:10.1007/BFb0052348
16. Rogaway, P., Bellare, M.: Robust computational secret sharing and a unified account of classical secret-sharing goals. In: Proceedings of the 14th ACM conference on Computer and communications security, pp. 172–184. ACM (2007)
17. Shamir, A.: How to share a secret. Commun. ACM **22**(11), 612–613 (1979)
18. Tompa, M., Woll, H.: How to share a secret with cheaters. J. Cryptology **1**(3), 133–138 (1989)

Side-Channel Analysis

Climbing Down the Hierarchy: Hierarchical Classification for Machine Learning Side-Channel Attacks

Stjepan Picek[1]([⊠]), Annelie Heuser[2], Alan Jovic[3], and Axel Legay[4]

[1] KU Leuven ESAT/COSIC and imec, Kasteelpark Arenberg 10,
3001 Leuven-Heverlee, Belgium
stjepan@computer.org
[2] IRISA/CNRS, Rennes, France
[3] University of Zagreb, Faculty of Electrical Engineering and Computing,
Zagreb, Croatia
[4] IRISA/Inria, Rennes, France

Abstract. Machine learning techniques represent a powerful paradigm in side-channel analysis, but they come with a price. Selecting the appropriate algorithm as well as the parameters can sometimes be a difficult task. Nevertheless, the results obtained usually justify such an effort. However, a large part of those results use simplification of the data relation and in fact do not consider all the available information. In this paper, we analyze the hierarchical relation between the data and propose a novel hierarchical classification approach for side-channel analysis. With this technique, we are able to introduce two new attacks for machine learning side-channel analysis: Hierarchical attack and Structured attack. Our results show that both attacks can outperform machine learning techniques using the traditional approach as well as the template attack regarding accuracy. To support our claims, we give extensive experimental results and discuss the necessary conditions to conduct such attacks.

Keywords: Side-channel attacks · Profiled scenario · Machine learning techniques · Hierarchical classification · Hierarchical attack · Structured attack

1 Introduction

Side-channel attacks (SCAs) are capable of revealing secret keys on cryptographic devices from unintentionally emitted information during computation, like power consumption or electromagnetic emanation. To evaluate the worst case security threat, so-called profiled attacks have to be conducted, which consist of an additional profiling phase. In this phase, one has the full control over the device to build additional advanced models, which are then exploited to extract key-dependent information in the attacking phase.

© Springer International Publishing AG 2017
M. Joye and A. Nitaj (Eds.): AFRICACRYPT 2017, LNCS 10239, pp. 61–78, 2017.
DOI: 10.1007/978-3-319-57339-7_4

Profiled attacks can be divided into two main approaches. First, there exist the traditional methods like the template attack [1] and stochastic approach [2], relying on the maximum-likelihood estimation. Recently, machine learning (ML) techniques, adapted as side-channel attacks, have been proposed, which can be beneficial in many scenarios. In particular, ML techniques have originally been introduced to classify between two classes (i.e., a bit is set (= 1) or not (= 0)) [3,4]. The first extension to 9 Hamming weight (HW) classes for Support Vector Machines (SVMs) has been given in [5]. Following this approach, recent works studied ML techniques mostly using 9 (e.g., [6,7]) or up to 16 classes [8].

However, in most real-world applications, the measured leakage does not follow the Hamming weight (HW) or the Hamming distance (HD) model. For instance, the authors in [9] showed that the leakage of the DPAcontest v2 traces [10] using a Xilinx FPGA VirtexTM-5 [11] is highly "non-linear", i.e., the HD does not apply. Instead, one should consider the output of the S-Box itself as a sensitive variable, resulting in 256 classes (considering AES with 8-bit words). Similarly, even when using an Atmel ATMega-163 smart-card as in the DPAcontest v4 [12], the leakage model does not exactly follow the Hamming weight. Thus, especially in a profiled scenario, to be able to capture more knowledge on the leakage model, it is more advisable to use the S-Box output with 256 classes as a sensitive variable directly. Even more, considering 256 classes yields direct information on the key as each class is related to only one key guess. This is naturally not the case when considering 9 Hamming weight classes. For instance, in the worst case when classifying into the Hamming weight class 4 we have 70 corresponding key guesses. Therefore, in a low noise scenario an attacker is clearly able to reveal the secret key within only one trace using 256 classes, whereas this is not possible (on average[1]) when using HW classes.

1.1 Idea and Contributions

The more classes we have, the more instances (measurements) are necessary to obtain high accuracies, i.e., high probability to classify each element to its class correctly. Generally speaking one has either $n+1$ HW classes or 2^n classes relating directly to the key guess where n is the number of bits of intermediate states

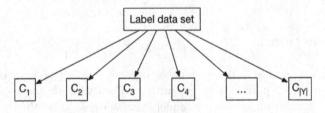

Fig. 1. Flat (standard) approach

---

[1] Note that, an attacker could reveal the secret key with only one trace if it corresponds to HW 0 or 8, which occurs with a probability of $\frac{2}{256}$.

and a key chunk[2]. Up to now in side-channel analysis, the class variables have been seen in a flat hierarchy, where each label directly results in the classified variables (see Fig. 1). Naturally, estimating 2^n instead of $n+1$ classes may bring statistical difficulties. When considering random classification, with 9 classes there is $1/9$ chance of a successful guess, while in the 256 classes scenario there is $1/256$ chance for a random hit. However, only considering the Hamming weight classes instead of the value itself brings two drawbacks as discussed before: first, it lowers the information about the secret key and second, it yields an imprecise estimation of the leakage model. On the other hand, with the increase in the number of classes, the computational complexity also rises. In general, for most ML techniques (with the exception of decision tree based techniques), the complexity for multi-class classification rises with $O(|\mathcal{Y}|)$, where $|\mathcal{Y}|$ is the number of classes. This is particularly critical when performing a proper tuning within the profiling phase, which may become very complex in terms of computation resources.

In this paper, in order to circumvent the problems arising when classifying $n+1$ or 2^n directly, we propose to adapt a divide-and-conquer strategy, which enables us to use as many classes as required from a side-channel leakage point of view with a much lower complexity of resources and with a higher accuracy compared to the standard approach. More precisely, our idea is to view the class variables in a tree structure with additional intermediate nodes which are given due to the natural clustering of the measured leakage.

A general illustration is given in Fig. 2, where we first divide into M nodes and then directly into the leafs. Note that compared to the flat approach in Fig. 1, the number of leaves does not change, but its depth does. Certainly, in some scenarios due to the given structure of the leakage, it may be suitable to include several layers of nodes resulting in a higher depth.

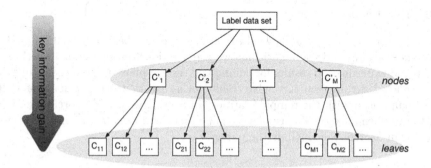

Fig. 2. Hierarchical approach

---

[2] For simplicity we assume that one key chunk is of the same size as one intermediate state chunk, however, this study can easily be extended for other scenarios as given e.g. in DES.

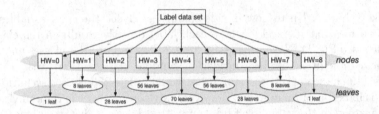

Fig. 3. Hierarchical approach considering HW classes as intermediate nodes

In the context of side-channel analysis, there is a simple hierarchy that one could follow, which we detail next. However, we stress that the hierarchical approach is not restricted to this specific scenario, but is rather a general concept. In particular, one could adapt the approach of clustering side-channel measurements by their similarity as introduced in [13] to build an appropriate tree structure. Note that here we use a priori knowledge to build more accurate classifiers, i.e., use the fact that we know (or can assume) the semantic hierarchy of the data.

First, we divide our measurements regarding the HW of the sensitive variable resulting in 9 classes and then each of those classes is further divided into the sensitive variable itself (leaf classes). Note that, since the HW of a 8-bit word forms a binomial distribution, two classes (HW 0 and HW 8) cannot be divided any further (i.e., these are already leaf classes), while the rest of the classes (HW 1 to HW 7) have various number of subclasses. In this case, using hierarchical approach, the largest number of classes we need to handle at once is for the class with the Hamming weight equal to 4, where there are 70 possible values. We give a depiction of hierarchical model with the Hamming weight/sensitive variable model in Fig. 3.

1.2   Road Map

This paper is organized as follows. In Sect. 2, we present basic information about machine learning and the algorithms we use. Section 3 presents our underlying setup and experimental results for tuning and testing phase for two powerful ML techniques. Next, in Sect. 4, we give results for a realistic testing scenario as well as a comparison with the template attack. In Sect. 5, we give a short discussion on the performance of ML as well as some possible future research directions. Finally, in Sect. 6, we conclude the paper.

2   Machine Learning Techniques

In this section, we briefly present machine learning techniques used in the paper. In our experimental setup, we use four algorithms, where two are relatively simple techniques, i.e., the Naive Bayes and the C4.5 decision tree, while the other two, Support Vector Machines and Rotation Forest, are more complex.

Remark 1. Note that we do not claim that these techniques are optimal, however, preliminary tests showed that both complex techniques had the highest accuracy out of a large pool of tested ML techniques. The reason why we additionally included two simple techniques will be discussed in Sect. 4.

2.1 Naive Bayes

The Naive Bayes classifier is a method based on the Bayesian rule which works under a simplifying assumption: it assumes that the predictor features (points in the measurement traces) are mutually independent given the target class. Existence of highly-correlated features in a dataset can thus influence the learning process and reduce the number of successful predictions. Additionally, Naive Bayes assumes normal distribution for predictor features. A Naive Bayes classifier outputs posterior probabilities as a result of the classification procedure [14]. Note that the space complexity for Naive Bayes algorithm for both training and testing phase is $O(|\mathcal{Y}|Dv)$, where $|\mathcal{Y}|$ is the number of classes, D is the number of features, and v is the average number of values for a features. On the other hand, time complexity for the training phase equals $O(ND)$ and for the testing phase is equal to $O(|\mathcal{Y}|D)$. Here, N is the number of training examples.

2.2 Decision Tree - C4.5

C4.5 is the landmark decision tree algorithm developed by Quinlan [15]. It is a divide-and-conquer algorithm that splits features at tree nodes using the information-based gain ratio criterion. The node splits on further branches if more information is gained (as measured by gain ratio) by the split than by keeping all the instances at the node. The runtime of the algorithm is $O(D) \times N \times \log N$ where D is the number of features and N is the number of instances [16]. The trees are first grown to full length and pruned afterwards in order to avoid data overfitting.

2.3 Rotation Forest

Rotation Forest (RF) is a more recent decision tree ensemble method proposed by Rodriguez et al. [17]. The ensemble is capable of both classification and regression, depending on the base classifier where in most applications, C4.5 algorithm is used as the base learner [15]. The algorithm focuses on presenting transformed data to the classifier by using a projection filter. The most common projection filter and the one that has been shown to be the main factor for the success of the ensemble is the principal component analysis (PCA) [18]. The running time is the same as for C4.5 multiplied with the number of iterations.

2.4 Support Vector Machines

Support Vector Machine (SVM) is a kernel based machine learning family of methods that are used to accurately classify both linearly separable and linearly inseparable data [19]. The basic idea when the data is not linearly separable is to transform them to a higher dimensional space by using a transformation kernel function. In this new space, the samples can usually be classified with a higher accuracy. Many types of kernel functions have been developed, with the most used ones being polynomial and radial-based. The computational complexity of SVM with radial kernel is between linear and quadratic in the number of instances. In this work, we investigate only the radial-based SVM. The most significant parameters are the cost of the margin C and the radial kernel parameter γ. As a learning method for SVM, sequential minimal optimization (SMO) type algorithm is used [20]. Because SMO is a binary classification algorithm, for multi-class classification purposes it is adapted to perform $N \times (N - 1)/2$ binary classifications, where N denotes the number of classes.

3   The Hierarchical Approach Under Test

3.1   Experimental Data

In order to ensure the reproducibility of our results, we use two publicly available data sets for our study.

DPAcontest v2 [10]. This version of the contest provides 1 000 000 measurements (in the template base) of an AES hardware implementation. Previous works showed that the most suitable leakage model (when attacking the last round of an unprotected hardware implementation) is the register writing in the last round, i.e.,

$$Y(k^*) = \underbrace{\mathbf{Sbox}^{-1}[C_{b_1} \oplus k^*]}_{\text{previous register value}} \oplus \underbrace{C_{b_2}}_{\text{ciphertext byte}}, \qquad (1)$$

where k^* denotes the secret key, $\mathbf{Sbox}^{-1}[\cdot]$ denotes the inverse Sbox operation, C_{b_1} and C_{b_2} are two ciphertext bytes, and the relation between b_1 and b_2 is given through the inverse ShiftRows operation of AES. In particular, we choose $b_1 = 12$ resulting in $b_2 = 8$ as it is one of the easiest bytes to attack[3]. For our study, we selected 50 points of interest with the highest correlation between $Y(k^*)$ and data set. Furthermore, we select 100 000 measurements randomly to conduct the subsequent experiments. Figure 4a shows the absolute correlation between $Y(k^*)$ and the measurements for our selected points. One can see that the measurements are relatively noisy and the resulting SNR (signal-to-noise ratio) lies between 0.0069 and 0.0096. To calculate the SNR, we use the model-based approach where we assume a leakage model and X is the measurement we calculate:

$$\frac{var(signal)}{var(noise)} = \frac{var(Y(k^*))}{var(X - Y(k^*))}. \qquad (2)$$

---

[3] See e.g., in the hall of fame on [10].

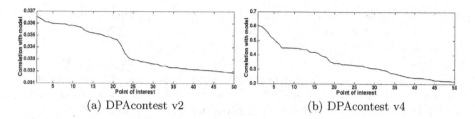

(a) DPAcontest v2   (b) DPAcontest v4

Fig. 4. Correlation between our model and the measurements

DPAcontest v4 [12]. The 4th version provides 100 000 measurements of a masked AES software implementation. However, as the mask is known, one can easily turn it into an unprotected scenario. Though, as it is a software implementation, the most leaking operation is not the register writing, but the processing of the S-box operation and we attack the first round. Accordingly, the leakage model changes to

$$Y(k^*) = \text{Sbox}[P_{b_1} \oplus k^*] \oplus \underbrace{M}_{\text{known mask}} \,, \tag{3}$$

where P_{b_1} is a plaintext byte and we choose $b_1 = 1$. Figure 4b shows the absolute correlation between $Y(k^*)$ and the measurements for our selected points. Compared to the measurements from version 2, there is much higher correlation and naturally also SNR, which is between 0.1188 and 5.8577.

3.2   Training Phase and Parameter Tuning

Tuning represents an important phase in order to properly use ML techniques, but unfortunately has often been ignored or underestimated in previous works on machine learning side-channel attacks. For parameter tuning, we first randomly selected the instances in a ratio of 2:1, where the total number of instances equals 20 000 instances. Those instances are randomly selected out of the datasets of 100 000 and 1 000 000 for DPAcontest v4 and DPAcontest v2, respectively. Then, we take the bigger set as the training set (the set with the 2/3 of the data) and the smaller set for testing (1/3 of the data). On the training set, we conduct 10-fold cross-validation with all parameters considered. In the 10-fold cross-validation, the original training sample is first randomly partitioned into 10 equal sized subsets. Then, a single subsample is selected to validate the data, while the remaining 9 subsets are used for training. The cross-validation process is repeated 10 times, where each of 10 subsamples is used once for validation. The obtained results are then averaged to produce an estimate. All the results in this section are presented as the percentage precision (accuracy) of the classifier. Here, accuracy is defined as the ratio between the sum of true positive and true negative records and sum of all records.

In our experiments we use Weka as the framework for conducting the ML analysis [21]. We do not give details about the tuning phase but we note that we made a grid search where for the Rotation Forest algorithm we investigated *Iteration* parameter in the range [10, 60] with a step of 10. For SVM, we experimented with γ and C parameters, where we tested γ values in the range [10, 70] with a step of 10 and C values in the range [0.1, 0.5] with a step of 0.1. Based on our experiments, we select as the best performing combinations Rotation Forest with 60 iterations and SVM with $C = 70, \gamma = 0.5$ for DPAcontest v4 scenario. For DPAcontest v2, we select Rotation Forest with 60 iterations and SVM with $C = 50, \gamma = 0.4$.

Considering our hierarchical approach and dividing first the measurements into 9 HW classes, we conduct a tuning phase with Rotation Forest and SVM for HW classes 1 to 7 (recall that it is not possible to divide HW classes 0 and 8 further into subclasses). We investigate the same parameter ranges as before, but again omit tuning details, since they are straightforward to obtain (the best obtained algorithms and parameters are given in the next section).

3.3   Testing Results

In this section we perform the testing on an independent set of traces to verify the performances for classifying into 9 and 256 classes (see Table 1) and the performance within each meta-class for the hierarchical approach (see Table 2). The results are given in Accuracy/F-measure/AUC form. Here, the area under the ROC curve is used to measure the accuracy and ROC curve is the ratio between true positive rate and false positive rate. AUC close to 1 represents a good test, while value close to 0.5 represents a random guessing. F-measure is the harmonic mean of the precision and recall, where precision is the ratio between true positive (TP - the number of examples predicted positive that are actually positive) and predicted positive, while recall is the ratio between true positives and actual positives [22]. All the values are given in percentages and in parenthesis, we give the parameter combinations reaching those values.

Note that this represents an ideal test scenario, where each meta-class only contains measurements from the correct HW class. Therefore, all the instances in a meta-class really belong to that meta-class. The next section discusses a more realistic attacking scenario where errors are propagated through the tree. For the testing results for the hierarchical approach (i.e., looking at each subclass), we only give the best obtained values. In addition to the best values, we give the parameter combinations used to obtain those values.

4   Realistic Testing

The goal in this section is to attack the implementation with our new approaches and assess their performance when compared to attacking immediately 256 classes. Here, we use 10 000 and 25 000 random measurements for all tests in order to have a fair comparison. The traces are divided uniformly at random in 2:1 ratio where we use 2/3 of measurements for profiling and 1/3 for the testing.

Table 1. Testing results for 9 and 256 classes (Accuracy/F-measure/AUC)

Algorithm	DPAcontest v4		DPAcontest v2	
	Rotation Forest	SVM	Rotation Forest	SVM
		9 classes		
Value	94.1/94.1/99.6	95.5/95.5/98.9 $(C = 70, \gamma = 0.5)$	25/19.8/50.2 $(Iter. = 60)$	23.67/19/50.1
		256 classes		
Value	26.7/24.3/50.9	27.8/28/96.9 $(C = 70, \gamma = 0.5)$	0.36/0.4/50.4	0.45/0.4/51 $(C = 50, \gamma = 0.4)$

Table 2. Testing results for the hierarchical approach (Accuracy/F-measure/AUC)

Set	DPAcontest v4	DPAcontest v2
HW1	69.6/68.9/91.1 (SVM, $C = 4, \gamma = 0.6$)	15.2/15.2/5.3 (RF, $Iter. = 50$)
HW2	67.8/57.5/96.3 (SVM, $C = 10, \gamma = 0.7$)	4.0/2.2/51.7 (SVM, $C = 1, \gamma = 0.1$)
HW3	49.5/49.4/97.3 (SVM, $C = 10, \gamma = 0.9$)	2.0/4/52.2 (SVM, $C = 1, \gamma = 0.1$)
HW4	46.4/46.4/97.5 (SVM, $C = 20, \gamma = 0.8$)	1.7/7/51.8 (SVM, $C = 1, \gamma = 0.4$)
HW5	49.9/50.1/97.3 (SVM, $C = 10, \gamma = 1$)	2.0/0.9/50.4 (SVM, $C = 1, \gamma = 0.4$)
HW6	57.7/58.0/96.3 (SVM, $C = 10, \gamma = 0.7$)	3.8/1.1/50.1 (SVM, $C = 1, \gamma = 0.1$)
HW7	74.5/74.6/92.2 (SVM, $C = 4, \gamma = 0.7$)	13.2/12.1/49 (SVM, $C = 1, \gamma = 0.2$)

The best results in all tables are highlighted with gray background color of a cell.

Therefore, we investigate a number of cases here that all fall within three categories:

1. Attacking directly all 256 classes.
2. Attacking 9 classes (i.e., the Hamming weight classes).
3. New attacks - Hierarchical attack and Structured attack.

4.1   Hierarchical Attack

In the Hierarchical attack, one first investigates how to classify measurements into a (relatively) small number of classes, i.e., into subclasses (which can be repeated several times) and then, in the second step, the obtained classification results are further classified into leaves. However, since not all the measurements are correctly classified in the first step, they need to be discarded (since they belong to subclasses) and the total number of available measurements will be consequently lower than the number of instances we begin with. Note that the number of leafs is the same as if one considers the flat approach, but the classification method in each independent step considers a smaller number of classes. With this approach, we are able not only to improve the accuracy, but also to lower the computational and space complexity for the classification process. Finally, since we are running independent experiments on each of the subclasses, it is also easy to parallelize the attack, which may not be an option when considering the flat approach. We give more algorithmic description of the hierarchical attack in the following listing:

1. Find a hierarchical relation to explore.
2. Run a classifier for each level of subclasses.
3. Consider all instances classified above some threshold value as correctly classified (e.g., all instances that have a probability of more than 90% to be correctly classified into certain subclass), otherwise discard.
4. For all instances kept in a subclass, run new classifier in order to find in which subclass they belong (repeat until leaf class is reached).

In our case study we exploit the HW of an intermediate value (see Eqs. (1) and (3)) and thus first divide into 9 HW classes. Then, we use the measurement predictions from that phase to conduct an attack on the intermediate value itself (leaf class).

4.2 Structured Attack

In addition to the Hierarchical attack, we introduce an attack combination of the standard flat approach and our hierarchical approach, which we denote as the Structured attack. Accordingly, we merge the information from both approaches and even further improve the accuracy. This is due to a fact that, when attacking with the flat approach, we expect that the final accuracy will be lower than for the hierarchical attack, but there will be instances where the flat classification classifies correctly, while hierarchical classification makes a wrong prediction. This combination is of particular interest when the computing power and the runtime complexity is of no importance, but only the accuracy of the attack. We give a more precise listing for the Structured attack in the following:

1. Run classifier with the flat approach.
2. Run the Hierarchical attack.
3. Assign weight factors for each of the two aforesaid steps.
4. Combine results from flat and hierarchical approach. Similarly as in the Hierarchical approach, set a threshold value that signifies which classification guesses to take as true.

4.3 Attack Results and Comparison with Template Attack

In order to facilitate a better understanding of the obtained results, we also use two simpler machine learning techniques - Naive Bayes and Decision Tree (C4.5). The Naive Bayes algorithm does not have parameters one could tune and the tuning phase for Decision Tree highlights that the default values of parameters are the best. Therefore, we use a minimum amount of two instances per leaf and a confidence factor for pruning equal to $C = 0.25$. It is important to state that these two methods are extremely fast (especially the Naive Bayes) and we consider it to be beneficially to run them always as a first indicator of what can be expected from an ML approach. Recall, the parameter combinations for RF and SVM algorithms used in this section are obtained in Sects. 3.2 and 3.3.

In Table 3, we present the results when working with 10 000 instances from the DPAcontest v2 and the DPAcontest v4 using 9 classes. Next, in Table 4, we present results for 10 000 instances using 256 classes. Note that the parameters for ML techniques are as presented in Table 1. As in the previous scenario, we can observe that SVM has the highest accuracy. Interestingly, Naive Bayes is more efficient than Decision Tree when working with 256 classes, but for 9 classes the situation is opposite. This is due to a jump of complexity appearing in the Decision Tree when the number of classes is increased. However, here we observe that Naive Bayes has the best accuracy when considering all 256 classes, which is a somewhat surprising result. We believe that a more extensive tuning phase for Rotation Forest and SVM would change this. However, we do not consider this completely justified when considering the huge difference in the runtime complexity between the Naive Bayes and these methods.

To rate the goodness of our achieved results, we additionally applied the template attack (TA) [1] to the same set of traces as used in the realistic attacking scenario and tested it for the standard approach of classifying 9 (see Tables 3 and 6) and 256 classes (see Tables 4 and 7) directly. TA is the most common and well-studied profiled side-channel attack and it is considered as the most powerful one from an information theoretic perspective given an infinite amount of measurements. However, compared to ML techniques, recent works showed its inferiority when, for example, the profiling set is not large enough or the attack is provided with too many useless (without information) points of interest (features) [23]. Actually, the attack principle is very close to the one of the Naive Bayes, where the main difference is the consideration of the features along with the measurements to be dependent. In particular, the Naive Bayes assumes independence and thus considers a univariate normal distribution. On the contrary, for TA the noise is considered dependent and thus a multivariate distribution is taken. For more details, we refer interested readers to [1]. To be more efficient and numerically stable, we applied the adaptation of using only one covariance matrix instead of 9 or 256 as described in [24]. Note that, for all the algorithms, we use the same datasets with the same feature selection process in order to make it as fair as possible (and to avoid seeing differences in results stemming from other than hierarchy causes, e.g. better feature selection).

The attack results of our new Hierarchical and Structured attack using 10 000 instances are given in Table 5. As both new attacks are considering the same leaf classes as in Table 4, we can directly compare their results. We note that in this set of experiments, we use the threshold value equal to 0.9, which represents that only the measurements with a high output probability of belonging to a certain class are taken as correctly classified. We can see that the Hierarchical SVM outperforms regular SVM, but the Structured SVM is by far the most efficient method for both the DPAcontest v4 and DPAcontest v2 scenario. We note that the training phases for the Hierarchical and Structured attacks consist of training

Table 3. Attack scenario with 9 classes, 10 000 instances

Algorithm	# classes	Parameters	Training	Testing
		DPAcontest v4		
Naive Bayes	9	-	66.8	65.9/66.0/91.2
Decision Tree	9	$c = 0.25, M = 2$	70.1	71.8/71.8/85.2
SVM	9	$C = 70, \gamma = 0.5$	90.9	91.39/91.4/98
TA	9	-	-	76.71
		DPAcontest v2		
Naive Bayes	9	-	11.78	11/10.2/50.1
Decision Tree	9	$c = 0.25, M = 2$	19.36	20.58/20.4/50.7
Rotation Forest	9	$Iter = 60$	24.69	25.12/19.6/51.2
TA	9	-	-	8.31

Table 4. Attack scenario with 256 classes, 10 000 instances

Algorithm	# classes	Parameters	Training	Testing
		DPAcontest v4		
Naive Bayes	256	-	18.5	17.0/16.3/93
Decision Tree	256	$c = 0.25, M = 2$	13.4	13.2/13.2/58.5
SVM	256	$C = 70, \gamma = 0.5$	30.4	27.8/28/96.9
TA	256	-	-	20.19
		DPAcontest v2		
Naive Bayes	256	-	0.42	0.58/0.1/51.3
Decision Tree	256	$c = 0.25, M = 2$	0.28	0.36/0.3/49.9
SVM	256	$C = 50, \gamma = 0.4$	0.43	0.45/0.4/51
TA	256	-	-	0.39

Table 5. Hierarchical and Structured attack, 10 000 instances

Algorithm	# classes	Parameters	Training	Testing
		DPAcontest v4		
Hierarchical attack	9/256	Table 2	38.23	31.36*
Structured attack	9/256	Tables 1 and 2	-	33.7
		DPAcontest v2		
Hierarchical ML	9/256	Table 2	2.95	1.32**
Structured ML	9/256	Tables 1 and 2	-	0.91

* From 3 016 correctly classified instances for 9 classes.
** From 829 correctly classified instances for 9 classes.

phases of the whole hierarchy (i.e., classes HW1 up to HW7) and therefore we give here the median value of the training accuracies. Note that we had only 6 700 instances to train for the Hierarchical attack, which gives on average 26 traces per class. When conducting the Hierarchical attack, we cannot use the

Table 6. Attack scenario with 9 classes, 25 000 instances

Algorithm	# classes	Parameters	Training	Testing (Acc./F-measure/AUC)
		DPAcontest v4		
Naive Bayes	9	-	70.01	67.85/67.9/91.7
Decision Tree	9	$c = 0.25, M = 2$	74.39	74.75/74.7/86.7
SVM	9	$C = 70, \gamma = 0.5$	93.83	94.32/94.3/98.6
TA	9	-	-	77.85
		DPAcontest v2		
Naive Bayes	9	-	8.21	8.1/10.2/50.3
Decision Tree	9	$c = 0.25, M = 2$	19.43	20.26/20.2/50.7
Rotation Forest	9	$Iter = 60$	25.15	24.71/19.5/50.4
TA	9	-	-	6.47

Table 7. Attack scenario with 256 classes, 25 000 instances

Algorithm	# classes	Parameters	Training (Acc.)	Testing (Acc./F-measure/AUC)
		DPAcontest v4		
Naive Bayes	256	-	20.44	20.27/18.4/94.5
Decision Tree	256	$c = 0.25, M = 2$	15.38	16.23/16.2/60.5
SVM	256	$C = 70, \gamma = 0.5$	35.54	35.02/35.1/98.1
TA	256	-	-	25.07
		DPAcontest v2		
Naive Bayes	256	-	0.65	0.5/0.1/50.8
Decision Tree	256	$c = 0.25, M = 2$	0.42	0.39/0.4/50
TA	256	-	-	0.4

whole test set (i.e., 3 300 instances), since some of them are wrongly classified when classifying into 9 classes. We give details about the available number of instances in notes below the table. For instance, the remark "From 3 016 correctly classified instances for 9 classes" means that after classifying into 9 classes, we have 3 016 correctly classified instances. Then, when classifying into subclasses we have only 3 016 instances in total and we see that in total 33.7% of those instances are correctly classified into intermediate values. In Tables 6, 7, and 8, we present results for 25 000 instances. When comparing the results achieved using 10 000 instances, we observe that the results of all attacks for the DPAcontest v4 are improving, whereas for the higher noise scenario from DPAcontest v2, the results are not changing noticeably when considering 9 and 256 classes (i.e., when working with standard ML techniques). Note that here we have on average 65 instances per class when training for hierarchical attack.

Table 8. Hierarchical and Structured attack, 25 000 instances

Algorithm	# classes	Parameters	Training	Testing
		DPAcontest v4		
Hierarchical attack	9/256	Table 2	44.01	40.74*
Structured attack	9/256	Tables 1 and 2	-	44.43
		DPAcontest v2		
Hierarchical ML	9/256	Table 2	2.92	1.69**
Structured ML	9/256	Tables 1 and 2	-	0.92

* From 7 844 correctly classified instances for 9 classes.
** From 2 066 correctly classified instances for 9 classes.

5   Discussion

In our examples, when conducting the hierarchical approach, we consider an extreme case: first dividing into 9 classes in accordance with the HW, and then dividing into all values for the corresponding HW. For some classes (HW 0 and 8) the hierarchical approach does not make a difference, since it is not possible to divide them any further. Contrary, for the Hamming weight class 4 contains 70 leaves, which is again a complex scenario. Therefore, one could instead use sets of two (or any other number) values that are mapped to the same class. For instance, in the Hamming weight class 4, values 23 and 27 can be grouped into a subclass. Then, in the next step, one uses a binary classification for those two values.

Using SVM with a flat classification, 256 classes, and measurements from the DPAcontest v4 (10 000 instances), we correctly classify 917 out of 3 300 instances (27.8%). When classifying into 9 classes, SVM reaches an accuracy of 91.4%, which translates into 3 016 correctly classified instances. The Hierarchical attack has an accuracy of 31.3%, which amounts to 945 instances that are correctly classified. Although the difference between the flat and hierarchical approach is small in this example, we note that we still improve the accuracy without using any extra information and with only a small overhead from the computational side. Even more so, when considering the Structured attack, the accuracy equals 33.7%, which is in total equal to 1 112 correctly classified instances. When compared to the flat approach, this amounts to 21% more instances that are correctly classified, which represents a significant improvement.

When considering the measurements of the DPAcontest v2, the classification is much more difficult. Indeed, when classifying into 9 classes, the accuracy is around 25% which results into 5 out of 9 classes having correctly classified instances. Therefore, to significantly improve the accuracy of the Hierarchical attack, we would need to use much more measurements. On the basis of the results from Table 5, it could seem that the Hierarchical attack has better accuracy (1.32%) than the Structured attack (0.91%) but that is actually not true. Indeed, when considering the Hierarchical attack, the accuracy can be calculated only from the number of instances that are correctly classified into subclasses

(829 instances), which results in around 11 correctly classified instances. On the other hand, the accuracy of 0.91% for the Structured attack must be taken on the whole test set (3 300) which equals 30 correctly classified instances. Therefore, we see that the Structured attack offers a significant improvement over all other considered methods.

We emphasize that in order to obtain a fair comparison the Hierarchical and Structured attack must be compared with the 256 classes scenario, and not with the results from 9 classes. With the increase in the number of instances to 25 000, the superior performance of the Hierarchical and Structured attacks becomes even more apparent. For instance, when considering the measurements of the DPAcontest v4, SVM with the flat approach and classifying 256 classes has the accuracy of 35%, TA of 25%, and Structured attack of 44%.

On a more general level, we present here two novel attacks that are able to significantly increase the efficiency of ML techniques when compared to related work. Naturally, conducting the Hierarchical attack is more computationally expensive than just attacking the Hamming weight classes, and the Structured attack is even more expensive since one needs to use flat approach on all 256 classes as well as the Hierarchical attack. However, the increase in the number of experiments is well compensated with the increased accuracy of those methods. On the other hand, the hierarchical approach for ML techniques is beneficial from the runtime complexity side, since using smaller number of classes decreases the runtime of ML, while dividing experiments enables one to easily use parallel computing. The process of making a hierarchy is here considered to be simple and therefore its complexity is negligible. Naturally, this does not need always to happen, which would make our attack more complex in accordance with the process of finding the hierarchy.

When considering realistic settings, one does not know whether a classifier correctly classified certain instances into subclasses. Therefore, it is necessary to use a threshold which serves as a cut-off for all measurements below it. Naturally, the value of such a threshold is a parameter that can be tuned and that differs with respect to the underlying setting. For instance, measurements with smaller levels of noise can have higher threshold values since it is expected that the classifier will be able to classify certain instance with high probability of success. However, we note that the threshold level is in the end to be set by the attacker, with regards to how reliable he considers the classifier to be.

In this paper, we considered the HW of the intermediate value as a first level in the hierarchy. However, we do not claim that this choice is optimal or should be generally taken. Another approach would be to use the values of each bit of the intermediate value as a level of hierarchy, e.g., all the measurements where the first bit equals to 1 goes into one class and where the first bit equal to 0 into other class. Then, each of those subclasses has 128 subclasses. A more general approach would be, if some hierarchical structure is not readily observable, to build a hierarchy with the automatic generation of subclasses, where algorithm groups leaf classes by their similarity [13].

Finally, we give several observations why the hierarchical attack might improve the accuracy in some cases. The first reason is because we use a priori knowledge about the dataset (i.e., we know the semantic hierarchy). Naturally, this can also be a source of mistake, where the question is how severe would a (slightly) wrong hierarchy influence the results. Since in our experiments, we use only two levels of hierarchy, then consequently, the propagation of error in the classification cannot go far. The second reason why hierarchical attacks improve the accuracy over flat approach is that they can limit the model complexity and constrain the expressiveness of a hypothetical class. We leave for future research the experiments showing which reason has more influence on success in these scenarios. Moreover, it would be interesting to explore how robust is the hierarchical classification when the hierarchy does not model the data completely. Still, we emphasize that the complexity of classification for each subclass and the corresponding subclasses is lower than in the case of the flat classification (since most of the algorithms have complexity increasing linearly with the number of classes).

As future work, we are interested in exploring how hierarchical and structured approaches behave when using a larger number of instances. Moreover, we observe that in the hierarchical approach, wrongly classified measurements often exhibit some structure (e.g., the measurements belonging to one class are dominantly classified as belonging to some other class) and we would like to investigate the automatic generation of classes (similar to [25]). With such an approach, we expect to find some new subclasses that can be used in the hierarchical approach.

6   Conclusions

In this paper, we introduced the concept of hierarchical machine learning classification for side-channel analysis. Instead of attacking immediately the sensitive variable or just the Hamming weight of it, we propose to use a divide-and-conquer approach in a form of class hierarchy. To show the practicability of our new approach, we conducted our analysis on two publicly available data sets from the DPAcontest with different SNRs and made a comparisons to machine learning techniques and the template attack using the standard (flat) approach. Our results show that, for both data sets, the Hierarchical and Structured attacks outperform other ML approaches as well as the template attack. Aside from the better accuracy with our hierarchical approach, an additional advantage is also the lower computational complexity for ML techniques, which renders more plausible such attacks when using realistic data sets with large number of measurements and points in time.

Acknowledgments. S. Picek was supported in part by Croatian Science Foundation under the project IP-2014-09-4882.

References

1. Chari, S., Rao, J.R., Rohatgi, P.: Template attacks. In: Kaliski, B.S., Koç, K., Paar, C. (eds.) CHES 2002. LNCS, vol. 2523, pp. 13–28. Springer, Heidelberg (2003). doi:10.1007/3-540-36400-5_3
2. Schindler, W., Lemke, K., Paar, C.: A stochastic model for differential side channel cryptanalysis. In: Rao, J.R., Sunar, B. (eds.) CHES 2005. LNCS, vol. 3659, pp. 30–46. Springer, Heidelberg (2005). doi:10.1007/11545262_3
3. Lerman, L., Bontempi, G., Markowitch, O.: Side channel attack: an approach based on machine learning. In: Second International Workshop on Constructive SideChannel Analysis and Secure Design, Center for Advanced Security Research Darmstadt, pp. 29–41 (2011)
4. Hospodar, G., Gierlichs, B., De Mulder, E., Verbauwhede, I., Vandewalle, J.: Machine learning in side-channel analysis: a first study. J. Cryptographic Eng. 1, 293–302 (2011). doi:10.1007/s13389-011-0023-x
5. Heuser, A., Zohner, M.: Intelligent machine homicide. In: Schindler, W., Huss, S.A. (eds.) COSADE 2012. LNCS, vol. 7275, pp. 249–264. Springer, Heidelberg (2012). doi:10.1007/978-3-642-29912-4_18
6. Lerman, L., Bontempi, G., Markowitch, O.: The bias-variance decomposition in profiled attacks. J. Cryptographic Eng. 5(4), 255–267 (2015)
7. Lerman, L., Bontempi, G., Markowitch, O.: Power analysis attack: an approach based on machine learning. IJACT 3(2), 97–115 (2014)
8. Lerman, L., Medeiros, S.F., Bontempi, G., Markowitch, O.: A machine learning approach against a masked AES. In: Francillon, A., Rohatgi, P. (eds.) CARDIS 2013. LNCS, vol. 8419, pp. 61–75. Springer, Cham (2014). doi:10.1007/978-3-319-08302-5_5
9. Heuser, A., Kasper, M., Schindler, W., Stöttinger, M.: A new difference method for side-channel analysis with high-dimensional leakage models. In: Dunkelman, O. (ed.) CT-RSA 2012. LNCS, vol. 7178, pp. 365–382. Springer, Heidelberg (2012). doi:10.1007/978-3-642-27954-6_23
10. TELECOM ParisTech SEN research group: DPA Contest. 2nd edn. (2009–2010). http://www.DPAcontest.org/v2/
11. Xilinx: Virtex-5 libraries guide for HDL designs. http://www.xilinx.com/support/documentation/sw_manuals/xilinx14_4/virtex5_hdl.pdf
12. TELECOM ParisTech SEN research group: DPA Contest. 4th edn. (2013–2014). http://www.DPAcontest.org/v4/
13. de Almendra Freitas, C.O., Oliveira, L.S., Aires, S.B.K., Bortolozzi, F.: Metaclasses and zoning mechanism applied to handwriting recognition. J. UCS 14(2), 211–223 (2008)
14. Friedman, N., Geiger, D., Goldszmidt, M.: Bayesian network classifiers. Mach. Learn. 29(2), 131–163 (1997)
15. Quinlan, J.R.: C4.5: Programs for Machine Learning. Morgan Kaufmann Publishers Inc., San Francisco (1993)
16. Frank, E., Witten, I.H.: Generating accurate rule sets without global optimization. In: Shavlik, J. (ed.) Fifteenth International Conference on Machine Learning, pp. 144–151. Morgan Kaufmann (1998)
17. Rodriguez, J.J., Kuncheva, L.I., Alonso, C.J.: Rotation forest: a new classifier ensemble method. IEEE Trans. Pattern Anal. Mach. Intell. 28(10), 1619–1630 (2006)

18. Kuncheva, L.I., Rodríguez, J.J.: An experimental study on rotation forest ensembles. In: Haindl, M., Kittler, J., Roli, F. (eds.) MCS 2007. LNCS, vol. 4472, pp. 459–468. Springer, Heidelberg (2007). doi:10.1007/978-3-540-72523-7_46

19. Vapnik, V.N.: The Nature of Statistical Learning Theory. Springer, New York (1995)

20. Platt, J.: Fast training of support vector machines using sequential minimal optimization. In: Schoelkopf, B., Burges, C., Smola, A. (eds.) Advances in Kernel Methods - Support Vector Learning. MIT Press (1998)

21. Hall, M., Frank, E., Holmes, G., Pfahringer, B., Reutemann, P., Witten, I.H.: The WEKA data mining software: an update. SIGKDD Explor. Newsl. **11**(1), 10–18 (2009)

22. Powers, D.M.W.: Evaluation: from precision, recall and F-factor to ROC, informedness, markedness and correlation (2007)

23. Lerman, L., Poussier, R., Bontempi, G., Markowitch, O., Standaert, F.-X.: Template attacks vs. machine learning revisited (and the curse of dimensionality in side-channel analysis). In: Mangard, S., Poschmann, A.Y. (eds.) COSADE 2014. LNCS, vol. 9064, pp. 20–33. Springer, Cham (2015). doi:10.1007/978-3-319-21476-4_2

24. Choudary, O., Kuhn, M.G.: Efficient template attacks. In: Francillon, A., Rohatgi, P. (eds.) CARDIS 2013. LNCS, vol. 8419, pp. 253–270. Springer, Cham (2014). doi:10.1007/978-3-319-08302-5_17

25. Whitnall, C., Oswald, E.: Robust profiling for DPA-style attacks. In: Güneysu, T., Handschuh, H. (eds.) CHES 2015. LNCS, vol. 9293, pp. 3–21. Springer, Heidelberg (2015). doi:10.1007/978-3-662-48324-4_1

Multivariate Analysis Exploiting Static Power on Nanoscale CMOS Circuits for Cryptographic Applications

Milena Djukanovic[1(✉)], Davide Bellizia[2],
Giuseppe Scotti[2], and Alessandro Trifiletti[2]

[1] Faculty of Electrical Engineering, University of Montenegro, Podgorica, Montenegro
milenadj@ac.me
[2] DIET, Università di Roma "La Sapienza", Rome, Italy
{bellizia,scotti,trifiletti}@diet.uniroma1.it

Abstract. Latest nanometer CMOS technology nodes have highlighted new issues in security of cryptographic hardware implementations. The constant growth of the static power consumption has led to a new class of side-channel attacks. Common attacks exploiting static power use an univariate approach to recover information from cryptographic engines. In our work, a multivariate approach based on information theoretic security metrics is presented. The temperature-dependence helps to exploit more information leakage from the hardware implementation. Starting from a univariate analysis, mutual information reveals that increasing the working temperature, the information leaked through the static power side channel is increased as well. In this work a multivariate analysis exploiting static power consumption is presented in which the temperature-domain is used to extract more information. The use of information theoretic approach allows to precisely quantify the amount of information that can be leaked from a cryptographic hardware implementation. The perceived information shows taking advantage of the use of more than one temperature, the security level can be decreased. The improvement achieved using the presented approach is demonstrated on a 40 nm CMOS implementation of the Present 80 crypto core.

Keywords: Side-channel attack · Static current · Cryptography · CMOS · Power analysis attack · Perceived information

1 Introduction

Side-channel attacks have proved to be one of the most important threats against modern cryptographic implementations in the last two decades. The security assessment of commercial products (such as smart cards) has implied new design and evaluation of countermeasures in order to withstand up-to-date physical attacks on hardware devices. Complex realization of cryptographic algorithms do not always relate to the robustness in terms of security of their physical

© Springer International Publishing AG 2017
M. Joye and A. Nitaj (Eds.): AFRICACRYPT 2017, LNCS 10239, pp. 79–94, 2017.
DOI: 10.1007/978-3-319-57339-7_5

implementations, as shown by Kocher in 1996, demonstrating the possibility to recover secret information by introducing a novel method for exploiting the leaked information from the device - Side Channel Attack (SCA) [1].

The scaling of the CMOS technology, that is the basis of most present micro-electronic devices, is a permanent trend since the introduction of integrated circuits. Consequences of this scaling lead to faster growth of static power consumption, having as a result new issues in hardware security. These issues refer to static current increase [2] and process variations influence [3]. In the area of modern cryptographic cores this leads to new expectations of being high-performance, area efficient and low power [4]. Well defined procedure for exploiting static current with "a-priori" model is given in [5,6] in the form of Leakage Power Analysis (LPA) using the Pearson's correlation coefficient as statistical distinguisher. So far, Differential Power Analysis (DPA) and Correlation Power Analysis (CPA) attacks have become a major threat to security of crypto cores [7]. Their effectiveness is based on the possibility of finding correlation between the dynamic power consumption and the data processed by a logic circuit.

The exploitability of the static power consumption as source of information leakage has gained an increasingly importance in the scientific community. First results of practical evaluation of the possibility to recover sensible information using static power as side channel have been presented in [8]. Author introduced experimental evaluation of static power consumption of three different Xilinx FPGAs, built in three different technologies (65 nm, 45 nm and 28 nm), demonstrating the feasibility of exploitability of static power in real devices. In [9], authors extended previous results in order to assess if the exploitability of static power consumption could provide a real benefits to the adversary, compared to standard procedures that use to exploit dynamic power consumption. The analysis carried out in [9] has pointed out that if the adversary has the control over the clock signal of the device under attack, the static power consumption is critical from a security point of view, since the adversary can strongly reduce the noise on measurements of static power samples. Authors in [10] presented a study of the vulnerability of several dual-rail pre-charged logic families in the context of attacks exploiting static power. Simulated results on 40 nm implementations show that countermeasures conceived to protect cryptographic hardware from dynamic power attack still exhibit a strong information leakage through static power. The importance of this consideration is critical, since standard-cell based gate-level (WDDL [11] and MDPL [12]) countermeasures are weaker in terms of security performance compared to CMOS if static consumption is used as side channel.

In this work, we evaluate the possibility to exploit information leaked through static power consumption executing univariate and multivariate analysis. For both analysis we use 4-bit PRESENT crypto-core and the full implementation of the PRESENT-80 block cipher in order to extract information from a hardware implementation of a cryptographic algorithm using static currents as source of information leakage, according to previous works [13,14]. The possibility to use more than one domain or dimension at a time (e.g. univariate case) leads us to more precise results (e.g. multivariate cases), even in the presence of process variations of simulated lightweight crypto core. Results are evaluated on the base of recently proposed information theoretic security metrics, that allow to precisely quantify the amount of information that can be extracted from hardware implementations. The multivariate approach that we propose, highlights the possibility to increase the actual information leakage, overcoming issues related to intra-die variations. Monte Carlo simulated chips are used to show the impact of intra-die variations on the information leakage, and how our method is suitable to precisely assess the security level of an implementation.

The remainder of the paper is organized as follows. In Sect. 2 the leakage phenomena is described along with some security metrics which have been used for validating the simulations of leakage data. Section 3 presents two case studies - 4-bit PRESENT crypto-core and full 40 nm CMOS lightweight implementation of PRESENT-80 block cipher. In Sects. 4 and 5 are presented univariate and multivariate analysis of information leakage of two adopted study cases, respectively. Suggestions for future work and conclusions are given in Sect. 6.

2 Background

The static current conducted by MOS transistors consists of three major sources: inverse junction current, gate tunnel current and sub-threshold current. In up-to-date technologies, the sub-threshold current is the most important leakage contribution in a MOS transistor [15]. More specifically, the sub-threshold current I_{leak} is given by:

$$I_{leak} = K \frac{W}{L} e^{\frac{V_{gs}-(V_{th}-\eta V_{ds}-\gamma V_{bs})}{nV_T}} \cdot \left(1 - e^{\frac{-V_{ds}}{V_T}} \right) \tag{1}$$

where V_{th} is the threshold voltage, V_T is the thermal voltage, V_{gs}, V_{ds} and V_{bs} are operating voltages of the device. In [10], authors show the data-dependence of the static current for basic cells in CMOS technology, comparing technologies from 90 to 28 nm process.

Under the perspective of static power consumption exploitation, an adversary measures the static power consumption of a device by storing the traces of the absorbed current, and exploiting the dependence of the samples on internal data. For assessing the resistance of a hardware implementation to the extraction of information through static power, several security metrics have been adopted in literature [9,10]. For our investigations we use both actual and information theoretic security metrics.

- **Signal-To-Noise Ratio (SNR).** According to [16], it quantifies the physical leakage of an hardware implementation, by means of the ratio of the variance of the data-dependent component of a power trace (σ_{data}^2), and the variance of the noise, due to switching noise of uncorrelated and non-cryptographic logics on-chip ($\sigma_{sw,noise}^2$) and to electronic noise ($\sigma_{el,noise}^2$).

$$SNR = \frac{\sigma_{data}^2}{\sigma_{el,noise}^2 + \sigma_{sw,noise}^2} \tag{2}$$

- **Measurements To Disclosure (MTD).** In order to give an estimation of the robustness of an implementation to a side-channel attack, it is useful to find the minimum number of measurements that are necessary to recover the secret key used by the hardware implementation [17, 18].
- **Mutual Information.** Introduced in [19], the mutual information quantifies how much information is leaked from a cryptographic implementation, by means of Shannon entropy of a random variable X and the observed leakage L:

$$MI(X; L) = H[X] - \sum_{x \in X} Pr(x) \sum_{l \in L} Pr_{chip}(l/x) \log_2 Pr_{chip}(x/l) \quad (3)$$

where $H[X]$ is the entropy of the variable X, $Pr(x)$ is the probability of $x \in X$, and $Pr_{chip}(x/l)$ is derived from $Pr_{chip}(l/x)$ thanks to Bayes' theorem. The underlying assumption is based on the fact that the adversary has a perfect knowledge about the distribution $Pr_{chip}(l/x)$. So, it represents the upper bound of the amount of information that can be extracted from a hardware implementation in a given scenario.
- **Perceived Information.** The perceived information, presented in [20] and defined in [21], captures the information about the secret variable X obtained when observing leakage L, generated according to the probability distribution $Pr_{chip}(L/x)$, and interpreted with the model $Pr_{model}(L/x)$:

$$\widehat{PI}(X; L) = H[X] - \sum_{x \in X} Pr(x) \sum_{l \in L} Pr_{chip}(l/x) \log_2 Pr_{model}(x/l) \quad (4)$$

The $\widehat{PI}(X; L)$ estimates how well the model $Pr_{model}(L/x)$ is suitable for the implementation under analysis, and how much information an adversary can extract with this model, considering it non perfect. The use of this metric allows to define the "average case" for the adversary, because assumptions used in this case describe well a realistic attack scenario. In a real scenario, we have:

$$\widehat{PI}(X; L) \leq MI(X; L) \quad (5)$$

3 Case Study

In this work, we have used simulated static power measurements collected on transistor-level designs of the PRESENT-80 block cipher [22]. The PRESENT-80 block cipher is an ultra-lightweight algorithm which uses 80-bit key and 64-bit plaintext. The encoding operation is made of 31 rounds of the following paradigm:

- **AddRoundKey.** The round plaintext at i-th round is XORed with the most significant 64-bit of round key at the i-th round.
- **SubsLayer.** 16 parallel 4×4 bit SBOXes perform non-linear function on the output of the *AddRoundKey* layer.
- **PermLayer.** A linear permutation is performed on the output of the previous layer.

The last round omits the *SubsLayer* and *PermLayer*, and only the *AddRoundKey* is performed. The 80-bit round key is computed at each round, and it is derived from the original round key, as shown in [22].

The analysis presented in this work is performed with two case studies, that use PRESENT-80 as reference. Both of them are designed with 40 nm CMOS technology, provided by STMicroelectronics, with standard threshold voltage option. The library used is provided with BSIM4 models, and with statistical parameters, in order to better evaluate intra-die variations.

3.1   4-Bit PRESENT Crypto-Core

The PRESENT-80 algorithm can be easily implemented in a slice architecture. The basic slice is composed by a 4-bit data-path. A 4-bit XOR combines input vectors (e.g. plaintext and secret key), and the output feeds a single PRESENT SBOX. The output register samples the 4-bit encrypted word. The block scheme of the data-path is depicted in Fig. 1a.

3.2   Full Implementation of PRESENT-80 Block Cipher

In addition to the 4-bit PRESENT crypto-core, our analysis is performed also on the full CMOS implementation. The design is intended to be a lightweight implementation of the PRESENT-80 block cipher. We have used the *iterative loop* architecture to design our cryptographic engine [22], as depicted in Fig. 1b, in order to get an area-efficient architecture. The final ciphertext is captured at the output of the *AddRoundKey* at the end of the last round. It has to be noticed that the *PermLayer* has no combinational logic inside, and it does not provide power consumption due to proper wiring to achieve the linear permutation.

3.3   Testbench

In this work, we present results based on simulated power traces collected on Cadence Virtuoso environment. In a real experiment, an adversary has to be able to measure the DC current after the target word is processed by the device under analysis. The static current can be measured in several ways. In order to retrieve the static consumption, a simple method is to sample the current drawn at V_{DD} pin of the cryptographic engine before the sampling edge of the clock signal, after the target operation is computed/executed. Similarly, we can sample the static current after stopping the clock signal. In this way, the measurement procedure can exclude the presence of dynamic phenomena, and only the static current is collected. We have chosen the latter way, in order to assess precisely the amount of information that is leaked from a hardware implementation of a cryptographic algorithm, considering only static phenomena. Measurements are collected in the same way for both case studies. Power supply voltage is kept constant at 1V, and clock frequency has been set to 10 MHz, according to previous work [23,24].

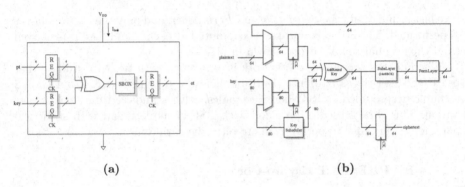

Fig. 1. Left: data path of the 4-bit PRESENT crypto-core. Right: RTL-level schematic of the PRESENT-80 block cipher implementation using iterative looping architecture.

4   Univariate Analysis of Information Leakage

The univariate analysis is the simplest way to evaluate the possibility to exploit information leaked through static power consumption. A univariate analysis is based on the use of a single current sample, which is used as relevant for the information leakage. From static currents point of view, all other samples within the same clock cycle are redundant. Moreover, samples from different clock periods are uncorrelated, and their cross-correlation is not useful to retrieve information, because they are generated from different process, and, more in general, from different data. If we consider a single sample, we have:

$$P_{leak}[j^*] = P_{data}[j^*] + P_{noise}[j^*] \tag{6}$$

So, each static current sample can be considered as the sum of a data-dependent component $P_{leak}[j^*]$ and the extraction of a noise process $P_{noise}[j^*]$, which can be assumed as uncorrelated and given by the combination of different sources, as electronic noise, switching noise and quantization noise. The model described by Eq. (6) is widely accepted in cryptographic literature, and it is reasonably to consider the noise process as Gaussian distributed [16], as shown in the following (Fig. 2):

$$P_{noise}[j^*] \sim N(0, \sigma_{noise}^2) \tag{7}$$

where σ_{noise}^2 is the variance of the noise process. Due to the presence of noise, a static current sample can be considered as a Gaussian distributed random variable. A large number of measurements have to be collected to recover information, and filtering out the noise. Since simulated power traces are given noise-free from Cadence Virtuoso, we added Gaussian white noise to simulation, accordingly to previous work [10,19–21]. If we assume $N(l|\mu_{x,N}, \sigma_{x,N}^2)$, the probability distribution of the normal variable L evaluated on input vector x, using the same formalism defined in [21], we can denote:

$$L(X, N) = L^{sim}(X, N) + N \tag{8}$$

with N has zero mean and variance $\sigma_{x,N}^2$.

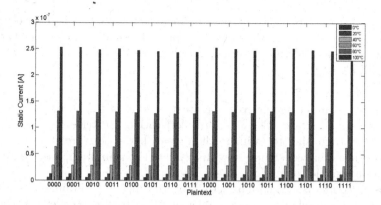

Fig. 2. Mean value of static currents measured on the 4-bit PRESENT crypto-core over all Monte Carlo simulations' chips. Results are shown regarding the input plaintext and the working temperature.

4.1 4-Bit PRESENT Crypto-Core

Our analysis on the 4-bit PRESENT crypto-core has been carried out firstly as univariate. The univariate analysis uses only one domain or dimension at a time. In this case, the only domain we have used is the current sample magnitude. In Eq. (1), we can notice a strong dependency of the static current from the temperature. To completely characterize the behavior of the crypto-core in the working temperature, we have analyzed its static power consumption in the range 0–100 °C, with steps of 20 °C. As we can notice, the behavior of the static current is in accordance to [15], and the simplified model shown in Eq. (1) holds also in the presence of intra-die variations. In Table 1, averaged sample means and standard deviations of static current measurements are shown.

Table 1. Averaged sample means and standard deviations results for the Monte Carlo simulations of the 4-bit PRESENT crypto-core.

Temperature	0 °C	20 °C	40 °C	60 °C	80 °C	100 °C
μ	5,02 nA	11,85 nA	28,06 nA	62,52 nA	129,01 nA	248,27 nA
σ	0,19 nA	0,52 nA	1,25 nA	2,64 nA	5,12 nA	9,50 nA

It is common to adopt linear power model for hardware implementations. Linear power models are based on the assumption that the power consumption of a CMOS circuit is linear with a weighting function of the processed data (e.g. Hamming Weight or Hamming Distance). In [6], a formulation of the simplest attack procedure exploiting static power consumption is given using a simple power model for m-bit slice implementation:

$$I_{stat,TOT} = w \cdot (I_H - I_L) + m \cdot I_L \qquad (9)$$

where I_L and I_H are the static current for a low level and high level in the corresponding bit. The model is able to capture the linear dependence between the static current and the Hamming Weight w of the target word x, which is given by Eq. (10):

$$x = f(I, k) \tag{10}$$

where I is the input vector (e.g. plaintext) and k is the secret key used by the device under attack. The Pearson's correlation coefficient is then used as statistical distinguisher to estimate the correct key. The correlation coefficient compares the actual consumption with estimated consumption, based on proper assumptions on the value of the secret key k.

This procedure is also used as actual security metric, to evaluate the robustness of a cryptographic hardware implementation to this class of side-channel attacks. To better assess the amount of information leaked through the static power consumption, we have used an information theoretic approach, as stated in Sect. 2. The information theoretic approach does not use *a-priori* power model of the device under analysis. In this context, the mutual information has been evaluated for each temperature on chip 0, as depicted in Fig. 3, according to Eq. (3). Looking at Fig. 3, it has to be noticed that increasing the working temperature mutual information curves shift to right, which means that the implementation leaks more if the temperature is increased. We should remind that the mutual information describes the amount of information that an adversary can extract from the chosen side-channel, assuming that he/she has a perfect knowledge of the chip's behavior. So, from the adversary point of view, this is the best-case, but it is only possible when the adversary has a perfect profile of the hardware implementation. This position does not stand in practice, because, in general, an adversary does not have this kind of knowledge. And, also with the possibility that he/she can build a perfect profile of the static power consumption of the device, this position will hold only if the profile is built and used to attack the same chip. As we can notice, under this assumption, the amount of information

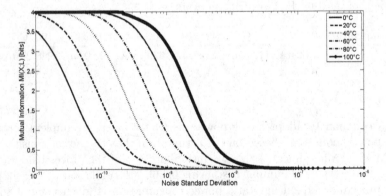

Fig. 3. Mutual information as function of noise standard deviation for the 4-bit PRESENT crypto-core in the temperature range 0–100 °C.

that can be extracted exploiting the static power consumption increases using higher temperatures. If we consider 3.5 bit for 0 °C and compare with 3.5 bit for 100 °C, the noise standard deviation is two orders of magnitude higher, which means that even in the presence of lower SNR, we can extract the same amount of information. This result is important considering that only changing the working temperature of the device under analysis, the amount of information leaked is strongly different. We should mention that also the temperature-dependence of static current has strong effects on the information leakage, and, as a consequence, on the possibility to get a successful attack.

4.2   Full Implementation of PRESENT-80 Block Cipher

To complete the univariate analysis, it is useful to analyze deeply the information that can be extracted from the full implementation of the PRESENT-80 block cipher. Also in this case, the same technology process has been used, with the same threshold option. For the full implementation, intra-die variations have not been taken into account. Static current samples are collected in the temperature range 0–100 °C, with steps of 25 °C. Mean values of static current measurements are plotted in Fig. 4. Similarly to the 4-bit case, it is necessary to evaluate the information that can be exploited through static power. The security metric used is again the mutual information. Results are shown in Fig. 5.

The temperature-dependence of the amount of the information leaked through the chosen side-channel is strong also in the full-implementation. In the range 25–100 °C, the mutual information exhibits a slower reduction between 4 and 3.5 bit, respect to 0 °C plot. At 0 °C, the slope of the mutual information is very similar to slopes of mutual information plots depicted in Fig. 3 for the 4-bit crypto-core.

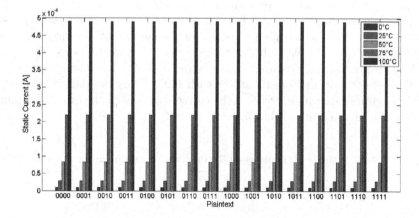

Fig. 4. Mean value of static currents measured on the full implementation of the PRESENT-80 crypto-core over all Monte Carlo simulations' chips. Results are shown regarding the input plaintext and the working temperature.

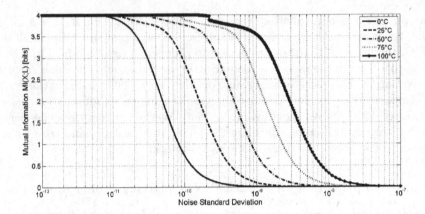

Fig. 5. Plot of mutual information of the full implementation of the PRESENT-80 block cipher in the temperature range 0–100 °C as function of the noise standard deviation.

In order to better assess the concept of "exploitability" and its relation with mutual information, the full implementation of the PRESENT-80 has been attacked using the procedure shown in [6]. To simulate a real experiment, Gaussian white noise has been added in Matlab. A SNR of −60 dB has been used to perform simulated attack. The most significant nibble at the output of the *AddRoundKey* at the first round has been chosen as target word, as shown in Eq. (11). The number of measurements used has been limited to 100 k, to better simulate a real scenario.

$$x_{63,62,61,60} = I_{63,62,61,60} \oplus k_{80,79,78,77} \tag{11}$$

Experiments have been repeated increasing the number of plaintext to get the Measurement To Disclosure (MTD) as function of the working temperature. Plots of correlation coefficients as function of the number of measurements collected are shown in Fig. 6a–e. It has to be noticed that in the range 50–100 °C the MTD is progressively reduced from 84 k to 19.1 k. For lower temperatures, it is not possible to get a successful attack with 100 k. Using the same SNR, for 0–25 °C, the procedure can not recover the correct key, within the given maximum number of measurements (Table 2).

Table 2. Results of the MTD analysis on the full implementation in the range 0–100 °C.

Temperature	0 °C	25 °C	50 °C	75 °C	100 °C
MTD	>100 k	>100 k	84 k	72 k	19.1 k

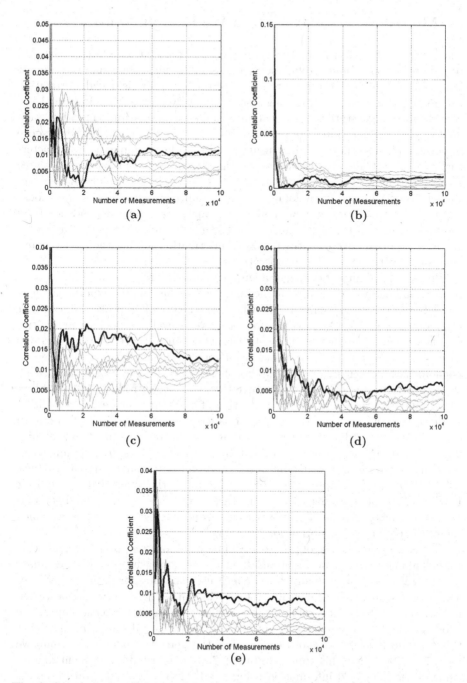

Fig. 6. Correlation coefficient as function of number of measurements used to perform the attack on the full implementation of PRESENT-80. Plots (a)–(e) are referred to working temperatures from 0 °C to 100 °C, respectively. Black bold lines are referred to correlation coefficients of the correct key.

5   Multivariate Analysis: Can We Exploit More?

The multivariate analysis in the dynamic power exploitation is mainly devoted to the use of two domains: dynamic current samples magnitude and the time. It has been proven that in hardware implementations, there is a high correlation (in time) between two neighboring points in a power trace [16]. In Sect. 3, we considered a single static current sample as a Gaussian distributed random variable for a given operation. This non-deterministic behavior is due to the presence of noise, and, if we now consider using more than one sample, each characterized by its own Gaussian distribution, we have to move to *multivariate analysis*. In our analysis we have used the static power consumption as side-channel, so, instead using time as other domain to get a multivariate approach, we choose to use the *temperature* as new domain. In Eq. (1), a strong dependence from the temperature of the static current is shown, and analysis reported in Sect. 4 confirm the possibility to extract "more" information forcing the device under analysis to work with higher temperatures.

Process variability of deep sub-micron technologies raises new issues in both security and analysis of cryptographic implementations. In [21], experiments performed on 20 prototype chips implemented in 65 nm CMOS process shown that common strategy like DPA are no more effective to evaluate the level of security of a device. As consequence of the strong dispersion of process parameters due to intra-die and inter-die variations in nanometer technologies, static current phenomena are even more variable. So, it is not easy to evaluate "how much the implementation leaks". To overcome this issue, we introduce the possibility to get a multivariate approach to evaluation of cryptographic devices exploiting static power consumption. Our approach is based on the possibility to perform experiments on the device under analysis at different temperatures and combining results to extract as much information as possible. To study the impact of intra-die variations on the amount of information that an adversary can extract, we use 100 circuit samples generated by Monte Carlo simulation of the 4-bit PRESENT crypto-core depicted in Fig. 1a. As reference, we have used the univariate analysis performed on chips with a working temperature of 100 °C, which shows the best mutual information behavior.

In this section, the use of the perceived information concept is necessary due to the presence of intra-die variations. In fact, this approach allows us to estimate the "average case" for the adversary, which uses one chip for profiling and uses this information to extract sensible information on a similar one. The perceived information is useful to evaluate how profiles that are built on a chip A are good (or fit) actual power consumption of a chip B. In [21], authors show that this scenario is perfectly captured by using the perceived information, and we have chosen to follow this approach. In Fig. 7a, the distribution of the maximum value of the perceived information using [60 °C; 100 °C] as temperature points over all sample chips is shown. It has to be noticed that the distribution is not symmetric, thus not normal, and this behavior can be demonstrated for each possible combination of temperature points. To better estimate the *typical outcome* of this distribution, we have chosen to use the *median* value instead of

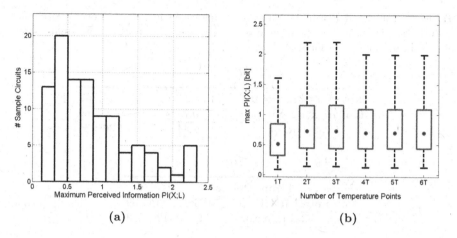

(a)   (b)

Fig. 7. Left: distribution of the maximum value of the perceived information over 100 chips using [60 °C; 100 °C] as temperature points. Right: box-plot of the maximum value of perceived information regarding the number of temperature points. Black dots represent the median value for each case. The bottom and top of the boxes are referred to first and third quartiles, respectively.

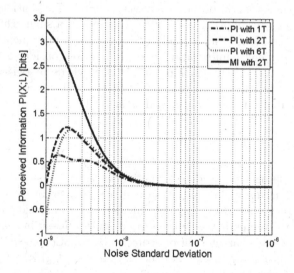

Fig. 8. Perceived information plot as function of the noise standard deviation for chip 46.

the mean value, due to positive skewness [25]. The analysis of the maximum value of the perceived information has been performed for each possible temperature points combination, over all Monte Carlo simulation's chips, and medians have been evaluated and reported for best combinations of temperature points, from 1-point to 6-points, in the range 0–100 °C.

Table 3. Top: median values of maximum perceived information over 100 circuits with intra-die variations. 1-point case is used as reference. Bottom: maximum values of perceived information for chip 46.

(a)

No. of Temp. Points	1	2	3	4	5	6
median(max(PI(X;L)))[bit]	0.51	0.73	0.72	0.69	0.69	0.69
Gain Factor	1	1.44	1.42	1.36	1.36	1.36

(b)

No. of Temp. Points	1	2	6
max(PI(X;L))[bit]	0.638	1.220	1.169
Gain Factor	1	1.912	1.832

Analyzing Table 3a, we could say that using more than one temperature, also in the presence of intra-die variations, the amount of information extracted using non perfect profiles can be 44% greater respect to the univariate analysis. This result confirms the strengths of the multivariate approach exploiting information from static power consumption. Moreover, focusing on a single chip, randomly extracted from the Monte Carlo simulation's set, the improvement in the maximum value of perceived information is of 91% if we consider 2 temperature points, as shown in Table 3b. Looking at Fig. 8, it yields to twice information extracted within the same noise standard deviation, which can assess a strong increase in probability to get a successful attack within the same scenario.

6   Conclusion

A multivariate approach to exploit information leakage using static power dissipation is presented in this work, and this is the first time in cryptographic literature that the temperature is used to expand dimensionality in static power analysis. The analysis of the temperature-dependency of static current in a chip has led to a complete univariate characterization using an information theoretic approach. The use of the mutual information has shown that using higher temperature, the possibility to get a successful experiment is increased significantly. This analysis has been performed on 40 nm CMOS 4-bit data-path of the PRESENT block cipher and on a full implementation of the PRESENT-80. Also the *Leakage Power Analysis* presented in [6] is used and simulated attacks have been performed on the full implementation to show the relationship between information theoretic and actual security metrics. The MTD value can be significantly reduced using higher temperature. With the aim of studying the effect of intra-die variations with an information theoretic approach, the multivariate analysis has been applied intensively on sample chips generated through Monte Carlo simulations. The use of perceived information led to a complete characterization of the actual possibility that an adversary can recover sensible information from a cryptographic implementation. With a population of 100 chips, the median of the distribution of the maximum value of the $\widehat{PI}(X; L)$ is increased by 44% compared to the univariate case. Focusing on a randomly picked chip, the increasing in the maximum value of perceived information is of 91% using 2 temperatures, comparing the best result obtained with 1-point analysis, which means that the multivariate approach can lead to a strong reduction in the security level. The multivariate approach remarks the possibility to recover sensible information using static power consumption as side-channel.

References

1. Kocher, P.C.: Timing attacks on implementations of Diffie-Hellman, RSA, DSS, and other systems. In: Koblitz, N. (ed.) CRYPTO 1996. LNCS, vol. 1109, pp. 104–113. Springer, Heidelberg (1996). doi:10.1007/3-540-68697-5_9

2. Roy, K., Mukhopadhyay, S., Mahmoodi-Meimand, H.: Leakage current mechanisms and leakage reduction techniques in deep-submicrometer CMOS circuits. Proc. IEEE **91**(2), 305–327 (2003)

3. Alioto, M., Bongiovanni, S., Djukanovic, M., Scotti, G., Trifiletti, A.: Effectiveness of leakage power analysis attacks on DPA-resistant logic styles under process variations. IEEE Trans. Circuits Syst. I Regul. Papers **61**(2), 429–442 (2014)

4. Eisenbarth, T., Kumar, S., Paar, C., Poschmann, A., Uhsadel, L.: A survey of lightweight-cryptography implementations. IEEE Des. Test **24**(6), 522–533 (2007)

5. Alioto, M., Giancane, L., Scotti, G., Trifiletti, A.: Leakage power analysis attacks: well-defined procedure and first experimental results. In: 2009 International Conference on Microelectronics - ICM, pp. 46–49 (2009)

6. Alioto, M., Giancane, L., Scotti, G., Trifiletti, A.: Leakage power analysis attacks: a novel class of attacks to nanometer cryptographic circuits. IEEE Trans. Circuits Syst. I **57**(2), 355–367 (2010)

7. Brier, E., Clavier, C., Olivier, F.: Correlation power analysis with a leakage model. In: Joye, M., Quisquater, J.-J. (eds.) CHES 2004. LNCS, vol. 3156, pp. 16–29. Springer, Heidelberg (2004). doi:10.1007/978-3-540-28632-5_2

8. Moradi, A.: Side-channel leakage through static power. In: Batina, L., Robshaw, M. (eds.) CHES 2014. LNCS, vol. 8731, pp. 562–579. Springer, Heidelberg (2014). doi:10.1007/978-3-662-44709-3_31

9. Pozo, S.M.D., Standaert, F., Kamel, D., Moradi, A.: Side-channel attacks from static power: when should we care? In: Proceedings of the 2015 Design, Automation & Test in Europe Conference & Exhibition. DATE 2015, Grenoble, pp. 145–150, 9–13 March 2015

10. Bellizia, D., Bongiovanni, S., Monsurro, P., Scotti, G., Trifiletti, A.: Univariate power analysis attacks exploiting static dissipation of nanometer CMOS VLSI circuits for cryptographic applications. IEEE Trans. Emerg. Topics Comput. **PP**(99), 1 (2016)

11. Tiri, K., Verbauwhede, I.: A logic level design methodology for a secure DPA resistant ASIC or FPGA implementation. In: Proceedings Design, Automation and Test in Europe Conference and Exhibition, vol. 1, pp. 246–251, February 2004

12. Popp, T., Mangard, S.: Masked dual-rail pre-charge logic: DPA-resistance without routing constraints. In: Rao, J.R., Sunar, B. (eds.) CHES 2005. LNCS, vol. 3659, pp. 172–186. Springer, Heidelberg (2005). doi:10.1007/11545262_13

13. Bellizia, D., Scotti, G., Trifiletti, A.: Implementation of the present-80 block cipher and analysis of its vulnerability to side channel attacks exploiting static power. In: 23rd International Conference Mixed Design of Integrated Circuits and Systems. MIXDES 2016, pp. 211–216, June 2016

14. Bellizia, D., Djukanovic, M., Scotti, G., Trifiletti, A.: Template attacks exploiting static power and application to CMOS lightweight crypto-hardware. Int. J. Circuit Theory Appl. 45(2), 229–241 (2016)

15. Chandrakasan, A.P., Bowhill, W.J., Fox, F.: Design of High-Performance Microprocessor Circuits, 1st edn. IEEE Press, New York (2000)

16. Mangard, S., Oswald, E., Popp, T.: Power Analysis Attacks: Revealing the Secrets of Smart Cards. Advances in Information Security. Springer, New York (2007)

17. Mangard, S.: Hardware countermeasures against DPA – a statistical analysis of their effectiveness. In: Okamoto, T. (ed.) CT-RSA 2004. LNCS, vol. 2964, pp. 222–235. Springer, Heidelberg (2004). doi:10.1007/978-3-540-24660-2_18

18. Tiri, K., Hwang, D., Hodjat, A., Lai, B.-C., Yang, S., Schaumont, P., Verbauwhede, I.: Prototype IC with WDDL and differential routing – DPA resistance assessment. In: Rao, J.R., Sunar, B. (eds.) CHES 2005. LNCS, vol. 3659, pp. 354–365. Springer, Heidelberg (2005). doi:10.1007/11545262_26

19. Macé, F., Standaert, F.-X., Quisquater, J.-J.: Information theoretic evaluation of side-channel resistant logic styles. In: Paillier, P., Verbauwhede, I. (eds.) CHES 2007. LNCS, vol. 4727, pp. 427–442. Springer, Heidelberg (2007). doi:10.1007/978-3-540-74735-2_29

20. Standaert, F.-X., Malkin, T.G., Yung, M.: A unified framework for the analysis of side-channel key recovery attacks. In: Joux, A. (ed.) EUROCRYPT 2009. LNCS, vol. 5479, pp. 443–461. Springer, Heidelberg (2009). doi:10.1007/978-3-642-01001-9_26

21. Renauld, M., Standaert, F.-X., Veyrat-Charvillon, N., Kamel, D., Flandre, D.: A formal study of power variability issues and side-channel attacks for nanoscale devices. In: Paterson, K.G. (ed.) EUROCRYPT 2011. LNCS, vol. 6632, pp. 109–128. Springer, Heidelberg (2011). doi:10.1007/978-3-642-20465-4_8

22. Bogdanov, A., et al.: PRESENT: an ultra-lightweight block cipher. In: Paillier, P., Verbauwhede, I. (eds.) CHES 2007. LNCS, vol. 4727, pp. 450–466. Springer, Heidelberg (2007). doi:10.1007/978-3-540-74735-2_31

23. Bongiovanni, S., Centurelli, F., Scotti, G., Trifiletti, A.: Design and validation through a frequency-based metric of a new countermeasure to protect nanometer ics from side-channel attacks. J. Cryptogr. Eng. 5(4), 269–288 (2015)

24. Rolfes, C., Poschmann, A., Leander, G., Paar, C.: Ultra-lightweight implementations for smart devices – security for 1000 gate equivalents. In: Grimaud, G., Standaert, F.-X. (eds.) CARDIS 2008. LNCS, vol. 5189, pp. 89–103. Springer, Heidelberg (2008). doi:10.1007/978-3-540-85893-5_7

25. Knight, K.: Mathematical Statistics. Texts in Statistical Science Series. Chapman & Hall/CRC Press, Boca Raton (2000)

Differential Bias Attack for Block Cipher Under Randomized Leakage with Key Enumeration

Haruhisa Kosuge[✉] and Hidema Tanaka

National Defense Academy of Japan, Yokosuka, Japan
{ed16005,hidema}@nda.ac.jp

Abstract. In the formal analysis of side-channel attacks, a theoretical model of side-channel information (*leakage model*) is supposed and dedicated attacks for the model are considered. In ASIACRYPT2015, a new leakage model for the analysis of block cipher was proposed by Bogdanov et al. The model assumes an adversary who has leaked values whose positions are unknown and randomly chosen from internal results (*random leakage model*). They also proposed an attack, *differential bias attack* for the model. This paper improves the security analysis on AES under the random leakage model. In the previous method, the adversary requires at least 2^{34} chosen plaintexts, therefore, it is infeasible to recover a secret key with a small number of data. However, there may be an adversary who can recover the secret key using his computing power. To consider the security against the adversary, we reestimate complexity for the adversary given a small number of data. We propose another hypothesis-testing method which can minimize the number of required data. The reestimation of complexity shows that the proposed method requires time complexity more than $T > 2^{60}$ because of time-data trade-off, however, some attacks are feasible under $T \leq 2^{80}$. In addition to the above method, we apply key enumeration to differential bias attack, and evaluate its efficiency by rank estimation. From the experimental evaluation, we show that the success rate of the attack can be practical if there is an advantageous restriction on the positions of leaked values.

Keywords: Block cipher · Side-channel attack · Formal security analysis · Leakage model · AES · Differential bias attack · Key enumeration · Rank estimation

1 Introduction

1.1 Background

Conventionally, the security of symmetric-key cryptographic primitives is evaluated under the assumption that the adversary is given black-box access to cryptosystems. However, such a model does not always ensure security of cryptographic implementations for side-channel attacks. In the last two decades, many attacks have been demonstrated by exploiting device-specific side channels.

© Springer International Publishing AG 2017
M. Joye and A. Nitaj (Eds.): AFRICACRYPT 2017, LNCS 10239, pp. 95–113, 2017.
DOI: 10.1007/978-3-319-57339-7_6

Due to such characteristic, the empirical evaluation of cryptographic devices is important. In addition, the formal analysis of side-channel attacks has been studied [6,11,18,19,21]. These works incorporate the side-channel information into the security analysis and develop countermeasures to defeat common side-channel attacks. Obviously, assumptions for the adversary are important in formal analysis. In addition to black-box access (plaintexts and corresponding ciphertexts), partial information about internal results of encryption process leaks following a certain model (*leakage model*).

In ASIACRYPT2015, a new leakage model was proposed by Bogdanov and Isobe [2]. The model assumes an adversary who has leaked values (bits or bytes in AES) emitting from internal operations, however, the positions where leaked bits/bytes come from are unknown and randomly chosen. We call the model *random leakage model*. Note that the leaked values are outputs of AddRoundkey or subkeys of internal rounds. Also, they impose restrictions on the round number and position of bits/bytes within the round (*leakage range*) to let the adversary have an additional advantage. Software implementation seems to be more relevant to the random leakage model than hardware implementation. In software operations, parts of the secret data were copied to a different medium such as swap space and the data remains undeleted after the operations. The adversary may have access to the data, e.g., by physical access to the computer. If he does not know where the data come from, the attack can be analyzed in the random leakage model.

The same authors also proposed *differential bias attack*, which is feasible under the random leakage model. The attack exploits truncated differential characteristics that can be obtained by correct subkey guess, and recover subkeys by applying hypothesis testing. Differential bias attack can break AES in the practical time complexity such as $T \leq 2^{60}$, however, the required data D is not practical ($D \geq 2^{34}$).

1.2 Contribution

This paper is intended as an improvement on the security analysis on AES (especially AES-128: 128-bit key length and 10-round version of AES) under the random leakage model. In the previous method [2], the adversary requires at least 2^{34} chosen plaintexts. Therefore, key recovery of AES-128 is infeasible for the adversary given a small number of data, therefore, restriction on the number of times to encrypt data by single key will easily prevent the attack. However, it is a wrong assumption, since there is an adversary with a computing power making up for the disadvantage. In order to evaluate the security against such adversary, we reestimate complexity for the adversary given a small number of data. We use a hypothesis-testing method that is also used in [2]. We consider time-data tradeoff and minimize the number of required data.

In addition, we assume an adversary given smaller number of data than the above attacks. For such adversary, we propose a new method which combines differential bias attack and key enumeration [20]. Though there are some variations

[3,5,15–17], we choose histogram-based key enumeration proposed by Poussier et al. [17] for its simplicity, efficiency and preciseness. Also, we evaluate complexity and success rate of the proposed attack by using histogram-based rank estimation [10]. Note that we can substitute any key enumeration and rank estimation algorithms for histogram-based ones.

We evaluate the results from both viewpoints of the adversary (of which the goal is to recover the secret key) and the evaluator (of which the goal is to analyze security of his cryptographic device). For the adversary (resp. evaluator), we assume that the adversary can check 2^{60} (resp. 2^{80}) key candidates of AES-128 [12], and summarize the results as follows.

1. By minimizing the number of required data D in differential bias attack, we show feasibility of the attacks under $D \leq 2^{30}$. Because of time-data tradeoff, the time complexity T increases to be $T > 2^{60}$. From the viewpoint of the evaluator, some attacks are feasible ($T \leq 2^{80}$) if leakages emit from advantageous leakage ranges.
2. Applying key enumeration to differential bias attack, we propose a new method for an adversary given a smaller number of data than the methods using hypothesis testing. We evaluate its success rate by increasing the number of samples when restriction on the leakage range is the most advantageous for the adversary. As a result, we confirm that the success rate is practical under $T \leq 2^{80}$ even if the number of samples q is less than the one used in hypothesis-testing methods. Hence the adversary given smaller number of data than the above methods is still a threat for the evaluator if the leakage range is advantageous for him.

The above results indicate that differential bias attack is a practical threat from the view point of the evaluator. Also, we should consider a situation that the attack is used as a complement of other side-channel attacks for the reduction of time complexity. As a countermeasure for differential bias attack, we insist on the importance of intensive protection of crucial rounds such as second round of AES (in addition to first round for other general side-channel attacks).

Notations. We use bold fonts for vectors, sans serif ones for functions, calligraphic ones for sets.

2   Previous Works

2.1   Leakage Model for Side-Channel Attacks

Recently, there is a trend to formally evaluate the security of cryptographic devices [6,11,18,19,21]. We assume an adversary who has access to a physical implementation of block ciphers and observes side channel leakage l emitting from encrypting process (*online phase*). Afterward, the adversary attempts to recover a secret key exploiting the data obtained in online phase with his computing power (*offline phase*). As a realistic assumption, the leakage of sensitive internal result v is expressed by a deterministic function δ (e.g. hamming weight) and a noise b (e.g. Gaussian noise), i.e., $l = \delta(v) + b$ [7]. In a different way, Dinur and Shamir defined a leakage model which assumes an adversary who is given a value of bit in internal operations where the position is fixed and the adversary knows it [6]. In the similar way, Bogdanov and Isobe defined another leakage model assuming that the position of leaked bit (or byte) is randomly-chosen and unknown for the adversary [2]. We focus on the last model [2] and call it *random leakage model*.

When a set of leakages $\mathcal{L}_D$ is observed in encryption processes of D plaintexts $\mathcal{X}_D$, we suppose a leaked variable $l \in \mathcal{L}_D$ is a bit in *bitwise random leakage model* or a byte in *bytewise random leakage model*, where the position of bit/byte is unknown and randomly chosen from intermediate results of encryptions. Note that the model resembles to one defined in probing attacks [11], however, the model is simpler and dedicated to analysis of block ciphers.

We assume n-bit secret key $\mathbf{k} = (k_0, k_1, ..., k_{N_s-1})$, where k_i is a n_s-bit subkey ($n_s = n/N_s$). Since AES-128 is under the consideration, the numbers are $N_s = 16$ and $n_s = 8$. Let $\mathbf{z}^r = (z_0^r, z_1^r, ..., z_{15}^r)$ be an output state of AddRoundkey in r-th round ($r \in [1 : 10]$), and $\mathbf{x} = (x_0, x_1, ..., x_{15})$ a plaintext. Also, we denote $\mathbf{k}^r = (k_0^r, k_1^r, ..., k_{15}^r)$ as subkeys (expanded keys from the secret key $\mathbf{k}$) added in r-th round and $\mathbf{k}^0(= \mathbf{k})$ is added in the initial AddRoundkey. In a byte λ_i, we denote j-th bit as $\lambda_{i,j}$. Assuming the bitwise (resp. bytewise) random leakage model, the adversary observes $x_{i,j}$ (resp. x_i), $z_{i,j}^r$ (resp. z_i^r) or $k_{i,j}^r$ (resp. k_i^r) from an encryption process. In [2], the authors impose restrictions on the round number r (*time*) and the bit/byte position (i, j) (*space*). Because of the restrictions, given data may become more advantageous for the adversary. We call two dimensional range (*time* & *space*) as *leakage range*.

2.2 Differential Bias Attack [2]

Differential bias attack uses *truncated differential characteristics* to recover the secret key as follows. Figures 1 and 2 show examples of truncated differential characteristics. We denote $\mathbf{y} = (y_0, y_1, ..., y_{15})$ as an output of MixColumns of 1st round. A colored-cell is a probability-one non-zero difference, a white-cell is a probability-one zero difference, and "?" denotes an unknown (zero/non-zero) difference. The adversary chooses two different 4 bytes of $(x_0, x_5, x_{10}, x_{15})$ which results in only one non-zero difference in (y_0, y_1, y_2, y_3) if a 32-bit subkey $(k_0, k_5, k_{10}, k_{15})$ is correctly guessed. Note that only one non-zero difference means there are two 4 bytes such that their difference includes 1 non-zero byte and 3 zero bytes. Then he inputs a pair of plaintexts $\mathbf{x}$ consisting of the above 4 bytes and 12 bytes with the same values. In the case of Fig. 1 (*correct case*), the 32-bit subkey is correctly guessed and there is only one non-zero difference in $\mathbf{y}$. In Fig. 2 (*wrong case*), the guessed subkey is wrong and there are four unknown differences in $\mathbf{y}$.

There is a bias between two distributions for differences of leakages for *correct* and *wrong* cases. Let $\Pr[l \oplus l'|\text{correct}]$ and $\Pr[l \oplus l'|\text{wrong}]$ be probabilities when the correct and wrong cases hold, where l and l' are values of leaked bits/bytes. As defined in Sect. 2.1, the values of l and l' are ones of bits/bytes in $\mathbf{x}$, $\mathbf{z}^r$

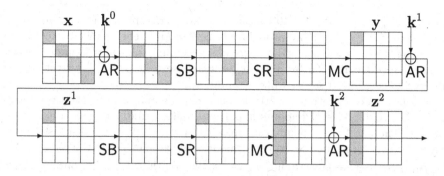

Fig. 1. Truncated differential characteristic in correct case.

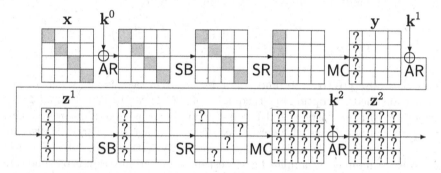

Fig. 2. Truncated differential characteristic in wrong case.

Algorithm 1. Previous method of differential bias attack with hypothesis testing [2].

---

input $\mathcal{X}_D$ and $\mathcal{L}_D$.
for $i = 0 \rightarrow 3$ do
   for $\mathbf{k}_i^* \in \mathbb{F}_2^{32}$ do
      Initialize a likelihood ratio $\Lambda_{\mathbf{k}_i^*} \leftarrow 1$.
      while $(1 - \alpha)/\beta < \Lambda_{\mathbf{k}_i^*} < \beta/(1 - \alpha)$ do
         Choose $\mathbf{y}_i$ and $\mathbf{y}_i'$ randomly where only 1 byte has a difference.
         Obtain $\mathbf{x}_i$ and $\mathbf{x}_i'$ which result in $\mathbf{y}_i$ and $\mathbf{y}_i'$ when $\mathbf{k}_i^*$ is correct.
         Choose two plaintexts $\mathbf{x}$ and $\mathbf{x}'$ which contain $\mathbf{x}_i$ and $\mathbf{x}_i'$, respectively.
         Acquire $\{l_j\}_{j=0}^{q_s-1}$ and $\{l_{j'}'\}_{j'=0}^{q_s-1}$ from $\mathcal{L}_D$ corresponding to $\mathbf{x}$ and $\mathbf{x}'$.
         Compute $\Lambda_{\mathbf{k}_i^*} = \Lambda_{\mathbf{k}_i^*} \times \prod_{j,j'} Pr[l_j \oplus l_{j'}'|\text{correct}]/Pr[l_j \oplus l_{j'}'|\text{wrong}]$.
      end while
      if $\beta/(1 - \alpha) \leq \Lambda_{\mathbf{k}_i^*}$ then
         $\mathbf{k}_i^*$ is added to a set of candidates $\mathcal{K}_i$.
      end if
   end for
end for
Brute-force search for the correct key $\mathbf{k}^* \in \mathcal{K}(= \mathcal{K}_0 \times \mathcal{K}_1 \times \mathcal{K}_2 \times \mathcal{K}_3)$.
return $\mathbf{k}^*$

($r \in [1:10]$) or $\mathbf{k}^r$ ($r \in [0:10]$) if there is no restriction on the leakage range. In $\mathbf{z}^1$ and $\mathbf{z}^2$, there are more probability-one zero difference in the correct case. Therefore, $\Pr[l \oplus l' = 0|\text{correct}] > \Pr[l \oplus l' = 0|\text{wrong}]$ holds for randomly chosen l and l'. If a distribution for a subkey guess close to the one of correct case is observed, the guessed key is a likely candidate.

In this way, differential bias attack takes a divide-and-conquer strategy on 32-bit subkeys. The paper [2] adopts hypothesis testing to distinguish subkey candidates. The number of required samples determines efficiency of a test. Information theory provides the expected number of samples [9]. Let $\mathsf{D}(p_0 \| p_1) = \sum_c p_0(c) \log \frac{p_0(c)}{p_1(c)}$ be a discrimination of two distributions p_0 and p_1 obtained by a sample. Suppose a test is to distinguish two distributions by obtaining q samples. The expected number of samples to distinguish p_0 and p_1 is:

$$\mathbb{E}[q] \geq \frac{\beta \cdot \log\left(\frac{\beta}{1-\alpha}\right) + (1-\beta) \cdot \log\left(\frac{1-\beta}{\alpha}\right)}{\mathsf{D}(p_0 \| p_1)}, \tag{1}$$

where α is false positive rate and β is false negative rate [9,14]. When we use information-theoretic optimal test, equality of Eq. (1) can be met. Formal distributions for $\Pr[l \oplus l'|\text{correct}]$ and $\Pr[l \oplus l'|\text{wrong}]$ can be computed by counting the number of bits/bytes with zero differences in leakage range and the discrimination of two distributions is easily calculated. By setting error rates α and β, the expected number of samples for the estimated discrimination is obtained. In [2], the authors supposed $\alpha = \beta = 2^{-32}$. We show an example of differential bias attack in Algorithm 1. Since there is no description on which test they use in [2], we use sequential probability ratio test [22]. Note that we denote $\mathbf{x}_i$, $\mathbf{k}_i$ and $\mathbf{y}_i$ as follows.

$$
\begin{aligned}
\mathbf{x}_0 &= (x_0, x_5, x_{10}, x_{15}), & \mathbf{k}_0 &= (k_0, k_5, k_{10}, k_{15}), & \mathbf{y}_0 &= (y_0, y_1, y_2, y_3) \\
\mathbf{x}_1 &= (x_4, x_9, x_{14}, x_3), & \mathbf{k}_1 &= (k_4, k_9, k_{14}, k_3), & \mathbf{y}_1 &= (y_4, y_5, y_6, y_7) \\
\mathbf{x}_2 &= (x_8, x_{13}, x_2, x_7), & \mathbf{k}_2 &= (k_8, k_{13}, k_2, k_7), & \mathbf{y}_2 &= (y_8, y_9, y_{10}, y_{11}) \\
\mathbf{x}_3 &= (x_{12}, x_1, x_6, x_{11}), & \mathbf{k}_3 &= (k_{12}, k_1, k_6, k_{11}), & \mathbf{y}_3 &= (y_{12}, y_{13}, y_{14}, y_{15})
\end{aligned} \tag{2}
$$

Differential bias attack is feasible under chosen-plaintext, known-plaintext and chosen-ciphertext settings. We only focus on chosen plaintext setting, since the same discussions can apply to the other settings. The time complexity T and the required data D are estimated by using the expected value of q as follows.

$$T = 4 \cdot 2^{32} \cdot \frac{\mathbb{E}[q]}{q_s^2} \cdot \frac{q_s}{10} + (1 + 2^{-32} \cdot (2^{32} - 1))^4 \approx 2^{30.68} \cdot \frac{\mathbb{E}[q]}{q_s}, \tag{3}$$

$$D = 4 \cdot 2^{32} \cdot q_s, \tag{4}$$

where q_s is the number of leakages measured by encrypting plaintexts with the same value. Since each encryption process leaks only one bit/byte, q_s bits/bytes are obtained by q_s times encryption of the same plaintext. Using q_s bits/bytes leaked from different positions, we compute q_s^2 times XOR operation and the complexity is estimated as $q_s/10$ times of encryptions of AES-128 in [2]. Since the expected number of samples is $\mathbb{E}[q]$, the operation iterates for $\mathbb{E}[q]/q_s^2$ different chosen plaintexts.

2.3   Key Enumeration and Rank Estimation

Conventionally, side-channel attacks have focused on the online phase and try to recover keys without much computation in the offline phase. However, we need to consider computing power in order to suppose an adversary who can recover the secret key with restricted side-channel information. Since the first algorithm, *optimal key enumeration*, was proposed in SAC2012 [20], several studies have been made on the algorithm [3,5,15–17]. Key enumeration is used in any side-channel attack if it takes divide-and-conquer strategy. When we can obtain lists of posterior probabilities for all N_s subkeys after the measurement (i.e. $\Pr[k_i = k_i^* | \mathcal{X}_D, \mathcal{L}_D]$, where k_i^* is a guessed key and $i \in [0 : N_s - 1]$), we can enumerate candidates of the secret key $\mathbf{k}$ with their joint probabilities $\prod_i \Pr[k_i = k_i^* | \mathcal{X}_D, \mathcal{L}_D]$ from the most likely one to the least likely one. In profiled attacks such as template attack [4], lists of probabilities $\Pr[k_i = k_i^* | \mathcal{X}_D, \mathcal{L}_D]$ are directly obtained, however in non-profiled attacks (e.g. differential power analysis [13]), one can make the probabilities by applying Bayesian extension [20].

By contrast, rank estimation is an evaluation tool for side-channel attacks enhanced by key enumeration [21]. Some variations are developed [1,8,10] after the first proposal in EUROCRYPT2013 [21], and there are tight connections between key enumeration and rank estimation algorithms [17]. The algorithm is to evaluate the rank of a correct key when the evaluator is given the posterior probability of the correct key. The rank directly determines the time complexity of the attack, since it indicates the number of key candidates to enumerate and verify.

Even if the correct key is in the last rank ($2^{128} - 1$ in AES-128), we can execute rank estimation in a realistic time. It is impossible for key enumeration to achieve this efficiency, since it enumerates a huge number of key candidates. However, an estimation error inevitably occurs. Each algorithm has a technique to estimate upper and lower bounds for the estimated rank. Therefore, we need to use a fast algorithm with tight bounds. Since histogram-based rank estimation proposed by Glowacz et al. [10] satisfies the above criteria, we use the algorithm to evaluate our proposed attack. Because of a connection to *histogram-based key enumeration* proposed by Poussier et al. [17], histogram-based rank estimation outputs time complexity of this key-enumeration algorithm. Therefore, we consider an application of the histogram-based key enumeration in Sect. 4.

Histogram-based Rank Estimation [10]. Algorithm 2 shows the histogram-based rank estimation. Note that the algorithm is also used in the histogram-based key enumeration. In this case, Algorithm 2 outputs a set of all histograms $\mathcal{H} = \{H_0, H_1, ..., H_{N_s-1}, H_{0:1}, H_{0:2}, ..., H_{0:N_s-1}\}$. We obtain lists of log posterior probabilities $LP_i = \log(\Pr[k_i = k_i^* | \mathcal{X}_D, \mathcal{L}_D])$. From the lists, histograms H_i with equally-sized N_{bin} bins are constructed. The number of subkeys in the bin b,

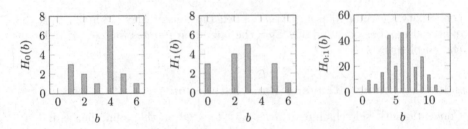

Fig. 3. Examples of histograms [17].

i.e., $H_i(b)$, is increased by 1 if there is a value LP_i in the list such that $b = \lfloor LP_i/S_{bin} \rfloor$, where S_{bin} is the binsize. Note that the binsize is the length of an interval of LP_i included in a bin. We show an example of histograms in Fig. 3, where $H_{0:1}$ is a histogram constructed by the convolution of H_0 and H_1 ($N_{bin} = 7$). Convoluting histograms from H_0 to H_i, we can construct $H_{0:i}$ (the number of bins is $i \cdot N_{bin} - (i-1)$). Histogram-based rank estimation outputs an estimated rank by using the last histogram $H_{0:N_s-1}$ and the index of bin in which the correct key may be included. Let $b^\dagger (\geq 0)$ be such index, and it is obtained by $b^\dagger = \sum_{i=0}^{N_s-1} H_i(b_i^\dagger)$, where $b_i^\dagger$ denotes an index of a bin in which a correct subkey k_i is included.

There is an estimation error caused by the convolution of histograms, however, the rank can be lower and upper-bounded as follows [10]. Let $\{LP_i\}_{i=0}^{N_s-1}$ be N_s lists of log posterior probabilities with their j-th elements denoted as $LP_i^{(j)}$, and a central value $m_i^{(j)}$. When the binsize is S_{bin}, we have (for $\forall j$):

$$|LP_i^{(j)} - m_i^{(j)}| \leq \frac{S_{bin}}{2},$$

$$\implies \left| \sum_{i=0}^{N_s-1} LP_i^{(j)} - \sum_{i=0}^{N_s-1} m_i^{(j)} \right| \leq \frac{S_{bin}}{2} \cdot N_s = \frac{N_s}{2} \cdot S_{bin}. \tag{5}$$

Therefore, the difference between the actual sum of log probabilities and the sum of central values is limited to $\lceil N_s/2 \rceil$ bins. Hence, we can limit the index of the bin in which the correct key is actually included as:

$$E_{lower} = \sum_{b=b^\dagger+\lceil N_s/2 \rceil}^{N_s \cdot N_{bin} - N_s} H_{0:N_s-1}(b) \quad \text{(lower bound)}, \tag{6}$$

$$E_{upper} = \sum_{b=b^\dagger-\lceil N_s/2 \rceil}^{N_s \cdot N_{bin} - N_s} H_{0:N_s-1}(b) \quad \text{(upper bound)}. \tag{7}$$

Histogram-based Key Enumeration [17]. We introduce the histogram-based key enumeration in Algorithm 3. Note that size_of(H) is the number of bins in H. The algorithm uses a set of histograms $\mathcal{H}$ which is obtained from Algorithm 2. Also, it requires a list S which connects subkeys and their bins. We denote

---

Algorithm 2. Histogram-based rank estimation [10, 17].

---

input $\{H_0, H_1, ..., H_{N_s-1}\}$.

Initialize a histogram $H_{0:0} \leftarrow H_0$.

for $i = 1 \rightarrow N_s - 1$ do

    for $b = 0 \rightarrow i \times N_{bin} - i$ do

        for $b' = 0 \rightarrow N_{bin} - 1$ do

            $H_{0:i}(b + b') \leftarrow H_{0:i}(b + b') + H_{0:i-1}(b) \times H_i(b')$

        end for

    end for

end for

(return $\mathcal{H} = \{H_0, H_1, ..., H_{N_s-1}, H_{0:1}, H_{0:2}, ..., H_{0:N_s-1}\}$)

Compute estimated rank $E \leftarrow \sum_{b=b^\dagger}^{N_s \times N_{bin} - N_s} H_{0:N_s-1}(b)$

return E

Algorithm 3. Histogram-based key enumeration [17].

---

input $\mathcal{H}$ and $\mathcal{S}$.

for $b_{sum} = b_{start} \rightarrow b_{stop}$ do

    Decompose_bin($N_s - 1, b_{sum}, \mathcal{K}$)

end for

function Decompose_bin($csh, b_{sum}, \mathcal{K}$)   ▷ Recursively called.

    if $csh = 1$ then

        Initialize the index of bin as $b \leftarrow$ size_of(H_0) $- 1$.

        while $(b \geq 0)$ & $(b + $ size_of(H_1) $> b_{sum})$ do

            if $(H_0(b) > 0)$ & $(H_1(b_{sum} - b) > 0)$ then

                Copy a set of subkeys in $H_0(b)$ to $\mathcal{K}_0$ and ones in $H_1(b_{bin} - b)$ to $\mathcal{K}_1$.

                for $\mathbf{k} \in \mathcal{K} = \mathcal{K}_0 \times \mathcal{K}_1 \times ... \times \mathcal{K}_{N_s-1}$ do

                    if $\mathbf{k}$ is compatible with plaintexts and ciphertexts then

                        return $\mathbf{k}$ (as the correct key)

                    end if

                end for

            end if

            Decrease the index of bin as $b \leftarrow b - 1$.

        end while

    else

        Initialize the index of bin as $b \leftarrow$ size_of(H_{csh}) $- 1$

        while $(b \geq 0)$ & $(b + $ size_of($H_{0:csh-1}$) $> b_{sum})$ do

            if $(H_{csh}(b) > 0)$ & $(H_{0:csh-1}(b_{sum} - b) > 0)$ then

                Copy a set of subkeys in the bin b of H_{csh} to $\mathcal{K}_{csh} \leftarrow \mathcal{S}_{csh}(b)$.

                Call Decompose_bin($csh - 1, b_{sum} - b, \mathcal{K}$) recursively.

            end if

            Decrease the index of bin as $b \leftarrow b - 1$.

        end while

    end if

    return "*failure*" (correct key is not found)

end function

$\mathcal{S} = \{\{\mathcal{S}_i(b)\}_{b=0}^{N_{bin}-1} | i \in [0 : N_s - 1]\}$, where $\{\mathcal{S}_i(b)\}_{b=0}^{N_{bin}-1}$ is a list of N_{bin} sets with their b-th element $\mathcal{S}_i(b)$ is a set of subkeys in the bin b of histogram H_i.

As a setting, the adversary chooses an index of starting bin b_{start} and stopping bin b_{stop} of the last histogram $H_{0:N_s-1}$. In most cases, he sets $b_{start} = $ size_of$(H_{0:N_s-1})$, since it is the bin in which the most likely candidates are included. Also, the value of b_{stop} is determined by his computing power (he can check $\sum_{b=b_{start}}^{b_{stop}} H_{0:N_s-1}(b)$ key candidates). The procedure consists of a recursive call of a function Decompose_bin and key validation inside the function. A histogram $H_{0:csh}$ is decomposed into $H_{0:csh-1}$ and H_{csh}, and a set of subkeys in a non-empty bin $\mathcal{S}_{csh}(b)$ is copied to $\mathcal{K}_{csh}$. From the last histogram $H_{0:N_s-1}$ to the first one $H_{0:1}$, the decomposition is recursively executed to obtain a set of key candidates $\mathcal{K}$.

Compared to the optimal key enumeration [20], the histogram-based key enumeration is suboptimal, since it can not enumerate candidates in the same bin and there is an estimation error caused by the convolution. However, the time complexity of the algorithm is negligible compared to the total key validation, since the number of candidates $|\mathcal{K}|$ in each key verification is much more than the number of times to call Decompose_bin. In average, it equals $|\mathcal{K}| = T/(N_s \cdot N_{bin} - (N_s - 1))$, where T is the total number of keys to enumerate. Also, the algorithm is easily parallelized.

3   Reestimation of Complexity by Time-Data Tradeoff

The previous method of differential bias attack [2] requires plenty of plaintexts with leakages. The adversary requires at least 2^{34} chosen plaintexts. Additionally, he queries multiple leakages from the same chosen plaintext to reduce the time complexity (see Eq. (3)). In this way, the previous method requires additional data. Therefore, we can not evaluate security against more practical adversary. We propose another method for differential bias attack which is feasible with a smaller number of data in Sect. 3.1. The method adopts hypothesis testing in the same manner as the previous one, however, there is a difference in the way to obtain samples. Also, we compare two methods by varying the settings of leakage range in Sect. 3.2.

3.1   New Hypothesis-Testing Method

To reconsider the number of data required to obtain $\mathbb{E}[q]$ samples (see Eq. (1)) for each key candidate, we exploit bijection of 32-bit functions composed of four Sboxes and one MDS matrix. Let F_i be such function, i.e., $\mathbf{y}_i = F_i(\mathbf{x}_i, \mathbf{k}_i)$. We consider 32-bit values of a pair of plaintexts $(\mathbf{x}_i, \mathbf{x}_i')$ and their outputs $(\mathbf{y}_i, \mathbf{y}_i')$ (see Eq. (2)). We define a function DF_i which outputs $(\mathbf{y}_i, \mathbf{y}_i') = \mathsf{DF}_i((\mathbf{x}_i, \mathbf{x}_i'), \mathbf{k}_i)$. When we regard $\mathbf{k}_i$ as constant, DF_i is a bijective function ($\mathsf{DF}_i : (\mathbb{F}_2^{32})^2 \rightarrow (\mathbb{F}_2^{32})^2$). From the fact, we derive a ratio of $(\mathbf{x}_i, \mathbf{x}_i')$ such that $\mathbf{y}_i \oplus \mathbf{y}_i'$ has only one non-zero byte (we call *condition of sample*) as follows.

First, we calculate the number of $(\mathbf{y}_i, \mathbf{y}'_i)$ such that $\mathbf{y}_i \oplus \mathbf{y}'_i$ has only one non-zero byte. There are $2^8 \cdot (2^8 - 1)$ possible values for a byte with non-zero difference and 2^{24} possible values for remaining 3 bytes with zero differences, and there are 4 ways to choose the byte with non-zero difference. Therefore, the number is calculated as $4 \cdot 2^{24} \cdot 2^8 \cdot (2^8 - 1)$. Next, we consider the number of $(\mathbf{x}_i, \mathbf{x}'_i)$ such that $(\mathbf{y}_i, \mathbf{y}'_i)$ has only one non-zero byte. Since DF_i is bijective when $\mathbf{k}_i$ is constant, there is a unique $(\mathbf{x}_i, \mathbf{x}'_i)$ for each $(\mathbf{y}_i, \mathbf{y}'_i)$. Therefore the number of target pairs of $(\mathbf{x}_i, \mathbf{x}'_i)$ satisfying the condition of sample equals one of $(\mathbf{y}_i, \mathbf{y}'_i)$. Hence, we can calculate the ratio as:

$$\frac{4 \cdot 2^{24} \cdot 2^8 \cdot (2^8 - 1)}{2^{32} \cdot (2^{32} - 1)} \approx 2^{-22}. \tag{8}$$

The above discussion is applicable to all block ciphers in which there are bijective components such as DF_i.

Using Eq. (8), we propose an algorithm which can minimize the number of required data in Algorithm 4. Let d be the total number of different plaintexts, there are $d \cdot (d-1)$ differences $\mathbf{x}_i \oplus \mathbf{x}'_i$ we can take. Among them, there are $d \cdot (d-1) \cdot 2^{-22}$ pairs in which $\mathbf{y}_i \oplus \mathbf{y}'_i$ has only one non-zero byte. Since it requires $\mathbb{E}[q]$ samples which satisfies the above to distinguish two distributions, d is estimated as follows.

$$\frac{d \cdot (d-1)}{2^{22}} = \frac{\mathbb{E}[q]}{q_s^2} \quad \Rightarrow \quad d \approx 2^{11} \cdot \frac{(\mathbb{E}[q])^{\frac{1}{2}}}{q_s} \tag{9}$$

The adversary chooses a pair of plaintexts from $\mathcal{X}_D$ and checks whether or not they satisfy the condition of sample for $\mathbf{k}_i$. Since the ratio is 2^{-22}, it requires 2^{22} times of execution of DF_i (equivalent to twice 1-round encryptions) to obtain a sample. When the number of required sample is $\mathbb{E}[q]$, the time complexity T and the number of required data D of Algorithm 4 are estimated as ($\alpha = \beta = 2^{32} - 1$):

$$T = 2^{32} \cdot \frac{\mathbb{E}[q]}{q_s^2} \cdot 2^{22} \cdot \frac{2}{10} \cdot 4 + (1 + 2^{-32} \cdot (2^{32} - 1))^4 \approx 2^{53.68} \cdot \frac{\mathbb{E}[q]}{q_s^2} \tag{10}$$

$$D = 2^{11} \cdot \frac{(\mathbb{E}[q])^{\frac{1}{2}}}{q_s} \cdot q_s \cdot 4 = 2^{13} \cdot (\mathbb{E}[q])^{\frac{1}{2}} \tag{11}$$

Contrary to Eqs. (4) and (11) shows the minimum number of data. However, there is a time-data tradeoff. The time complexity T increases because of redundant computation of DF_i (equivalent to 2 times 1-round encryption) to check whether a pair of plaintexts satisfies the condition of sample.

3.2   Comparison to the Previous Method

We explain the difference between the previous and proposed methods. The previous method requires chosen plaintexts which take all possible values in $\mathbf{x}_i$ to obtain all possible pairs of $(\mathbf{y}_i, \mathbf{y}'_i) = \mathsf{DF}_i((\mathbf{x}_i, \mathbf{x}'_i), \mathbf{k}_i)$ for any $\mathbf{k}_i$. Then, a pair of plaintexts for any $(\mathbf{y}_i, \mathbf{y}'_i)$ and $\mathbf{k}_i$ can be deterministically found in the

---

Algorithm 4. New method of differential bias attack with hypothesis testing.

input $\mathcal{X}_D$ and $\mathcal{L}_D$.
for $i = 0 \to 3$ do
    for $\mathbf{k}_i^* \in \mathbb{F}_2^{32}$ do
        Initialize a likelihood ratio $\Lambda_{\mathbf{k}_i^*} \leftarrow 1$.
        for $(\mathbf{x}, \mathbf{x}') \in \mathcal{X}_D \times \mathcal{X}_D$ do
            if $(\mathbf{y}_i, \mathbf{y}_i') = \mathsf{DF}_i((\mathbf{x}_i, \mathbf{x}_i'), \mathbf{k}_i^*)$ has only one non-zero difference then
                Acquire $\{l_j\}_{j=0}^{q_s-1}$ and $\{l_{j'}'\}_{j'=0}^{q_s-1}$ from $\mathcal{L}_D$ corresponding to $\mathbf{x}$ and $\mathbf{x}'$.
                $\Lambda_{\mathbf{k}_i^*} = \Lambda_{\mathbf{k}_i^*} \times \prod_{j,j'} Pr[l_j \oplus l_{j'}' | \text{correct}]/Pr[l_j \oplus l_{j'}' | \text{wrong}]$.
                if $\Lambda_{\mathbf{k}_i^*} \leq (1-\alpha)/\beta \cup \beta/(1-\alpha) \leq \Lambda_{\mathbf{k}_i^*}$ then
                    Terminate inner-most loop (**for** $\{\mathbf{x}, \mathbf{x}'\} \in \mathcal{X}_D \times \mathcal{X}_D$ **do**).
                end if
            end if
        end for
        if $\beta/(1-\alpha) \leq \Lambda_{\mathbf{k}_i^*}$ then
            $\mathbf{k}_i^*$ is added to a set of candidates $\mathcal{K}_i$.
        end if
    end for
end for
Brute-force search for the correct key $\mathbf{k}^* \in \mathcal{K}(= \mathcal{K}_0 \times \mathcal{K}_1 \times \mathcal{K}_2 \times \mathcal{K}_3)$.
return $\mathbf{k}^*$

---

list $\mathcal{X}_D$. However, the adversary has approximately 2^{44} samples for each key candidate from Eq. (8), and this is much more than the number required for the attack (see Tables 1 and 2). As opposed to the previous method, the proposed method aims to query the minimum number of chosen plaintexts sufficient for the attack. Since the adversary does not have all possible pairs of $(\mathbf{y}_i, \mathbf{y}_i')$, he should find pairs which satisfy the condition of sample and disregard the others. This procedure increases time complexity.

Tables 1 and 2 show complexity of the previous and proposed methods in the bitwise and bytewise random leakage models. We use the same leakage ranges as described in [2] for the comparison. Obviously, there is a time-data tradeoff

Table 1. Complexity of bitwise random leakage model.

Leakage range			$E[q]$	q_s	Current work		This work	
Randomized parameters	$z_{i,j}^r$							
	Constant	Variable			T	D	T	D
Space	$r = 1$	$\forall(i,j)$	$2^{28.84}$	256	$2^{51.84}$	$2^{42.00}$	$2^{74.52}$	$2^{27.42}$
Space	$r = 2$	$\forall(i,j)$	$2^{24.84}$	256	$2^{47.84}$	$2^{42.00}$	$2^{70.52}$	$2^{25.42}$
Space	$r = 3$	$\forall(i,j)$	$2^{39.98}$	256	$2^{62.98}$	$2^{42.00}$	$2^{85.66}$	$2^{32.99}$
Time	(i,j)	r	$2^{19.90}$	11	$2^{47.44}$	$2^{37.46}$	$2^{65.58}$	$2^{22.95}$
Time	(i,j)	r $(r \neq 9)$	$2^{19.36}$	10	$2^{47.04}$	$2^{37.32}$	$2^{65.04}$	$2^{22.68}$
Time	(i,j)	r $(r \neq 1,2,8,9)$	$2^{33.22}$	7	$2^{60.41}$	$2^{36.81}$	$2^{78.90}$	$2^{29.61}$
Time & space	None	(r,i,j)	$2^{38.04}$	256×11	$2^{57.58}$	$2^{45.46}$	$2^{83.72}$	$2^{32.02}$
Time & space	None	$\forall(r,i,j)$ $(r \neq 9)$	$2^{37.49}$	256×10	$2^{57.17}$	$2^{45.32}$	$2^{83.17}$	$2^{31.75}$
Time & space	None	$\forall(r,i,j)$ $(r \neq 1,2,8,9)$	$2^{51.22}$	256×7	$2^{71.41}$	$2^{44.81}$	$2^{96.90}$	$2^{38.61}$

Table 2. Complexity of bytewise random leakage model.

Leakage range			$\mathbb{E}[q]$	q_s	Current work		This work	
Randomized parameters	$z_{i,j}^r$							
	Constant	Variable			T	D	T	D
Space	$r = 1$	$\forall i$	$2^{19.84}$	32	$2^{45.84}$	$2^{39.00}$	$2^{65.52}$	$2^{22.92}$
Space	$r = 2$	$\forall i$	$2^{14.48}$	32	$2^{40.48}$	$2^{39.00}$	$2^{60.16}$	$2^{20.24}$
Space	$r = 3$	$\forall i$	$2^{29.64}$	32	$2^{55.64}$	$2^{42.00}$	$2^{75.32}$	$2^{27.82}$
Time	i	r	$2^{15.62}$	11	$2^{43.16}$	$2^{37.46}$	$2^{61.30}$	$2^{20.81}$
Time	i	r $(r \neq 9)$	$2^{15.52}$	10	$2^{43.19}$	$2^{37.32}$	$2^{61.20}$	$2^{20.76}$
Time	i	r $(r \neq 1,2,8,9)$	$2^{24.22}$	7	$2^{52.41}$	$2^{36.81}$	$2^{69.90}$	$2^{25.11}$
Time & space	None	(r,i)	$2^{24.04}$	16 × 11	$2^{46.58}$	$2^{42.45}$	$2^{69.72}$	$2^{25.02}$
Time & space	None	$\forall(r,i)$ $(r \neq 9)$	$2^{23.58}$	16 × 10	$2^{46.26}$	$2^{42.32}$	$2^{69.26}$	$2^{24.79}$
Time & space	None	$\forall(r,i)$ $(r \neq 1,2,8,9)$	$2^{38.04}$	16 × 7	$2^{61.23}$	$2^{41.80}$	$2^{83.72}$	$2^{32.02}$

between them. We see the tables from both viewpoints of the adversary and evaluator (see Sect. 1.2 for their definition). For the adversary (resp. evaluator), we assume that the adversary can check 2^{60} (resp. 2^{80}) key candidates [12].

In Tables 1 and 2, the proposed method is infeasible under $T \leq 2^{60}$. Therefore, the adversary should query as much data as possible to apply the previous method. On the contrary, some attacks of the proposed method are feasible under $T \leq 2^{80}$, therefore, the evaluator needs to consider some countermeasure in addition to restrict online phase against the adversary. As mentioned in [2], noise addition can be an effective countermeasure. When a probability that an observed bit/byte is correct (not a noise) is π, $\mathbb{E}[q]$ increases as $\mathbb{E}[q]' = \mathbb{E}[q] \cdot (\pi^2)^{-1}$. Therefore, time complexity and the required data for the proposed method increases as $T' = T \cdot (\pi^2)^{-1}$ and $D' = D \cdot (\pi)^{-1}$.

4 Application of Key Enumeration and Rank Estimation

Differential bias attacks with hypothesis testing shown in Sects. 2.2 and 3.1 can restrict key-candidate space into $|\mathcal{K}| = (1 + \alpha \cdot (2^{32} - 1))^4$, and the success rate of the attack is $(1 - \beta)^4$. The methods are based on the assumption that the adversary is given data sufficient to have small error rates α and β (e.g. $\alpha = \beta = 2^{-32}$). Therefore, the adversary achieve the key-candidate space close to 16 and the success rate close to 1. On this point, we also consider an attack which may be feasible even when given data are not sufficient to have small error rates. We apply histogram-based key enumeration [17] to differential bias attack and evaluate its efficiency by histogram-based rank estimation [10]. We show the specification of the algorithm in Sect. 4.1 and experimentally evaluate its efficiency in Sect. 4.2.

4.1 Differential Bias Attack with Key Enumeration

First, the adversary makes a list of log posterior probabilities for each subkey from random leakages and construct histograms $\{H_0, H_1, ..., H_{N_s-1}\}$

(see Sect. 2.3). Let $\mathsf{I}_{\mathsf{DF}_i}$ be a function which outputs 1 if a difference of $(\mathbf{y}_i, \mathbf{y}'_i) = \mathsf{DF}_i((\mathbf{x}_i, \mathbf{x}'_i), \mathbf{k}_i)$ has only one non-zero byte, and outputs 0 otherwise ($\mathsf{I}_{\mathsf{DF}_i}$ checks if $(\mathbf{x}_i, \mathbf{x}'_i)$ satisfy the condition of sample for $\mathbf{k}_i$). Using Bayes' theorem, we can compute a posterior probability of subkey candidate $\mathbf{k}_i^*$ as follows.

$$
\begin{aligned}
\Pr[\mathbf{k}_i = \mathbf{k}_i^* | \mathcal{X}_D, \mathcal{L}_D \times \mathcal{L}_D] &= \frac{\Pr[\mathcal{X}_D, \mathcal{L}_D \times \mathcal{L}_D | \mathbf{k}_i = \mathbf{k}_i^*] \cdot \Pr[\mathbf{k}_i = \mathbf{k}_i^*]}{\sum_{\mathbf{k}'_i} \Pr[\mathcal{X}_D, \mathcal{L}_D \times \mathcal{L}_D | \mathbf{k}_i = \mathbf{k}'_i] \cdot \Pr[\mathbf{k}_i = \mathbf{k}'_i]} \\
&= \frac{\Pr[\mathcal{X}_D, \mathcal{L}_D \times \mathcal{L}_D | \mathbf{k}_i = \mathbf{k}_i^*]}{\sum_{\mathbf{k}'_i} \Pr[\mathcal{X}_D, \mathcal{L}_D \times \mathcal{L}_D | \mathbf{k}_i = \mathbf{k}'_i]}
\end{aligned} \tag{12}
$$

Note that we assume that $\Pr[\mathbf{k}_i = \mathbf{k}_i^*]$ is a uniform distribution (uniform prior). Since the denominator is a normalizing constant, we only consider likelihood of $\mathbf{k}_i = \mathbf{k}_i^*$.

$$
\begin{aligned}
\Pr[\mathcal{X}_D, \mathcal{L}_D \times \mathcal{L}_D | \mathbf{k}_i = \mathbf{k}_i^*] &= \prod_{(l,l')} \Pr[l \oplus l' | \mathbf{k}_i = \mathbf{k}_i^*] \\
&= \left(\prod_{(l,l') | \mathsf{I}_{\mathsf{DF}_i}(\mathbf{x}_i, \mathbf{x}'_i, \mathbf{k}_i^*) = 0} \Pr[l \oplus l' | \text{ wrong}] \right) \cdot \left(\prod_{(l,l') | \mathsf{I}_{\mathsf{DF}_i}(\mathbf{x}_i, \mathbf{x}'_i, \mathbf{k}_i^*) = 1} \Pr[l \oplus l' | \text{ correct}] \right) \\
&= \left(\prod_{(l,l')} \Pr[l \oplus l' | \text{ wrong}] \right) \cdot \left(\prod_{(l,l') | \mathsf{I}_{\mathsf{DF}_i}(\mathbf{x}_i, \mathbf{x}'_i, \mathbf{k}_i^*) = 1} \frac{\Pr[l \oplus l' | \text{ correct}]}{\Pr[l \oplus l' | \text{ wrong}]} \right).
\end{aligned} \tag{13}
$$

Since $\prod_{(l,l')} \Pr[l \oplus l' | \text{ wrong}]$ is constant for all subkey candidates, key-dependent variable is $\prod_{(l,l') | \mathsf{I}_{\mathsf{DF}_i}(\mathbf{x}_i, \mathbf{x}'_i, \mathbf{k}_i^*) = 1} \frac{\Pr[l \oplus l' | \text{ correct}]}{\Pr[l \oplus l' | \text{ wrong}]}$. From Eqs. (12) and (13), the posterior probability can be expressed as follows.

$$
\Pr[\mathbf{k}_i = \mathbf{k}_i^* | \mathcal{X}_D, \mathcal{L}_D \times \mathcal{L}_D] \propto \prod_{(l,l') | \mathsf{I}_{\mathsf{DF}_i}(\mathbf{x}_i, \mathbf{x}'_i, \mathbf{k}_i^*) = 1} \frac{\Pr[l \oplus l' | \text{ correct}]}{\Pr[l \oplus l' | \text{ wrong}]} \tag{14}
$$

Hence, the posterior probability is obtained by likelihood ratio. From the posterior probability, we obtain a list of log posterior probabilities:

$$
\log \left(\Pr[\mathbf{k}_i = \mathbf{k}_i^* | \mathcal{X}_D, \mathcal{L}_D \times \mathcal{L}_D] \right) = \\
\sum_{(l,l') | \mathsf{I}_{\mathsf{DF}_i}(\mathbf{x}_i, \mathbf{x}'_i, \mathbf{k}_i^*) = 1} \log \left(\frac{\Pr[l \oplus l' | \text{ correct}]}{\Pr[l \oplus l' | \text{ wrong}]} \right) + const. \tag{15}
$$

Note that we can ignore the constant value in the construction of histograms.

Algorithm 5 shows an application of the histogram-based key enumeration to differential bias attack. Computation of log likelihood ratios for all subkeys requires the highest cost. The time complexity of the operation is calculated as $2^{32} \cdot d \cdot (d-1) \cdot 2/10 \cdot 4$, where d is the number of different chosen plaintexts. When we assume an adversary who can check 2^{60} (resp. 2^{80}) key candidates, the computation is feasible if $d \leq 2^{14.66}$ (resp. $d \leq 2^{24.66}$).

Algorithm 5. Differential bias attack with histogram-based key enumeration.

---

input $\mathcal{X}_D$ and $\mathcal{L}_D$.
for $i = 0 \rightarrow 3$ **do**
   for $\mathbf{k}_i^* \in \mathbb{F}_2^{32}$ **do**
      Initialize a log likelihood ratio $log(\Lambda_{\mathbf{k}_i^*}) \leftarrow 0$.
      for $(\mathbf{x}, \mathbf{x}') \in \mathcal{X}_D \times \mathcal{X}_D$ **do**
         if $(\mathbf{y}_i, \mathbf{y}_i') = \mathrm{DF}_i((\mathbf{x}_i, \mathbf{x}_i'), \mathbf{k}_i^*)$ has only one non-zero difference **then**
            Acquire $\{l_i\}_{i=0}^{q_s-1}$ and $\{l_j'\}_{j=0}^{q_s-1}$ from $\mathcal{L}_D$ corresponding to $\mathbf{x}$ and $\mathbf{x}'$.
            $log(\Lambda_{\mathbf{k}_i^*}) \leftarrow log(\Lambda_{\mathbf{k}_i^*}) + \sum_{i,j} log(Pr[l_i \oplus l_j' | \mathrm{correct}]/Pr[l_i \oplus l_j' | \mathrm{wrong}])$.
         end if
      end for
      Calculate a bin index $b \leftarrow \lfloor log(\Lambda_{\mathbf{k}_i^*})/S_{bin} \rfloor$.
      Add $\mathbf{k}_i^*$ to the bin b of H_i as $\mathcal{S}_i(b) \leftarrow \mathcal{S}_i(b) \cup \mathbf{k}_i^*$.
      Increase $H_i(b) \leftarrow H_i(b) + 1$.
   end for
end for
Call Algorithm 2 $\mathcal{H} \leftarrow \mathrm{rank\_estimation}(H_0, H_1, H_2, H_3)$.
Call Algorithm 3 $\mathbf{k}^* \leftarrow \mathrm{key\_enumeration}(\mathcal{H}, \mathcal{S})$.
return $\mathbf{k}^*$

Memory requirements for Algorithm 5 become cumbersome, since it requires a list of all 32-bit subkeys and their index of bins. We show the way to practically implement the algorithm as follows. Using Eq. (15), a list of log posterior probabilities for 32-bit subkey is obtained, and it requires $\mathcal{S}$ whose size is more than $2^{32} \times 4 \times 4$ [byte]. Therefore, the adversary may decompose a list for 32-bit subkey list into four lists for 8-bit subkeys. For example, we decompose four 32-bit lists to sixteen 8-bit lists by marginalization, i.e., $Pr[k_j^* | \mathcal{X}_D, \mathcal{L}_D \times \mathcal{L}_D] = \sum_{\mathbf{k}_i | k_j = k_j^*} Pr[\mathbf{k}_i = \mathbf{k}_i^* | \mathcal{X}_D, \mathcal{L}_D \times \mathcal{L}_D]$. When we decompose a list of N_s log posterior probabilities $\{LP_i^{(j)}\}_{i=1}^{N_s}$ (the binsize is S_{bin}) to one of $N_s' = N_s \times N_p$ lists $\{dLP_{i'}^{(j')}\}_{i'=1}^{N_s'}$ (the binsize is dS_{bin}), there may be an error caused by the decomposition. Supposing that $LP_i^{(j)} = \sum_{i'=0}^{N_p-1} dLP_{i'}^{(j')}$, we have:

$$\left| LP_i^{(j)} - m_i^{(j)} \right| \leq \frac{S_{bin}}{2}, \quad \left| dLP_{i'}^{(j')} - dm_{i'}^{(j')} \right| \leq \frac{dS_{bin}}{2},$$

$$\implies \left| LP_i^{(j)} - m_i^{(j)} \right| = \left| \sum_{i'=0}^{N_p-1} dLP_{i'}^{(j')} - \sum_{i'=0}^{N_p-1} dm_{i'}^{(j')} \right| \leq \frac{N_p \times dS_{bin}}{2}. \quad (16)$$

Therefore, there is no error caused by the decomposition if an inequality $S_{bin} \leq (N_p \times dS_{bin})$ holds. In this case, we only consider the error caused by the convolution of histograms shown in Eq. (5). When we execute rank estimation, $\mathcal{S}$ is not required, therefore, the above decomposition is not necessary.

Algorithm 6. Parallelized rank estimation for differential bias attack.

---

for $i = 0 \to 3$ do
    for $\mathbf{k}_i^* \in \mathbb{F}_2^{32}$ in parallel do
        for $j = 0 \to q - 1$ do
            Initialize a log likelihood ratio $log(\Lambda_{\mathbf{k}_i^*}) \leftarrow 0$.
            Choose $\mathbf{y}_i$ and $\mathbf{y}_i'$ randomly where only 1 byte has a difference.
            Obtain $\mathbf{x}_i$ and $\mathbf{x}_i'$ s.t. $(\mathbf{y}_i, \mathbf{y}_i') = \mathsf{DF}_i((\mathbf{x}_i, \mathbf{x}_i'), \mathbf{k}_i^*)$.
            Choose two plaintexts $\mathbf{x}$ and $\mathbf{x}'$ which contain $\mathbf{x}_i$ and $\mathbf{x}_i'$, respectively.
            Choose l and l' randomly corresponding to $\mathbf{x}$ and $\mathbf{x}'$.
            $log(\Lambda_{\mathbf{k}_i^*}) \leftarrow log(\Lambda_{\mathbf{k}_i^*}) + log(Pr[l \oplus l'|\text{correct}]/Pr[l \oplus l'|\text{wrong}])$.
        end for
        Calculate a bin index $b \leftarrow \lfloor log(\Lambda_{\mathbf{k}_i^*})/S_{bin} \rfloor$.
        Increase $H_i(b) \leftarrow H_i(b) + 1$.
    end for
end for
Call Algorithm 2 $E \leftarrow \mathsf{rank\_estimation}(H_0, H_1, H_2, H_3)$.
return E.

4.2 Experimental Evaluation

Contrary to the methods shown in Sects. 2.2 and 3.1, we can not estimate success rate using error rates. Therefore, we should estimate it experimentally. Algorithm 6 shows a procedure of the experiment and it is performed by CUDA (version 6.00) implementation on GPGPU (NVIDIA Tesla K20Xm) platform in order to parallelize the computation ("**for** $\mathbf{k}_i^* \in \mathbb{F}_2^{32}$ **in parallel do**" in Algorithm 6). Note that we simulate differential bias attack without actual data. For all $\mathbf{k}_i^*$, we encrypt q pairs of plaintexts, s.t., $(\mathbf{y}_i, \mathbf{y}_i') = \mathsf{DF}_i((\mathbf{x}_i, \mathbf{x}_i'), \mathbf{k}_i^*)$, and extract leakages from a leakage range randomly. After making four histograms H_0, H_1, H_2 and H_3, we execute histogram-based rank estimation [10].

We estimate success rate changing the number of samples q. Because of constraints on execution time (time complexity is $2^{32} \cdot q \cdot 4$), we set $q = 500 \cdot i$, $i \in \{1 : 5\}$. Since $q = 2500$ at most, it is much smaller than all expected number of samples $\mathbb{E}[q]$ in Tables 1 and 2. Therefore, we choose a leakage range which requires less samples. From ones of the bytewise random leakage model shown in Table 2, we choose one of the leakage ranges, space randomization of $r = 2$ (second line). Note that a leaked byte is randomly chosen from z_i^2 or k_i^2 ($i \in [0 : 15]$). This setting is the most advantageous one for the adversary. For all q, we obtain the upper bound of the estimated rank to estimate success rate for key-recovery under $T \leq 2^{80}$ and $T \leq 2^{60}$.

First, we make a histogram $H_{0:3}$ in condition that the correct key is unknown. Next, we obtain a bin index of the correct key $b^\dagger$ for 2^{32} times by changing plaintexts and position of leaked bytes randomly. Last, we count the number of candidates in $H_{0:3}$ such that $\sum_{b=b^\dagger}^{4 \cdot q - 4} H_{0:3}(b) \leq 2^{60}$ or 2^{80} (we set $N_{bin} = q$) and calculate its ratio (success rate). As mentioned in Sect. 3.2, we consider both viewpoints of the evaluator ($T \leq 2^{80}$) and the adversary ($T \leq 2^{60}$). Figure 4 shows the results. Success rate is practical under $T \leq 2^{80}$ (right graph) and

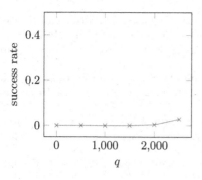 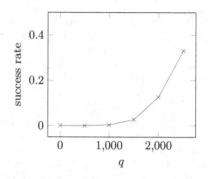

Fig. 4. Success rate of the proposal attack under the leakage range $r = 2$ and $T \leq 2^{60}$ (left) under $T \leq 2^{80}$ (right).

$q \geq 1500$ even if the number of samples q is smaller than $\mathbb{E}[q] = 2^{14.48}$ used in hypothesis-testing methods. Hence, the adversary given approximately $\mathbb{E}[q]/2^4$ samples for all key candidates ($2^{14.48}/1500 \approx 2^4$) is still a threat for the evaluator.

We confirm that success rate for the adversary to recover the secret key is approximately equal to 0 under $T \leq 2^{60}$, however, the attack can be practical under $T \leq 2^{80}$. The evaluator can have an effective security margin by evaluating cryptographic devices by assuming the adversary with such computing power.

5   Conclusion

This paper improves the security analysis of AES-128 under the random leakage model. More specifically, we generalize differential bias attack to be applicable under less advantageous condition for the adversary. First, we reestimate complexity for the adversary given smaller number of data by using new hypothesis-testing method. Second, we apply key enumeration to differential bias attack. Both results show that it is possible to recover the secret key under $T \leq 2^{80}$ with much smaller number of data. It indicates that differential bias attack is a practical threat from the view point of the evaluator. Also, the attack can complement other side-channel attacks to reduce time complexity. In addition to noise addition, intensive prevention of leakages from important rounds such as second round of AES can be an effective countermeasure.

References

1. Bernstein, D.J., Lange, T., van Vredendaal, C.: Tighter, faster, simpler side-channel security evaluations beyond computing power. Cryptology ePrint Archive, report 2015/221 (2015)
2. Bogdanov, A., Isobe, T.: How secure is AES under leakage. In: Iwata, T., Cheon, J.H. (eds.) ASIACRYPT 2015. LNCS, vol. 9453, pp. 361–385. Springer, Heidelberg (2015). doi:10.1007/978-3-662-48800-3_15

3. Bogdanov, A., Kizhvatov, I., Manzoor, K., Tischhauser, E., Witteman, M.: Fast and memory-efficient key recovery in side-channel attacks. In: Dunkelman, O., Keliher, L. (eds.) SAC 2015. LNCS, vol. 9566, pp. 310–327. Springer, Cham (2016). doi:10.1007/978-3-319-31301-6_19

4. Chari, S., Rao, J.R., Rohatgi, P.: Template attacks. In: Kaliski, B.S., Koç, K., Paar, C. (eds.) CHES 2002. LNCS, vol. 2523, pp. 13–28. Springer, Heidelberg (2003). doi:10.1007/3-540-36400-5_3

5. David, L., Wool, A.: A bounded-space near-optimal key enumeration algorithm for multi-dimensional side-channel attacks. Cryptology ePrint Archive, report 2015/1236 (2015)

6. Dinur, I., Shamir, A.: Side channel cube attacks on block ciphers. Cryptology ePrint Archive, report 2009/127 (2009)

7. Doget, J., Prouff, E., Rivain, M., Standaert, F.X.: Univariate side channel attacks and leakage modeling. J. Cryptogr. Eng. 1(2), 123–144 (2011)

8. Duc, A., Faust, S., Standaert, F.-X.: Making masking security proofs concrete. In: Oswald, E., Fischlin, M. (eds.) EUROCRYPT 2015. LNCS, vol. 9056, pp. 401–429. Springer, Heidelberg (2015). doi:10.1007/978-3-662-46800-5_16

9. Fluhrer, S.R., McGrew, D.A.: Statistical analysis of the alleged RC4 keystream generator. In: Goos, G., Hartmanis, J., Leeuwen, J., Schneier, B. (eds.) FSE 2000. LNCS, vol. 1978, pp. 19–30. Springer, Heidelberg (2001). doi:10.1007/3-540-44706-7_2

10. Glowacz, C., Grosso, V., Poussier, R., Schüth, J., Standaert, F.-X.: Simpler and more efficient rank estimation for side-channel security assessment. In: Leander, G. (ed.) FSE 2015. LNCS, vol. 9054, pp. 117–129. Springer, Heidelberg (2015). doi:10.1007/978-3-662-48116-5_6

11. Ishai, Y., Sahai, A., Wagner, D.: Private circuits: securing hardware against probing attacks. In: Boneh, D. (ed.) CRYPTO 2003. LNCS, vol. 2729, pp. 463–481. Springer, Heidelberg (2003). doi:10.1007/978-3-540-45146-4_27

12. Kleinjung, T., Lenstra, A.K., Page, D., Smart, N.P.: Using the cloud to determine key strengths. In: Galbraith, S., Nandi, M. (eds.) INDOCRYPT 2012. LNCS, vol. 7668, pp. 17–39. Springer, Heidelberg (2012). doi:10.1007/978-3-642-34931-7_3

13. Kocher, P., Jaffe, J., Jun, B.: Differential power analysis. In: Wiener, M. (ed.) CRYPTO 1999. LNCS, vol. 1666, pp. 388–397. Springer, Heidelberg (1999). doi:10.1007/3-540-48405-1_25

14. Mantin, I.: Predicting and distinguishing attacks on RC4 keystream generator. In: Cramer, R. (ed.) EUROCRYPT 2005. LNCS, vol. 3494, pp. 491–506. Springer, Heidelberg (2005). doi:10.1007/11426639_29

15. Manzoor, K., et al.: Efficient practical key recovery for side-channel attacks. Master's thesis, Aalto University, June 2014

16. Martin, D.P., O'Connell, J.F., Oswald, E., Stam, M.: Counting keys in parallel after a side channel attack. In: Iwata, T., Cheon, J.H. (eds.) ASIACRYPT 2015. LNCS, vol. 9453, pp. 313–337. Springer, Heidelberg (2015). doi:10.1007/978-3-662-48800-3_13

17. Poussier, R., Standaert, F.-X., Grosso, V.: Simple key enumeration (and rank estimation) using histograms: an integrated approach. In: Gierlichs, B., Poschmann, A.Y. (eds.) CHES 2016. LNCS, vol. 9813, pp. 61–81. Springer, Heidelberg (2016). doi:10.1007/978-3-662-53140-2_4

18. Renauld, M., Standaert, F.-X., Veyrat-Charvillon, N.: Algebraic side-channel attacks on the AES: why time also matters in DPA. In: Clavier, C., Gaj, K. (eds.) CHES 2009. LNCS, vol. 5747, pp. 97–111. Springer, Heidelberg (2009). doi:10.1007/978-3-642-04138-9_8

19. Schramm, K., Wollinger, T., Paar, C.: A new class of collision attacks and its application to DES. In: Johansson, T. (ed.) FSE 2003. LNCS, vol. 2887, pp. 206–222. Springer, Heidelberg (2003). doi:10.1007/978-3-540-39887-5_16

20. Veyrat-Charvillon, N., Gérard, B., Renauld, M., Standaert, F.-X.: An optimal key enumeration algorithm and its application to side-channel attacks. In: Knudsen, L.R., Wu, H. (eds.) SAC 2012. LNCS, vol. 7707, pp. 390–406. Springer, Heidelberg (2013). doi:10.1007/978-3-642-35999-6_25

21. Veyrat-Charvillon, N., Gérard, B., Standaert, F.-X.: Security evaluations beyond computing power. In: Johansson, T., Nguyen, P.Q. (eds.) EUROCRYPT 2013. LNCS, vol. 7881, pp. 126–141. Springer, Heidelberg (2013). doi:10.1007/978-3-642-38348-9_8

22. Wald, A.: Sequential tests of statistical hypotheses. In: Kotz, S., Johnson, N.L. (eds.) Breakthroughs in Statistics, pp. 256–298. Springer, Heidelberg (1992)

Differential Cryptanalysis

Impossible Differential Cryptanalysis of Reduced-Round SKINNY

Mohamed Tolba, Ahmed Abdelkhalek, and Amr M. Youssef$^{(\boxtimes)}$

Concordia Institute for Information Systems Engineering,
Concordia University, Montréal, QC, Canada
youssef@ciise.concordia.ca

Abstract. SKINNY is a new lightweight tweakable block cipher family proposed by Beierle *et al.* at CRYPTO 2016. SKINNY has 6 main variants where SKINNY-n-t is a block cipher that operates on n-bit blocks using t-bit tweakey (key and tweak) where $n = 64$ or 128 and $t = n, 2n$, or $3n$. In this paper, we present impossible differential attacks against reduced-round versions of all the 6 members of the SKINNY family in the single-tweakey model. More precisely, using an 11-round impossible differential distinguisher, we present impossible differential attacks against 18-round SKINNY-n-n, 20-round SKINNY-n-$2n$ and 22-round SKINNY-n-$3n$ ($n = 64$ or 128). To the best of our knowledge, these are the best attacks against these 6 variants in the single-tweakey model.

Keywords: Cryptanalysis · Impossible differential attacks · Tweakable · Block ciphers · SKINNY

1 Introduction

SKINNY [3] is a Substitution Permutation Network (SPN) family of tweakable lightweight block ciphers proposed at CRYPTO 2016 by Beierle *et al.* It supports two block lengths $n = 64$ or 128 and for each of them, the tweakey t can be either $n, 2n$ or $3n$. This family of ciphers inherits the recent design trend of having an SPN cipher with suboptimal internal components. More precisely, SKINNY uses a light tweakey schedule along with a round function that consists of a compact S-box and a sparse diffusion layer. However, these suboptimal components are arranged such that tight security bounds are guaranteed. Indeed, using Mixed Integer Linear Programming (MILP), the designers of SKINNY provide high security bounds against differential/linear attacks for all the SKINNY versions in both the single-tweakey and related-tweakey models. Furthermore, SKINNY has a good performance for round-based ASIC implementation as it requires a very small area using serial ASIC. Moreover, the designers of SKINNY show that its ASIC threshold implementation is very favorable to AES-128 threshold implementation [5]. Providing compact implementation and a high level of security with the existence of the tweakey was feasible by generalizing the Superposition TWEAKEY (STK) construction [7]. Lastly, being a tweakable block cipher allows SKINNY to be employed into a higher level of operating modes such as SCT [11].

© Springer International Publishing AG 2017
M. Joye and A. Nitaj (Eds.): AFRICACRYPT 2017, LNCS 10239, pp. 117–134, 2017.
DOI: 10.1007/978-3-319-57339-7_7

The designers of SKINNY presented 16-round attacks against SKINNY-n-n ($n = 64$ or 128) in the single-tweakey model utilizing 11-round impossible differential distinguisher. To provoke public cryptanalysis of SKINNY, they have announced a competition [2] against two particular variants of SKINNY, namely, SKINNY-64-128 and SKINNY-128-128, in which they indicated that the best known attack against SKINNY-64-128, in the single-tweakey model, is 18 rounds. As a result, a handful of third-party analysis have been published [1, 10, 12]. However, these attacks are in the arguably weaker attack model, the related-tweakey model, in which the attacker is assumed to have the ability to query the encryption oracle with keys that have specific relations.

In this paper, we present impossible differential attacks against reduced-round versions of all the 6 variants of SKINNY, namely, SKINNY-n-n, SKINNY-n-$2n$ and SKINNY-n-$3n$ ($n = 64$ or 128). All these attacks utilize the same 11-round impossible differential distinguisher. Then, we exploit the fact that the tweakey additions are only performed on the first two rows of the state, along with the MixColumns operation properties and the tweakey schedule relations, to extend this distinguisher by 7, 9, 11 rounds to launch key-recovery attacks in the single-tweakey model against 18, 20, 22 rounds of SKINNY-n-n, SKINNY-n-$2n$ and SKINNY-n-$3n$ ($n = 64$ or 128), respectively. Specifically, we extend the designers' 11-round impossible differential distinguisher by 3, 3 and 3 rounds above it and 4, 6 and 8 rounds below it to launch 18, 20 and 22 rounds attacks against SKINNY-n-n, SKINNY-n-$2n$ and SKINNY-n-$3n$ ($n = 64$ or 128), respectively. The time, data and memory complexities of our attacks are presented in Table 1.

Table 1. The time, data and memory complexities of our attacks.

Block cipher version	# of rounds	Time	Data	Memory
SKINNY-64-64	18	$2^{57.1}$	$2^{47.52}$	$2^{58.52}$
SKINNY-128-128	18	$2^{116.94}$	$2^{92.42}$	$2^{115.42}$
SKINNY-64-128	20	$2^{121.08}$	$2^{47.69}$	$2^{74.69}$
SKINNY-128-256	20	$2^{245.72}$	$2^{92.1}$	$2^{147.1}$
SKINNY-64-192	22	$2^{183.97}$	$2^{47.84}$	$2^{74.84}$
SKINNY-128-384	22	$2^{373.48}$	$2^{92.22}$	$2^{147.22}$

The rest of the paper is organized as follows. Section 2 provides the notations used throughout the paper and a brief description of SKINNY. In Sect. 3, we present the impossible differential distinguisher used in our attacks. The details of our attacks are presented in Sects. 4, 5 and 6, respectively. Finally, the paper is concluded in Sect. 7.

2   Specifications of SKINNY

The following notations are used throughout the rest of the paper:

- TK_i: The round tweakey used in round i.
- ETK_i: The equivalent round tweakey used in round i.
- x_i: The input to the SubCells (SC) operation at round i.
- y_i: The input to the AddRoundConstantTweakey (AK) operation at round i.
- y_i': The input to the AddRoundConstantEquivlantTweakey (AEK) operation at round i.
- z_i: The input to the ShiftRows (SR) operation at round i.
- w_i: The input to the MixColumns (MC) operation at round i.
- $x_i[j]$: The jth cell of x_i, where $0 \le j < 16$.
- $x_i[j \cdots l]$: The cells from j to l of x_i, where $j < l$.
- $x_i[j, l]$: The cells j and l of x_i.
- $x_i[j][k]$: The kth bit of the jth cell of x_i.
- $x_i[j]\{k, l, m\}$: The XOR of bits k, l, m of cell j of x_i.
- $x_i[col : j]$: The four cells in column j, e.g., $x_i[col : 0] = x_i[0, 4, 8, 12]$.
- $x_i[SR^{-1}[col : j]]$: The four cells in column j after the SR^{-1} operation is applied, e.g., $x_i[SR^{-1}[col : 0]] = x_i[0, 7, 10, 13]$.
- $x_i[col : j][k, l]$: The jth and l^{th} cells of column j of x_i, e.g., $x_i[col : 0][0, 1] = x_i[0, 4]$.
- $\Delta x_i, \Delta x_i[j]$: The difference at state x_i and cell $x_i[j]$, respectively.

The SKINNY family supports two block lengths of $n = 64$ and 128 bits. In both versions, the internal state IS is represented as a 4×4 array of cells such that one cell represents a nibble (when the block length $n = 64$) and a byte (when the block length $n = 128$). While classical block ciphers have two inputs, namely the plaintext and the key, and output the ciphertext, SKINNY is a tweakable block cipher [7,9] that uses an input called the tweakey instead of the key. Then, the user has the freedom to choose which part of the tweakey to be assigned to the key and which part to be assigned to the tweak. This family of block ciphers with block length n deploys three main tweakeys of lengths $t = n$ bits, $t = 2n$ bits and $t = 3n$ bits. Similar to the state, the tweakey state can be represented as z 4×4 arrays of cells, i.e., we have arrays $TK1$ (in case $z = 1$), $TK1$ and $TK2$ (in case $z = 2$), $TK1$, $TK2$, and $TK3$ (in case $z = 3$).

The encryption operation proceeds as follows. First, the plaintext $m = m_0 \| m_1 \| \cdots \| m_{14} \| m_{15}$ (where $|m_i| = n/16 = s$-bit) is loaded into the internal state IS row-wise as depicted in Fig. 1. Then, the tweakey input $tk = tk_0 \| tk_1 \| \cdots \| tk_{16z-1}$ (where $|tk_i|$ is s-bit as in the internal state) is loaded row-wise such that $TK1[i] = tk_i$ for $0 \le i \le 15$ (in case $z = 1$), $TK1[i] = tk_i$, $TK2[i] = tk_{16+i}$ for $0 \le i \le 15$ (in case $z = 2$) or $TK1[i] = tk_i$, $TK2[i] = tk_{16+i}$, $TK3[i] = tk_{32+i}$ for $0 \le i \le 15$ (in case $z = 3$). Finally, the internal state is updated by applying the round function r times, where the number of rounds r depends on the block length and the tweakey size as shown in Table 2.

As shown in Fig. 1, in each round, SKINNY applies five different operations, namely, SubCells, AddConstants, AddRoundTweakey, ShiftRows and MixColumns. The cipher does not apply whitening tweakeys. Consequently, parts of the first and last rounds do not add any security. In what follows, we describe the five different operations that are employed in each round:

Table 2. Number of rounds for SKINNY-n-t, with n-bit state and t-bit tweakey state.

Block size n	Tweakey size t		
	n	$2n$	$3n$
64	32	36	40
128	40	48	56

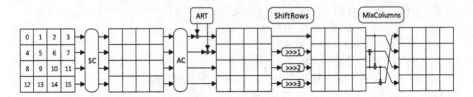

Fig. 1. The SKINNY round function

- SubCells (*SC*): A nonlinear bijective mapping applied on every cell of the internal state, where 4-bit (in case $n = 64$) or 8-bit (in case $n = 128$) S-boxes are applied.
- AddConstants (*AC*): A 4×4 round constant is XORed to the state. These round constants are generated using a 6-bit affine LFSR. The details of generating the round constants can be found in [3].
- AddRoundTweakey (*ART*): The first and second rows of all the tweakey arrays are XORed to the state. More precisely, for $0 \le i \le 7$, we have:
  - $IS[i] = IS[i] \oplus TK1[i]$, when $z = 1$,
  - $IS[i] = IS[i] \oplus TK1[i] \oplus TK2[i]$, when $z = 2$,
  - $IS[i] = IS[i] \oplus TK1[i] \oplus TK2[i] \oplus TK3[i]$, when $z = 3$.
- ShiftRows (*SR*): The rows of the state are rotated as in AES but to the right, i.e., the following permutation $P = [0, 1, 2, 3, 7, 4, 5, 6, 10, 11, 8, 9, 13, 14, 15, 12]$ is applied.
- MixColumns (*MC*): Each column in the state is multiplied by a binary matrix M, where M and its inverse M^{-1} are given as follows:

$$M = \begin{pmatrix} 1 & 0 & 1 & 1 \\ 1 & 0 & 0 & 0 \\ 0 & 1 & 1 & 0 \\ 1 & 0 & 1 & 0 \end{pmatrix}, M^{-1} = \begin{pmatrix} 0 & 1 & 0 & 0 \\ 0 & 1 & 1 & 1 \\ 0 & 1 & 0 & 1 \\ 1 & 0 & 0 & 1 \end{pmatrix}.$$

Tweakey Schedule. As depicted in Fig. 2, the tweakey arrays are updated through tweakey schedule as follows. First all the tweakey arrays, i.e., *TK1* (when $z = 1$), *TK1, TK2* (when $z = 2$), or *TK1, TK2, TK3* (when $z = 3$) are permuted using a permutation P_T such that $P_T = [9, 15, 8, 13, 10, 14, 12, 11, 0, 1, 2, 3, 4, 5, 6, 7]$. Finally, each cell in the first and second rows of *TK2, TK3* (when $z = 2$ or $z = 3$) is updated using the LFSR operations shown in Table 3, where x_0 is the LSB of the cell.

Table 3. The SKINNY LFSR used in the tweakey schedule, where s denotes the cell size in bits.

TK	s	LFSR
$TK2$	4	$(x_3 \parallel x_2 \parallel x_1 \parallel x_0) \rightarrow (x_2 \parallel x_1 \parallel x_0 \parallel x_3 \oplus x_2)$
	8	$(x_7 \parallel x_6 \parallel x_5 \parallel x_4 \parallel x_3 \parallel x_2 \parallel x_1 \parallel x_0) \rightarrow (x_6 \parallel x_5 \parallel x_4 \parallel x_3 \parallel x_2 \parallel x_1 \parallel x_0 \parallel x_7 \oplus x_5)$
$TK3$	4	$(x_3 \parallel x_2 \parallel x_1 \parallel x_0) \rightarrow (x_0 \oplus x_3 \parallel x_3 \parallel x_2 \parallel x_1)$
	8	$(x_7 \parallel x_6 \parallel x_5 \parallel x_4 \parallel x_3 \parallel x_2 \parallel x_1 \parallel x_0) \rightarrow (x_0 \oplus x_6 \parallel x_7 \parallel x_6 \parallel x_5 \parallel x_4 \parallel x_3 \parallel x_2 \parallel x_1)$

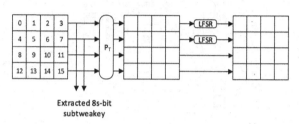

Extracted 8s-bit
subtweakey

Fig. 2. The tweakey schedule

In our attack, we use AddKey (AK) operation which compromises the AC and ART operations. Moreover, we swap the linear operations AK, $MC \circ SR$, and hence we use the equivalent subtweakey ETK instead of the subtweakey TK such that $ETK_{r+1} = MC \circ SR(TK_r)$.

3   An Impossible Differential Distinguisher of SKINNY

Impossible differential cryptanalysis was proposed independently by Biham, Biryukov and Shamir [4] and Knudsen [8]. It exploits a (truncated) differential characteristic of probability exactly 0 and thus acts as a distinguisher. Then, this distinguisher is turned into a key-recovery attack by prepending and/or appending additional rounds, which are usually referred to as the analysis rounds. The keys involved in the analysis rounds which lead to the impossible differential are wrong keys and thus are excluded. Miss-in-the-Middle is the general technique used to construct impossible differentials, where a cipher E is split such that $E = E_2 \circ E_1$, and we try to find two deterministic differentials, the first one covers E_1 and has the form $\Delta\delta \rightarrow \Delta\gamma$, and the second covers E_2^{-1}, and has the form $\Delta\beta \rightarrow \Delta\zeta$. When the intermediate differences $\Delta\gamma, \Delta\zeta$ do not match, the differential $\Delta\delta \rightarrow \Delta\beta$ that covers the whole cipher E holds with zero probability.

The designers of SKINNY exhaustively searched for the longest truncated impossible differential that has one active cell in both $\Delta\delta$ and $\Delta\beta$. They found 16 such truncated impossible differentials where each one covers 11 rounds. They exploited one of these 16 impossible differentials, illustrated in Fig. 3, to attack 16-round SKINNY-n-n ($n = 64$ or 128). This distinguisher, which we reuse in our attacks, states that a pair of messages that has only one active cell at $x_3[12]$ cannot have only one active cell at $x_{14}[8]$. The reason is that the active cell

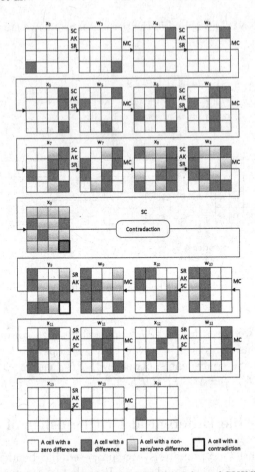

Fig. 3. Impossible differential distinguisher of SKINNY

$\Delta x_3[12]$ results in 4 active cells and 12 unknown cells after 6 rounds, i.e., at state x_9. From the other side, the active cell $\Delta x_{14}[8]$ results in 4 inactive cells, 5 unknown cells and 7 active cells at state Y_9 contradicting with the forward differential at $\Delta y_9[15]$.

Our attacks depend on the following proposition:

Proposition 1 *(Differential Property of the S-box). Given two nonzero differences Δi and Δo in $\mathbb{F}16$ or $\mathbb{F}256$, the equation: $S(x) + S(x + \Delta i) = \Delta o$ has one solution on average. This property also applies to S^{-1}.*

All our attacks use the same 11-round distinguisher, have 3 analysis rounds on its top. They, however, differ in the analysis rounds appended below it. In what follows, we describe our attack against SKINNY-64-128 in details and then mention only the main differences for the other attacks.

4   Impossible Differential Key-Recovery Attack on 20-Round SKINNY-n-2n (n = 64 or 128)

4.1   Impossible Differential Key-Recovery Attack on SKINNY-64-128

In this section, we present the first published attack on 20-round SKINNY-64-128 in the single-tweakey model. We use the notion of data structures to generate enough pairs of messages to launch the attack. In the first three rounds, we use the equivalent tweakey ETK instead of the tweakey TK. Therefore, the first round has no tweakey, and hence we can build our structures at y'_1. Then, we propagate it backward linearly through MC^{-1}, SR^{-1}, and SC^{-1} to obtain the corresponding plaintexts. Our utilized structure takes all the possible values in 7 nibbles $y'_1[3, 4, 5, 6, 9, 11, 14]$ while the remaining nibbles take a fixed value. Thus, one structure generates $2^{4 \times 7} \times (2^{4 \times 7} - 1)/2 \approx 2^{55}$ possible pairs. Hence, we have 2^{55} possible pairs of messages satisfying the plaintext differences. In addition, we utilize the following pre-computation tables in order to efficiently extract/filter the (equivalent) tweakey nibbles corresponding to the active state nibbles involved in the analysis rounds, where the table $H_l\{(E)TK_i[\mathbb{S}]\}$ (also referred to as H_l) is used to extract/filter the (equivalent) tweakey used in round i at cells belonging to the set $\mathbb{S}$ and H^* is computed once and used to extract all the tweakey nibbles of the last analysis round and those corresponding to column 1 in round 18.

$H_1\{TK_{18}[2, 6]\}$: For all the 2^{24} possible values of $\Delta z_{17}[SR^{-1}[col : 2][0, 1]]$, $z_{17}[SR^{-1}[col : 2]]$, compute $\Delta y_{18}[col : 2], y_{18}[col : 2]$. Then, store $\Delta z_{17}[SR^{-1}[col : 2][0, 1]]$, $z_{17}[SR^{-1}[col : 2]], y_{18}[col : 2][0, 1]$ in H_1 indexed by $\Delta y_{18}[col : 2], y_{18}[col : 2][2, 3]$. H_1 has 2^{24} rows and on average about $2^{24}/2^{24} = 1$ value in each row.

$H_2\{TK_{18}[0, 4]\}$: For all the 2^{28} possible values of $\Delta z_{17}[SR^{-1}[col : 0][0, 2, 3]]$, $z_{17}[SR^{-1}[col : 0]]$, compute $\Delta y_{18}[col : 0], y_{18}[col : 0]$. Then, store $\Delta z_{17}[SR^{-1}[col : 0][0, 2, 3]]$, $z_{17}[SR^{-1}[col : 0]], y_{18}[col : 0][0, 1]$ in H_2 indexed by $\Delta y_{18}[col : 0]$, $y_{18}[col : 0][2, 3]$. H_2 has 2^{24} rows and on average about $2^{28}/2^{24} = 2^4$ values in each row.

$H_3\{TK_{18}[3, 7]\}$: For all the 2^{28} possible values of $\Delta z_{17}[SR^{-1}[col : 3][0, 1, 3]]$, $z_{17}[SR^{-1}[col : 3]]$, compute $\Delta y_{18}[col : 3], y_{18}[col : 3]$. Then, store $\Delta z_{17}[SR^{-1}[col : 3][0, 1, 3]]$, $z_{17}[SR^{-1}[col : 3]], y_{18}[col : 3][0, 1]$ in H_3 indexed by $\Delta y_{18}[col : 3]$, $y_{18}[col : 3][2, 3]$. H_3 has 2^{24} rows and on average about $2^{28}/2^{24} = 2^4$ values in each row.

$H_4\{TK_{17}[0, 4]\}$: For all the 2^{20} possible values of $\Delta z_{16}[SR^{-1}[col : 0][0]]$, $z_{16}[SR^{-1}[col : 0]]$, compute $\Delta y_{17}[col : 0][0, 1, 3], y_{17}[col : 0]$. Then, store $\Delta z_{16} [SR^{-1}[col : 0][0]], z_{16}[SR^{-1}[col : 0]], y_{17}[col : 0][0, 1]$ in H_4 indexed by $\Delta y_{17}[col : 0][0, 1, 3], y_{17}[col : 0][2, 3]$. H_4 has 2^{20} rows and on average about $2^{20}/2^{20} = 1$ value in each row.

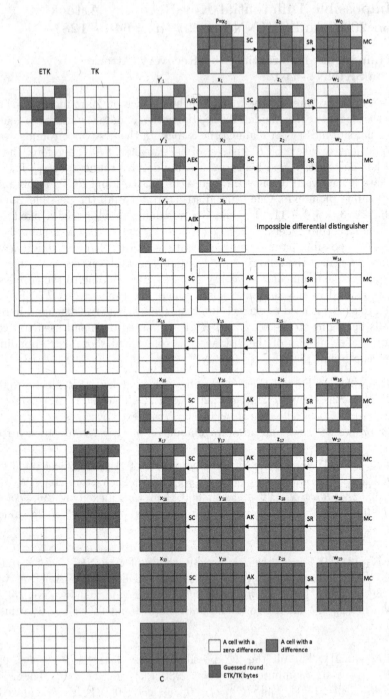

Fig. 4. Impossible differential attack on 20-round SKINNY-n-$2n$

$H_5\{TK_{17}[2,3,6]\}$: From the properties of the MixColumns, we have $\Delta x_{16}[0] = \Delta x_{16}[8] = \Delta x_{16}[12] = \Delta w_{15}[8]$. Therefore, for all the 2^{40} possible values for $\Delta x_{16}[8]$, $x_{16}[8,12], \Delta w_{16}[2,7], w_{16}[2,6,14], x_{17}[3,11]$, compute $w_{16}[10,15], \Delta y_{17}[2,3,6,10,11,14], y_{17}[2,3,6,10,11,14,15]$ such that $y_{17}[15] = SC([w_{16}[15] \oplus x_{17}[3])$, from the MixColumns operation. Then, store $\Delta z_{16}[SR^{-1}[col:2][0,2]], \Delta z_{16}[SR^{-1}[col:3][1,3]], z_{16}[SR^{-1}[col:2]], z_{16}[SR^{-1}[col:3][3]], y_{17}[2,3,6]$ in H_5 indexed by $\Delta y_{17}[2,3,6,10,11,14], y_{17}[10,11,14,15]$. H_5 has 2^{40} rows and on average about $2^{40}/2^{40} = 1$ value in each row.

$H_6\{TK_{17}[1,5]\}$: For all the 2^{24} possible values of $\Delta z_{16}[SR^{-1}[col:1][0,3]], z_{16}[SR^{-1}[col:1]]$, compute $\Delta y_{17}[col:1][0,1,3], y_{17}[col:1]$. Then, store $\Delta z_{16}[SR^{-1}[col:1][0,3]], z_{16}[SR^{-1}[col:1]], y_{17}[col:1][0,1]$ in H_6 indexed by $\Delta y_{17}[col:1][0,1,3], y_{17}[col:1][2,3]$. H_6 has 2^{20} rows and on average about $2^{24}/2^{20} = 2^4$ values in each row.

$H_7\{TK_{16}[0]\}$: For all the 2^{20} possible values of $\Delta z_{15}[SR^{-1}[col:0][2]], z_{15}[SR^{-1}[col:0]]$, compute $\Delta y_{16}[col:0][0,2,3], y_{16}[col:0]$. Then, store $\Delta z_{15}[SR^{-1}[col:0][2]], z_{15}[SR^{-1}[col:0]], y_{16}[col:0][0]$ in H_7 indexed by $\Delta y_{16}[col:0][0,2,3], y_{16}[col:0][2,3]$. H_7 has 2^{20} rows and on average about $2^{20}/2^{20} = 1$ value in each row.

$H_8\{TK_{16}[2]\}$: For all the 2^{20} possible values of $\Delta z_{15}[SR^{-1}[col:2][0]], z_{15}[SR^{-1}[col:2]]$, compute $\Delta y_{16}[col:2][0,1,3], y_{16}[col:2]$. Then, store $\Delta z_{15}[SR^{-1}[col:2][0]], z_{15}[SR^{-1}[col:2]], y_{16}[col:2][0,1]$ in H_8 indexed by $\Delta y_{16}[col:2][0,1,3], y_{16}[col:2][2,3]$. H_8 has 2^{20} rows and on average about $2^{20}/2^{20} = 1$ value in each row.

$H_9\{TK_{15}[2]\}$: From the properties of the MixColumns, we have $\Delta x_{15}[2] = \Delta x_{15}[10] = \Delta x_{15}[14] = \Delta w_{14}[10]$. Therefore, for all the 2^4 possible differences for $\Delta x_{15}[2,10]$, 2^8 possible values of $x_{15}[2,10]$ and 2^4 possible values of $TK_{15}[2]$, compute $\Delta z_{15}[2,10], z_{15}[2,10]$. Then, store $\Delta z_{15}[2]$ in H_9 indexed by $\Delta z_{15}[2,10], z_{15}[2,10], TK_{15}[2]$. H_9 has 2^{20} rows and on average about $2^{16}/2^{20} = 2^{-4}$ values in each row.

$H_{10}\{ETK_1[4,11,14]\}$: For all the 2^{12} possible differences of $\Delta w_1[5,9,13]$, we have only 2^4 valid differences that have exactly one difference in $\Delta y_2'[13]$ and 3 zero differences in $\Delta y_2'[1,5,9]$. Therefore, for all the 2^4 possible differences of $\Delta w_1[5,9,13]$, 2^{12} possible values of $w_1[5,9,13]$ and 2^8 possible values of $ETK_1[4,14]$, compute $\Delta y_1'[4,14], y_1'[4,14], \Delta x_1[11], x_1[11]$. Then, store $\Delta w_1[5,9,13], w_1[5,9,13], x_1[11]$ in H_{10} indexed by $\Delta y_1'[4,14], y_1'[4,14], \Delta x_1[11], ETK_1[4,14]$. H_{10} has 2^{28} rows and on average about $2^{24}/2^{28} = 2^{-4}$ values in each row.

$H_{11}\{ETK_1[3,6,9]\}$: For all the 2^{12} possible differences of $\Delta w_1[3,7,11]$, we have only 2^4 valid differences that have exactly one difference in $\Delta y_2'[7]$ and 3 zero differences in $\Delta y_2'[3,11,15]$. Therefore, for all the 2^4 possible differences of $\Delta w_1[3,7,11]$, 2^{12} possible values of $w_1[3,7,11]$ and 2^4 possible values of $ETK_1[6]$, compute $\Delta y_1'[6], y_1'[6], \Delta x_1[3,9], x_1[3,9]$. Then, store $\Delta w_1[3,7,11], w_1[3,7,11], x_1[3,9]$ in

H_{11} indexed by $\Delta x_1[3,9], \Delta y_1'[6], y_1'[6], ETK_1[6]$. H_{11} has 2^{20} rows and on average about $2^{20}/2^{20} = 1$ value in each row.

$\mathbf{H_{12}\{TK_{16}[1]\}}$: For all the 2^8 possible values of $\Delta x_{16}[1], x_{16}[1]$, compute $\Delta y_{16}[1], y_{16}[1]$. Then, store $y_{16}[1]$ in H_{12} indexed by $\Delta y_{16}[1]$. H_{12} has 2^4 rows and on average about $2^8/2^4 = 2^4$ values in each row.

$\mathbf{H_{13}\{ETK_1[1,5]\}}$: For all the 2^{16} possible values of $\Delta w_1[6], w_1[1,6], ETK_1[1,5]$ ($ETK_1[1] = ETK_1[5]$, see Appendix A in the full version of this paper [13]), compute $\Delta y_1'[5], y_1'[1,5]$. Then, store $\Delta w_1[6], w_1[1,6]$ in H_{13} indexed by $\Delta y_1'[5], y_1'[1,5]$, $ETK_1[1]$. H_{13} has 2^{16} rows and on average about $2^{16}/2^{16} = 1$ value in each row.

$\mathbf{H_{14}\{ETK_2[7,10,13]\}}$: From the properties of the MixColumns, we have $\Delta w_2[4] = \Delta w_2[8] = \Delta w_2[12] = \Delta y_3'[12]$. Therefore, for all the 2^4 possible differences for $\Delta w_2[4,8,12]$, 2^{12} possible values of $w_2[4,8,12]$ and 2^{12} possible values of $ETK_2[7,10,13]$, compute $\Delta y_2'[7,10,13], y_2'[7,10,13]$. Then, store $\Delta y_2'[10]$ in H_{14} indexed by $\Delta y_2'[7,10,13], y_2'[7,13], ETK_2[7,10,13]$. H_{14} has 2^{32} rows and on average about $2^{28}/2^{32} = 2^{-4}$ value in each row.

$\mathbf{H^*}$: For all the 2^{32} possible values of $\Delta z_i[SR^{-1}[col:j]], z_i[SR^{-1}[col:j]]$, compute $\Delta y_{i+1}[col:j], y_{i+1}[col:j]$. Then, store $\Delta z_i[SR^{-1}[col:j]], z_i[SR^{-1}[col:j]], y_{i+1}[col:j][0,1]$ in H^* indexed by $\Delta y_{i+1}[col:j], y_{i+1}[col:j][2,3]$. H^* has 2^{24} rows and on average about $2^{32}/2^{24} = 2^8$ values in each row.

Instead of guessing the tweakey nibbles involved in the analysis rounds as in the general approach of impossible differential attacks, we use the above mentioned pre-computation tables to deduce the tweakey nibbles that lead a specific pair of plaintext/ciphertext to the impossible differential and thus should be excluded. The details of our attack are as follows:

1. Generate 2^m structures as described above. Therefore, we have 2^{m+55} pairs of messages generated using 2^{m+28} messages. Then, ask the encryption oracle for their corresponding ciphertexts and decrypt them partially over MC^{-1}, SR^{-1} to compute z_{19}.
2. Determine the number of possible values of $TK_{19}[0:7]$ that satisfy the last round by performing the following steps for all the message pairs:

   (a) Access H^* for $i = 18, j = 0$ and compute $TK_{19}[0,4]$ such that $TK_{19}[0,4] = y_{19}[0,4] \oplus z_{19}[0,4]$[1]. Therefore, we have 2^8 possible tweakeys for $TK_{19}[0,4]$.
   (b) Access H^* for $i = 18, j = 1$ and compute $TK_{19}[1,5]$ such that $TK_{19}[1,5] = y_{19}[1,5] \oplus z_{19}[1,5]$. Therefore, we have $2^{8+8=16}$ possible tweakeys for $TK_{19}[0,1,4,5]$.

---

[1] $TK_{19}[0,4] = y_{19}[0,4] \oplus z_{19}[0,4]$ means that $TK_{19}[0] = y_{19}[0] \oplus z_{19}[0], TK_{19}[4] = y_{19}[4] \oplus z_{19}[4]$.

(c) Access H^* for $i = 18, j = 2$ and compute $TK_{19}[2,6]$ such that $TK_{19}[2,6] = y_{19}[2,6] \oplus z_{19}[2,6]$. Therefore, we have $2^{16+8=24}$ possible tweakeys for $TK_{19}[0,1,2,4,5,6]$.

(d) Access H^* for $i = 18, j = 3$ and compute $TK_{19}[3,7]$ such that $TK_{19}[3,7] = y_{19}[3,7] \oplus z_{19}[3,7]$. Therefore, we have $2^{24+8=32}$ possible tweakeys for $TK_{19}[0:7]$.

3. Determine the number of possible values of $TK_{18}[0:7]$ that satisfy the next to last round by performing the following steps for all the message pairs and remaining tweakeys that satisfy the path until now:

(a) Access H_1 and compute $TK_{18}[2,6]$ such that $TK_{18}[2,6] = y_{18}[2,6] \oplus z_{18}[2,6]$. Therefore, we have 2^{32} possible tweakeys for $TK_{19}[0:7], TK_{18}[2,6]$.

(b) Access H_2 and compute $TK_{18}[0,4]$ such that $TK_{18}[0,4] = y_{18}[0,4] \oplus z_{18}[0,4]$. Therefore, we have $2^{32+4=36}$ possible tweakeys for $TK_{19}[0:7]$, $TK_{18}[0,2,4,6]$.

(c) Access H_3 and compute $TK_{18}[3,7]$ such that $TK_{18}[3,7] = y_{18}[3,7] \oplus z_{18}[3,7]$. Therefore, we have $2^{36+4=40}$ possible tweakeys for $TK_{19}[0:7]$, $TK_{18}[0,2,3,4,6,7]$.

(d) Access H^* for $i = 17, j = 1$ and compute $TK_{18}[1,5]$ such that $TK_{18}[1,5] = y_{18}[1,5] \oplus z_{18}[1,5]$. Therefore, we have $2^{40+8=48}$ possible tweakeys for $TK_{19}[0:7], TK_{18}[0:7]$.

4. Determine the number of possible values of $TK_{17}[0:6]$ that satisfy the eighteenth round by performing the following steps for all the message pairs and remaining tweakeys that satisfy the path until now:

(a) Access H_4 and compute $TK_{17}[0,4]$ such that $TK_{17}[0,4] = y_{17}[0,4] \oplus z_{17}[0,4]$. Therefore, we have 2^{48} possible tweakeys for $TK_{19}[0:7], TK_{18}[0:7], TK_{17}[0,4]$.

(b) Access H_5 and compute $TK_{17}[2,3,6]$ such that $TK_{17}[2,3,6] = y_{17}[2,3,6] \oplus z_{17}[2,3,6]$. Therefore, we have 2^{48} possible tweakeys for $TK_{19}[0:7], TK_{18}[0:7], TK_{17}[0,2,3,4,6]$.

(c) Access H_6 and compute $TK_{17}[1,5]$ such that $TK_{17}[1,5] = y_{17}[1,5] \oplus z_{17}[1,5]$. Therefore, we have $2^{48+4=52}$ possible tweakeys for $TK_{19}[0:7]$, $TK_{18}[0:7], TK_{17}[0:6]$.

5. Determine the number of possible values of $TK_{16}[0,2]$ that satisfy the seventeenth round by performing the following steps for all the message pairs and remaining tweakeys that satisfy the path until now:

(a) Access H_7 and compute $TK_{16}[0]$ such that $TK_{16}[0] = y_{16}[0] \oplus z_{16}[0]$. Therefore, we have 2^{52} possible tweakeys for $TK_{19}[0:7], TK_{18}[0:7], TK_{17}[0:6], TK_{16}[0]$.

(b) Access H_8 and compute $TK_{16}[2]$ such that $TK_{16}[2] = y_{16}[2] \oplus z_{16}[2]$. Therefore, we have 2^{52} possible tweakeys for $TK_{19}[0:7], TK_{18}[0:7], TK_{17}[0:6], TK_{16}[0,2]^2$.

---

[2] Note that instead of having $TK_{16}[6]$ that lead to the impossible differential distinguisher, we have $x_{16}[6]$ that result in the same impossible differential distinguisher.

6. The knowledge of $TK_{19}[6]$ and $TK_{17}[4]$ enables us to deduce $TK_{15}[2]$ (see Appendix A in [13]). Hence, we determine the number of possible tweakey values that satisfy the sixteenth round by performing the following steps for all the message pairs and remaining tweakeys that satisfy the path until now:

   (a) Access H_9; and we will find 2^{-4} possible values in each row, i.e., we have 4-bit filter on the remaining tweakeys. Therefore, we have $2^{52-4=48}$ possible tweakeys for $TK_{19}[0:7]$, $TK_{18}[0:7]$, $TK_{17}[0:6]$, $TK_{16}[0,2]$ $TK_{15}[2]$.

7. The knowledge of $TK_{18}[2,4]$ and $TK_{16}[0,2]$ enables us to deduce $ETK_1[4,6, 14]$[3] (see Appendix A in [13]). Hence, we determine the number of possible values for $ETK_1[3,9,11]$ that satisfy the second round by performing the following steps for all the message pairs and remaining tweakeys that satisfy the path until now:

   (a) Access H_{10} and compute $ETK_1[11]$ such that $ETK_1[11] = y_1'[11] \oplus x_1[11]$; we will find 2^{-4} possible values in each row, i.e., we have 4-bit filter on the remaining tweakeys. Therefore, we have $2^{48-4=44}$ possible tweakeys for $TK_{19}[0:7]$, $TK_{18}[0:7]$, $TK_{17}[0:6]$, $TK_{16}[0,2]$, $TK_{15}[2]$, $ETK_1[4,6,11, 14]$.

   (b) Access H_{11} and compute $ETK_1[3,9]$ such that $ETK_1[3,9] = y_1'[3,9] \oplus x_1[3,9]$. Therefore, we have 2^{44} possible tweakeys for $TK_{19}[0:7]$, $TK_{18}[0:7]$, $TK_{17}[0:6]$, $TK_{16}[0,2]$, $TK_{15}[2]$, $ETK_1[3,4,6,9,11,14]$.

8. Determine the number of possible values for $TK_{16}[1]$ that satisfy the seventeenth round by performing the following steps for all the message pairs and remaining tweakeys that satisfy the path until now:

   (a) Access H_{12} and compute $TK_{16}[1]$ such that $TK_{16} = y_{16}[1] \oplus z_{16}[1]$. Therefore, we have $2^{44+4=48}$ possible tweakeys for $TK_{19}[0:7]$, $TK_{18}[0:7]$, $TK_{17}[0:6]$, $TK_{16}[0,1,2]$, $TK_{15}[2]$, $ETK_1[3,4,6,9,11,14]$.

9. The knowledge of $TK_{18}[0]$ and $TK_{16}[1]$ enables us to deduce $ETK_1[1,5]$ (see footnote 3) (see Appendix A in [13]). Hence, we determine the number of possible tweakey values that satisfy the second round by performing the following steps for all the message pairs and remaining tweakeys that satisfy the path until now:

   (a) Access H_{13} and we will find 1 possible value in each row. Therefore, we have 2^{48} possible tweakeys for $TK_{19}[0:7]$, $TK_{18}[0:7]$, $TK_{17}[0:6]$, $TK_{16}[0,1,2]$, $TK_{15}[2]$, $ETK_1[1,3,4,5,6,9,11,14]$,.

10. The knowledge of $TK_{19}[0,3,7]$ and $TK_{17}[1,3,5]$ enables us to deduce $ETK_2[7,10,13]$ (see Appendix A in [13]). Hence, we determine the number of possible tweakey values that satisfy the third round by performing the following steps for all the message pairs and remaining tweakeys that satisfy the path until now:

---

[3] Note that $ETK_1[6] = ETK_1[14]$ and $ETK_1[1] = ETK_1[5]$.

(a) Access H_{14} and we will find 2^{-4} possible values in each row. Therefore, we have $2^{48-4=44}$ possible tweakeys for $TK_{19}[0:7]$, $TK_{18}[0:7]$, $TK_{17}[0:6]$, $TK_{16}[0,1,2]$, $TK_{15}[2]$, $ETK_1[1,3,4,5,6,9,11,14]$, $ETK_2[7,10,13]$.

Attack Complexity. As depicted in Fig. 4, we have 38 tweakey nibbles that are involved in the analysis rounds. Thanks to the tweakey schedule, these 38 nibbles take only 2^{116} possible values (see Appendix A in [13]). For each of the 2^{m+55} message pairs, we remove, on average, 2^{44} out of 2^{116} possible values of these tweakey nibbles. Therefore, the probability that a wrong tweakey is not discarded with one pair is $1 - 2^{44-116} = 1 - 2^{-72}$. Hence, after processing all the 2^{m+55} pairs, we have $2^{116}(1 - 2^{-72})^{2^{m+55}} \approx 2^{116} \times (e^{-1})^{2^{m+55-72}} \approx 2^{116} \times 2^{-1.4 \times 2^{m-17}}$ remaining candidates for 116-bit of the tweakey. In order to determine the optimal value of m that leads to the best computational complexity, we evaluate the computational complexity of the attack as a function of m, as illustrated in Table 4. Similar to AES [6], the SKINNY round function can be implemented using 16 table lookups. As seen from Table 4, steps 5(a), 5(b) and 6(a) dominate the time complexity of the attack, and hence in order to optimize the time complexity of the attack we choose $m = 19.69$. Consequently, we have 2^{107} remaining tweakey candidates for the 116-bit of the tweakey. Therefore, the tweakey can be recovered by exhaustively searching the 2^{107} remaining tweakey candidates with 2^{12} remaining tweakey bits, that are not involved in the attack, using 2 plaintext/ciphertext pairs. Therefore, the total time complexity of the attack is $2 \times 2^{107} \times 2^{12} + 2^{120.15} = 2^{121.08}$ encryptions. The data complexity of the attack can be determined from step 1 in which we generate $2^{m=19.69}$ structures. Hence, the data complexity of the attack is $2^{19.69+28=47.69}$ chosen plaintexts. The memory complexity of the attack is dominated by the memory that is required to store $2^{m+55=74.69}$ pairs to exclude the wrong tweakeys, hence, it is $2^{74.69}$.

4.2 Impossible Differential Key-Recovery Attack on SKINNY-128-256

The only difference between SKINNY-64-128 and SKINNY-128-256 is the tweakey schedule, more precisely, the LFSR operation. The above attack on SKINNY-64-128 can be applied on SKINNY-128-256 while only considering that the cell size $s = 8$. Therefore, one structure can generate 2^{111} pairs with 2^{56} chosen plaintexts. According to the tweakey schedule, the 38 bytes involved in the attack have 2^{232} possible values (see Appendix B in the full version of this paper [13]). In this attack, we exclude, on overage, 2^{88} out of 2^{232} possible values of the involved tweakey bytes for every message pair. Hence, the probability that one wrong tweakey is not discarded is $1 - 2^{88-232} = 1 - 2^{-144}$. Therefore, we have $2^{232} \times (1 - 2^{-144})^{2^{m+111}} \approx 2^{232} \times (e^{-1})^{2^{m+111-144}} \approx 2^{232} \times 2^{-1.4 \times 2^{m-33}}$ remaining candidates for 232-bit of the tweakey bytes, after processing all the message pairs. In order to optimize the time complexity of the attack, we choose $m = 36.1$. Consequently, we have 2^{220} remaining candidates for 232-bit of the tweakey, and hence the tweakey can be recovered by exhaustively searching the remaining candidates with 2^{24} possible values, for the 24 bits of the tweakey that

Table 4. Time complexity of the different steps of the attack on 20-round SKINNY-64-128, where NT denotes the number of tweakeys to be excluded.

Step	Time complexity (in 20-round encryptions)	NT	$m = 19.69$
1	2^{m+28}	-	$2^{47.69}$
2(a)	$2^{m+55} \times \dfrac{1}{16 \times 20} \approx 2^{m+46.68}$	2^8	$2^{66.37}$
2(b)	$2^{m+55} \times 2^8 \times \dfrac{1}{16 \times 20} \approx 2^{m+54.68}$	2^{16}	$2^{74.37}$
2(c)	$2^{m+55} \times 2^{16} \times \dfrac{1}{16 \times 20} \approx 2^{m+62.68}$	2^{24}	$2^{82.37}$
2(d)	$2^{m+55} \times 2^{24} \times \dfrac{1}{16 \times 20} \approx 2^{m+70.68}$	2^{32}	$2^{90.37}$
3(a)	$2^{m+55} \times 2^{32} \times \dfrac{1}{16 \times 20} \approx 2^{m+78.68}$	2^{32}	$2^{98.37}$
3(b)	$2^{m+55} \times 2^{32} \times \dfrac{1}{16 \times 20} \approx 2^{m+78.68}$	2^{36}	$2^{98.37}$
3(c)	$2^{m+55} \times 2^{36} \times \dfrac{1}{16 \times 20} \approx 2^{m+82.68}$	2^{40}	$2^{102.37}$
3(d)	$2^{m+55} \times 2^{40} \times \dfrac{1}{16 \times 20} \approx 2^{m+86.68}$	2^{48}	$2^{106.37}$
4(a)	$2^{m+55} \times 2^{48} \times \dfrac{1}{16 \times 20} \approx 2^{m+94.68}$	2^{48}	$2^{114.37}$
4(b)	$2^{m+55} \times 2^{48} \times \dfrac{2}{16 \times 20} \approx 2^{m+95.68}$	2^{48}	$2^{115.37}$
4(c)	$2^{m+55} \times 2^{48} \times \dfrac{1}{16 \times 20} \approx 2^{m+94.68}$	2^{52}	$2^{114.37}$
5(a)	$2^{m+55} \times 2^{52} \times \dfrac{1}{16 \times 20} \approx 2^{m+98.68}$	2^{52}	$2^{118.37}$
5(b)	$2^{m+55} \times 2^{52} \times \dfrac{1}{16 \times 20} \approx 2^{m+98.68}$	2^{52}	$2^{118.37}$
6(a)	$2^{m+55} \times 2^{52} \times \dfrac{1}{16 \times 20} \approx 2^{m+98.68}$	2^{48}	$2^{118.37}$
7(a)	$2^{m+55} \times 2^{48} \times \dfrac{1}{16 \times 20} \approx 2^{m+94.68}$	2^{44}	$2^{114.37}$
7(b)	$2^{m+55} \times 2^{44} \times \dfrac{1}{16 \times 20} \approx 2^{m+90.68}$	2^{44}	$2^{110.37}$
8(a)	$2^{m+55} \times 2^{44} \times \dfrac{1}{16 \times 20} \approx 2^{m+90.68}$	2^{48}	$2^{110.37}$
9(a)	$2^{m+55} \times 2^{48} \times \dfrac{1}{16 \times 20} \approx 2^{m+94.68}$	2^{48}	$2^{114.37}$
10(a)	$2^{m+55} \times 2^{48} \times \dfrac{1}{16 \times 20} \approx 2^{m+94.68}$	2^{44}	$2^{114.37}$

are not involved in the attack, using 2 plaintext/ciphertext pairs. Therefore, the total time complexity of the attack is $2 \times 2^{220} \times 2^{24} + 2^{36.1+111} \times 2^{104} \times \frac{3}{16 \times 20}^4 = 2^{245} + 2^{244.36} = 2^{245.72}$. The data complexity of the attack is $2^{m+56=92.1}$ chosen plaintexts; and the memory complexity is dominated by storing $2^{m+111=147.1}$ message pairs.

---

[4] The second term is computed from step 5(a), 5(b) and 6(a).

5   Impossible Differential Key-Recovery Attack on 18-Round SKINNY-n-n ($n = 64$ or 128)

The only difference between SKINNY-64-64 and SKINNY-128-128 is the cell size s, where $s = 4$ (resp. $s = 8$) in case of SKINNY-64-64 (resp. SKINNY-128-128). Therefore, we present the steps of the two attacks concurrently as a function of s. This attack is applicable to the first 18 rounds of the 20-round attack on SKINNY-n-2n, i.e., the ciphertext $c = x_{18}$. Therefore, we use the same steps used in the previous attack from step 4 to the end and the same precomputation tables from H_4 to the end with the following modifications:

- Each structure can generate $2^{7 \times s} \times 2^{7 \times s - 1} = 2^{14 \times s - 1}$ with $2^{7 \times s}$ chosen plaintexts. Then, to apply the attack we take 2^m structures to generate $2^{m + 14 \times s - 1}$ pairs, but we have 4 s-bit filter in the transition over MC^{-1} from the ciphertext to w_{17}. Therefore, we have $2^{m + 14 \times s - 1 - 4 \times s = m + 10 \times s - 1}$ remaining pairs to launch the attack.
- The number of rows and entries in each table will be represented as a function of s. For example, H_6 has $2^{5 \times s}$ rows; and in each row, we have 2^s entries.
- The modifications of the number of tweakeys to be excluded from step 4 to the end are presented in Table 5.
- The relation of the tweakey cells can be found in Appendix C in the full version of this paper [13].

Attack Complexity. We have 22 tweakey cells that are involved in the analysis rounds where these 22 tweakey cells have only $2^{13 \times s}$ possible values (see Appendix C in [13]). The probability that one wrong tweakey is not discarded with one pair is $1 - 2^{-s - 13 \times s} = 1 - 2^{-14 \times s}$. Hence, after processing all the $2^{m + 10 \times s - 1}$ pairs, we have $2^{13 \times s}(1 - 2^{-14 \times s})^{2^{m + 10 \times s - 1}} \approx 2^{13 \times s} \times (e^{-1})^{2^{m + 10 \times s - 1 - 14 \times s}} \approx 2^{13 \times s} \times 2^{-1.4 \times 2^{m - 4 \times s - 1}}$ remaining candidates for $13 \times s$-bit of the tweakey. Steps 5(a), 5(b) and 6(a) dominate the time complexity of the attack, as seen from Table 5, and hence in order to optimize the time complexity of the attack we choose $m = 19.52$ (resp. $m = 36.42$) in case of SKINNY-64-64 (resp. SKINNY-128-128). Consequently, we have 2^{44} (resp. 2^{89}) remaining tweakey candidates for the 52-bit (resp. 104-bit) of the tweakey. Therefore, the tweakey can be recovered by exhaustively searching the 2^{44} (resp. 2^{89}) remaining tweakey candidates with 2^{12} (resp. 2^{24}) for the other tweakey bits, that are not involved in the attack, using 1 plaintext/ciphertext pair. Therefore, the total time complexity of the attack is $2^{44} \times 2^{12} + 2^{56.14} = 2^{57.1}$ (resp. $2^{89} \times 2^{24} + 2^{116.84} = 2^{116.94}$) encryptions in case of SKINNY-64-64 (resp. SKINNY-128-128). The data complexity of the attack can be determined from step 1 in which we generate $2^{m = 19.52}$ (resp. $2^{m = 36.42}$) structures. Hence, the data complexity of the attack is $2^{19.52 + 28 = 47.52}$ (resp. $2^{36.42 + 56 = 92.42}$) chosen plaintexts in case of SKINNY-64-64 (resp. SKINNY-128-128). The memory complexity is dominated by the memory required to store the $2^{58.52}$ (resp. $2^{115.42}$) pairs after the ciphertext filtration and is estimated to be $2^{58.52}$ (resp. $2^{115.42}$) in case of SKINNY-64-64 (resp. SKINNY-128-128).

6   Impossible Differential Key-Recovery Attack on 22-Round SKINNY-n-3n (n = 64 or 128)

SKINNY-64-192 differs from SKINNY-128-384 in the cell size s and the tweakey schedule. As the tweakey schedule does not influence the attack procedure, we present the two attacks as a function of s. The 20-round attack on SKINNY-n-2n (n = 64 or 128) can be extended to 22-round attack on SKINNY-n-3n (n = 64 or 128) by appending 2 rounds, i.e., the ciphertext $c = x_{22}$. Therefore, we can use the same attack procedures of SKINNY-n-2n (n = 64 or 128) to attack SKINNY-n-3n (n = 64 or 128) by repeating step 2 three times to extract the tweakey cells $TK_{19}[0:7], TK_{20}[0:7], TK_{21}[0:7]$. The details of the tweakey schedule can be found in Appendix D in the full version of this paper [13]. Moreover, as in the previous attack on 18-round SKINNY-n-n (n = 64 or 128), each structure can generate $2^{7 \times s} \times 2^{7 \times s - 1} = 2^{14 \times s - 1}$ with $2^{7 \times s}$ chosen plaintexts. Then, we take 2^m structures to generate $2^{m+14 \times s - 1}$ pairs using $2^{m+7 \times s}$ chosen plaintexts.

Attack Complexity. The 54 tweakey cells that are involved in the analysis rounds have only $2^{45 \times s}$ possible values. The probability that a wrong tweakey is not discarded with one pair is $1 - 2^{27 \times s - 45 \times s} = 1 - 2^{-18 \times s}$. Hence, after processing all the $2^{m+14 \times s - 1}$ pairs, we have $2^{45 \times s}(1 - 2^{-18 \times s})^{2^{m+14 \times s - 1}} \approx 2^{45 \times s} \times (e^{-1})^{2^{m+14 \times s - 1 - 18 \times s}} \approx 2^{45 \times s} \times 2^{-1.4 \times 2^{m - 4 \times s - 1}}$ remaining candidates for $45 \times s$-bit of the tweakey. In order to optimize the time complexity of the

Table 5. Time complexity of the different steps of the attack on 18-round SKINNY-64-64 and SKINNY-128-128, where NT denotes the number of tweakeys to be excluded.

Step	Time Complexity (in 18-round encryptions)	NT	$s = 4, m = 19.52$	$s = 8, m = 36.42$
1	$2^{m+7 \times s}$	-	$2^{47.52}$	$2^{92.42}$
4(a)	$2^{m+10 \times s - 1} \times \dfrac{1}{16 \times 18} \approx 2^{m+10 \times s - 9.17}$	1	$2^{50.35}$	$2^{107.25}$
4(b)	$2^{m+10 \times s - 1} \times \dfrac{2}{16 \times 18} \approx 2^{m+10 \times s - 8.17}$	1	$2^{51.35}$	$2^{108.25}$
4(c)	$2^{m+10 \times s - 1} \times \dfrac{1}{16 \times 18} \approx 2^{m+10 \times s - 9.17}$	2^s	$2^{50.35}$	$2^{107.25}$
5(a)	$2^{m+10 \times s - 1} \times 2^s \times \dfrac{1}{16 \times 18} \approx 2^{m+11 \times s - 9.17}$	2^s	$2^{54.35}$	$2^{115.25}$
5(b)	$2^{m+10 \times s - 1} \times 2^s \times \dfrac{1}{16 \times 18} \approx 2^{m+11 \times s - 9.17}$	2^s	$2^{54.35}$	$2^{115.25}$
6(a)	$2^{m+10 \times s - 1} \times 2^s \times \dfrac{1}{16 \times 18} \approx 2^{m+11 \times s - 9.17}$	1	$2^{54.35}$	$2^{115.25}$
7(a)	$2^{m+10 \times s - 1} \times \dfrac{1}{16 \times 18} \approx 2^{m+10 \times s - 9.17}$	2^{-s}	$2^{50.35}$	$2^{107.25}$
7(b)	$2^{m+10 \times s - 1} \times 2^{-s} \times \dfrac{1}{16 \times 18} \approx 2^{m+9 \times s - 9.17}$	2^{-s}	$2^{46.35}$	$2^{99.25}$
8(a)	$2^{m+10 \times s - 1} \times 2^{-s} \times \dfrac{1}{16 \times 18} \approx 2^{m+9 \times s - 9.17}$	1	$2^{46.35}$	$2^{99.25}$
9(a)	$2^{m+10 \times s - 1} \times \dfrac{1}{16 \times 18} \approx 2^{m+10 \times s - 9.17}$	1	$2^{50.35}$	$2^{107.25}$
10(a)	$2^{m+10 \times s - 1} \times \dfrac{1}{16 \times 18} \approx 2^{m+10 \times s - 9.17}$	2^{-s} [a]	$2^{50.35}$	$2^{107.25}$

[a] After this step, we have 2^{-s} tweakeys to be excluded for each message pair, i.e., we exclude 1 tweakey after processing 2^s pairs.

attack, we choose $m = 19.84$ (resp. $m = 36.22$) in case of SKINNY-64-192 (resp. SKINNY-128-384). Consequently, we have 2^{170} (resp. 2^{347}) remaining tweakey candidates for the 180-bit (resp. 360-bit) of the tweakey. Therefore, the tweakey can be recovered by exhaustively searching the 2^{170} (resp. 2^{347}) remaining tweakey candidates with 2^{12} (resp. 2^{24}) for the other tweakey bits, that are not involved in the attack, using 3 (calculated from the unicity distance) plaintext/ciphertext pairs. Therefore, the total time complexity of the attack is $3 \times 2^{170} \times 2^{12} + 2^{183.97} = 2^{184.79}$ (resp. $3 \times 2^{347} \times 2^{24} + 2^{372.35} = 2^{373.48}$) encryptions in case of SKINNY-64-192 (resp. SKINNY-128-384). The data complexity of the attack is $2^{19.84+28=47.84}$ (resp. $2^{36.22+56=92.22}$) chosen plaintexts in case of SKINNY-64-192 (resp. SKINNY-128-384). The memory complexity of the attack is $2^{74.84}$ (resp. $2^{147.22}$) in case of SKINNY-64-64 (resp. SKINNY-128-384).

7 Conclusion

In this work, we presented impossible differential attacks against reduced-round versions of all the 6 SKINNY's variants. All of these attacks use the same impossible differential distinguisher that covers 11-round. We extended this 11-round distinguisher by 7, 9 and 11 rounds to attack 18, 20 and 22 rounds of SKINNY-n-n, SKINNY-n-$2n$ and SKINNY-n-$3n$ ($n = 64$ or 128), respectively, exploiting the properties of the MixColumns operation, the simple tweakey schedule and the fact that the tweakey is only added to the first two rows of the state. The presented attacks are currently the best known ones on all the variants of SKINNY in the single-tweakey model.

References

1. Ankele, R., Banik, S., Chakraborti, A., List, E., Mendel, F., Sim, S. M., Wang, G.: Related-key impossible-differential attack on reduced-round SKINNY. Cryptology ePrint Archive, Report 2016/1127 (2016). http://eprint.iacr.org/2016/1127
2. Beierle, C., Jean, J., Klbl, S., Leander, G., Moradi, A., Peyrin, T., Sasaki, Y., Sasdrich, P., Sim, S.M.: Skinny family of block ciphers: cryptanalysis competition (2016)
3. Beierle, C., Jean, J., Kölbl, S., Leander, G., Moradi, A., Peyrin, T., Sasaki, Y., Sasdrich, P., Sim, S.M.: The SKINNY family of block ciphers and its low-latency variant MANTIS. In: Robshaw, M., Katz, J. (eds.) CRYPTO 2016. LNCS, vol. 9815, pp. 123–153. Springer, Heidelberg (2016). doi:10.1007/978-3-662-53008-5_5
4. Biham, E., Biryukov, A., Shamir, A.: Cryptanalysis of Skipjack reduced to 31 rounds using impossible differentials. In: Stern, J. (ed.) EUROCRYPT 1999. LNCS, vol. 1592, pp. 12–23. Springer, Heidelberg (1999). doi:10.1007/3-540-48910-X_2
5. Bilgin, B., Gierlichs, B., Nikova, S., Nikov, V., Rijmen, V.: A more efficient AES threshold implementation. In: Pointcheval, D., Vergnaud, D. (eds.) AFRICACRYPT 2014. LNCS, vol. 8469, pp. 267–284. Springer, Cham (2014). doi:10.1007/978-3-319-06734-6_17
6. Daemen, J., Rijmen, V.: The Design of Rijndael. Springer, Heidelberg (2002)

7. Jean, J., Nikolić, I., Peyrin, T.: Tweaks and keys for block ciphers: the TWEAKEY framework. In: Sarkar, P., Iwata, T. (eds.) ASIACRYPT 2014. LNCS, vol. 8874, pp. 274–288. Springer, Heidelberg (2014). doi:10.1007/978-3-662-45608-8_15
8. Knudsen, L.: A 128-bit block cipher. Complexity **258**(2), 216 (1998). NIST AES Proposal
9. Liskov, M., Rivest, R.L., Wagner, D.: Tweakable block ciphers. J. Cryptol. **24**(3), 588–613 (2011)
10. Liu, G., Ghosh, M., Song, L.: Security analysis of SKINNY under related-tweakey settings. Cryptology ePrint Archive, Report 2016/1108 (2016). http://eprint.iacr. org/2016/1108
11. Peyrin, T., Seurin, Y.: Counter-in-tweak: authenticated encryption modes for tweakable block ciphers. In: Robshaw, M., Katz, J. (eds.) CRYPTO 2016. LNCS, vol. 9814, pp. 33–63. Springer, Heidelberg (2016). doi:10.1007/978-3-662-53018-4_2
12. Sadeghi, S., Mohammadi, T., Bagheri, N.: Cryptanalysis of reduced round SKINNY block cipher. Cryptology ePrint Archive, Report 2016/1120 (2016). http://eprint. iacr.org/2016/1120
13. Tolba, M., Abdelkhalek, A., Youssef, A.M.: Impossible differential cryptanalysis of reduced-round skinny. Cryptology ePrint Archive, Report 2016/1115 (2016). http://eprint.iacr.org/2016/1115

Impossible Differential Attack
on Reduced Round SPARX-64/128

Ahmed Abdelkhalek, Mohamed Tolba, and Amr M. Youssef[(⊠)]

Concordia Institute for Information Systems Engineering,
Concordia University, Montréal, QC, Canada
youssef@ciise.concordia.ca

Abstract. SPARX-64/128 is an ARX-based block cipher with 64-bit block size and 128-bit key. It was published in ASIACRYPT 2016 as one of the instantiations of a family of ARX-based block ciphers with provable security against single-characteristic differential and linear cryptanalysis. In this work, we present 12 and 13-round impossible distinguishers on SPARX-64/128 that can be used to attack 15 and 16-round SPARX-64/128 with post-whitening keys, respectively. While the 15-round attack starts from round 0, the 16-round one, exploiting the key schedule, has to start from round 2.

Keywords: Block ciphers · Impossible differential · Miss-in-the-middle · SPARX

1 Introduction

SPARX is a family of ARX-based block ciphers that was published in ASIACRYPT 2016 [6]. It was designed with the goal of putting forward a general strategy for designing ARX-based symmetric-key primitives with provable security against single-characteristic differential and linear cryptanalysis. As a dual to the wide trail strategy [4] adopted by many S-box based block ciphers, the designers proposed the long trail strategy. This strategy promotes the use of a rather weak but large S-box, i.e., an ARX-based S-box, along with a very light linear layer. Fostering the existence of long trails, that involve an uninterrupted sequence of calls to the S-box interleaved with key additions, rather than having maximum diffusion in each linear layer is at the core of this proposed strategy. The long trail strategy allowed the designers to bound the maximum differential and linear probabilities for any number of rounds of a block cipher designed following such strategy. SPARX-64/128 is a member of this family of block ciphers following the long trail strategy with 64-bit block size and 128-bit key. The only cryptanalysis of SPARX was done by its designers as they presented a 13-round bit-based division property distinguisher that they used to launch an integral attack against 15-round SPARX-64/128 [5]. No other attacks were given in the short/full versions of the design paper.

© Springer International Publishing AG 2017
M. Joye and A. Nitaj (Eds.): AFRICACRYPT 2017, LNCS 10239, pp. 135–146, 2017.
DOI: 10.1007/978-3-319-57339-7_8

Impossible differential cryptanalysis that was independently proposed by Biham *et al.* [3] and Knudsen [9] is one of the most powerful cryptanalytic techniques. Firstly, we try to find a certain input difference that propagates to a specific output difference with zero probability resulting in an impossible differential distinguisher. In general, the input and output differences can be truncated. Then, after finding the longest possible impossible differential, it is used in a key recovery attack by prepending and/or appending a few additional rounds which are usually called the analysis rounds. The attack proceeds as follows: first, we collect pairs with certain plaintext and ciphertext differences. Then, we guess some bits of the key material involved in the analysis rounds and if one of the pairs satisfies the input and output differences of the impossible differential under some subkey bits, then these subkey bits must be wrong. Thus, we discard as many wrong keys as possible and do an exhaustive search on the surviving ones along with the rest of the key. The early abort technique [10] allows us to guess the involved key material on steps to discard the undesired pairs as early as possible and therefore reduce the time complexity of the attack.

In this paper, we present a 12-round truncated impossible differential on SPARX-64/128 that can be extended to a 13-round impossible differential with a specific input difference and a truncated output difference. We use the 12-round impossible differential to launch an impossible differential attack against 15-round SPARX-64/128 including the post-whitening key with data complexity of 2^{51} chosen plaintexts, time complexity of $2^{94.1}$ 15-round encryptions and memory complexity of $2^{43.5}$ 64-bit blocks. Then, we use the 13-round impossible differential to attack 16-round SPARX-64/128, including the post-whitening key, starting from round 2 with data, time and memory complexities of $2^{61.5}$ known plaintexts, 2^{94} 16-round encryptions, and $2^{61.5}$ 64-bit blocks, respectively.

The remainder of the paper is organized as follows. In Sect. 2, the notations used throughout the paper are given followed by the specification of SPARX-64/128. Our impossible differentials are presented in Sect. 3. Afterwards, in Sect. 4, we provide a detailed description of our impossible differential attacks on SPARX-64/128. Finally, Sect. 5 concludes the paper.

2   Description of SPARX-64/128

Notations. The following notations are used throughout the paper:

- K: The master key.
- k_i: The ith 16-bit of the key state, where $0 \leq i \leq 7$.
- k_i^j: The ith 16-bit of the key state after applying the key schedule permutation j times, where $0 \leq i \leq 7$ and $0 \leq j \leq 17$ for SPARX-64/128.
- $RK_{(a,i)}$: The 32-bit round key used at branch a of round i where $0 \leq i \leq 24$ and $a = 0$ (1) denotes the left (right) branch of SPARX-64/128.
- $X_{(a,i)}$ $(Y_{(a,i)})$: The left (right) 16-bit input at branch a of round i where $0 \leq i \leq 24$, $a = 0$ (1) denotes the left (right) branch of SPARX-64/128, and the LSB of either $X_{(a,i)}$ or $Y_{(a,i)}$ is on the right.

- w: The number of 32-bit words, i.e., $w = 2$ for a 64-bit block and $w = 4$ for a 128-bit master key.
- R^3: The iteration of 3 rounds of SPECKEY with their corresponding key additions.
- L_w: Linear mixing layer used in SPARX with w-word block size, thus L_2 represents the linear mixing layer used in SPARX-64/128.
- $\boxplus$: Addition mod 2^{16}.
- $\oplus$: Bitwise XOR.
- $\lll q$ ($\ggg q$): Rotation of a word by q bits to the left (right).
- $\|$: Concatenation of bits.
- $0xabcd$: A 16-bit number in hexadecimal representation.

2.1 Specifications of SPARX-64/128

SPARX [5, 6] is a family of ARX-based Substitution-Permutation Network (SPN) block ciphers. It follows the SPN design construction while using ARX-based S-boxes instead of S-boxes based on look-up tables. ARX-based S-boxes form a specific category of S-boxes that rely solely on addition, rotation and XOR operations to provide both non-linearity and diffusion. The SPARX family adopts the 32-bit SPECKEY ARX-based S-box, shown in Fig. 1, which resembles one round of SPECK-32 [1,2] with only one difference, that is, the key is added to the whole 32-bit state instead of just half the state as in SPECK-32.

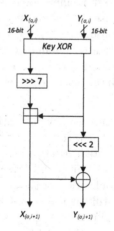

Fig. 1. The SPECKEY ARX-based S-box used in the SPARX family.

For a given member of the SPARX family whose block size is n bits, the plaintext is divided into $w = n/32$ words of 32 bits each. Then, the SPECKEY S-box (S), being applied to w words in parallel, is iterated r times interleaved by the addition of independent subkeys. Then, a linear mixing layer (L_w) is applied to ensure diffusion between the words. The structure made of a key addition

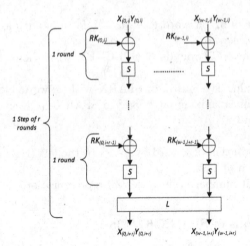

Fig. 2. SPARX structure

followed by S is called a round while the structure made of r rounds followed by L_w is called a step, as depicted in Fig. 2. Thus, the ciphertext corresponding to a given plaintext is generated by iterating such steps. The number of steps and the number of rounds in each step depend on both the block size of the cipher and the size of the key it utilizes.

SPARX-64/128 is the lightest member of this family operating on 64-bit blocks using 128-bit keys. It uses 3 rounds in each step and iterates over 8 steps, i.e., the total number of rounds is 24. More precisely, in SPARX-64/128, 2 SPECKEY S-boxes (S) are iterated simultaneously 3-times, while being interleaved by the addition of the round keys and then a linear mixing layer (L_2) is applied, as shown in Fig. 3a. The structure of L_2 is depicted in the dotted square in Fig. 3b.

Key Schedule. The 128-bit master key instantiates the key state, denoted by $k_0^0\|k_1^0\|k_2^0\|k_3^0\|k_4^0\|k_5^0\|k_6^0\|k_7^0$. Then, the 3×32-bit round keys used in the left branch of the first step are extracted. Afterwards, the permutation illustrated in Fig. 4 is applied and then the 3×32-bit round keys used in the right branch of the first step are extracted. The application of the permutation and the extraction of the keys are interleaved untill all the round keys encompassing the post-whitening ones are generated. This means that, first, the round keys of a branch of a given step j are generated and then the key state is updated. The following observation on the key schedule is exploited in our attacks.

Observation: The last round key of a given step and the first round key of the subsequent step can be deduced from one another. To clarify this point, we consider the last round key of step 0 and the first round key of step 1. The 64-bit round key of the third round is $k_4^0\|k_5^0, k_4^1\|k_5^1$ and the 64-bit round key of the fourth round is $k_0^2\|k_1^2, k_0^3\|k_1^3$. According to the key schedule: $k_0^2 = k_6^0 = k_4^0$, $k_1^2 = k_7^1 \boxplus 2 = k_5^0 \boxplus 2$, $k_0^3 = k_6^2 = k_4^1$ and $k_0^3 = k_7^1 \boxplus 3 = k_5^1 \boxplus 3$.

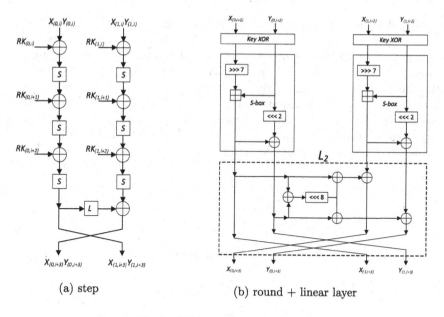

(a) step

(b) round + linear layer

Fig. 3. SPARX-64/128 structure

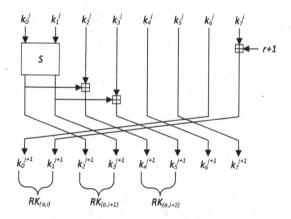

Fig. 4. SPARX-64/128 key schedule permutation, where the counter r is initialized to 0.

Finally, it is to be noted that we measure the memory complexity of our attacks in number of 64-bit blocks and the time complexity in terms of the equivalent number of round-reduced encryptions.

3   Impossible Differentials of SPARX-64/128

A 12-round impossible differential is readily noticeable when considering SPARX-64/128 to be a twisted variant of a Feistel construction where the two halves undergo a keyed function before getting mixed and swapped. Indeed, as depicted in Fig. 5, if the left branch of SPARX-64/128 at round i has a zero difference while the right half has a nonzero difference, then after 2 steps (6 rounds), the input at the left branch must have a nonzero difference. From the other direction, if the input of the right branch of round $i+12$ has a nonzero difference, i.e., Γ and the input of the left branch at that round has a difference $L_2(\Gamma)$, then after the linear transformation, the right branch will have a zero difference which propagates unaltered for 2 complete steps (6 rounds) and contradicts with the forward differential at the left branch.

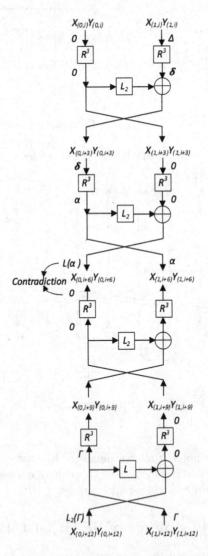

Fig. 5. 12-round impossible differential SPARX-64/128

This 12-round truncated impossible differential can be extended to a 13-round distinguisher with a specific input difference and truncated output difference. This is feasible by exploiting the fact that there exist differentials with probability 1 for one SPECKEY round and one of these differentials is a fixed point of L_2. Particularly, if the input difference of the distinguisher is chosen to be $0x8000\ 0x8000$ then by propagating it backward through L_2 we have the same difference at both the right and left branches as an output for the S-box and this output difference corresponds to the input difference $0x0040\ 0x0000$ with probability 1. Hence, the input of the 13-round distinguisher is $0x0040\ 0x0000$ and $0x0040\ 0x0000$ while the output is still truncated in the form of $L_2(\Gamma)$ and Γ.

4   Impossible Differential Cryptanalysis of SPARX-64/128

The 12 and 13-round impossible distinguishers described above can be used to attack 15 and 16-round SPARX-64/128, respectively. Both attacks include the post-whitening key, however, the 16-round attack starts at round 2.

4.1   15-Round Impossible Differential Attack on SPARX-64/128

In this attack, we have chosen to place the 12-round distinguisher at the top, end it with a specific difference that meets the constraint of $L_2(\Gamma)$ and Γ, and then append 3 rounds that have a high probability as shown in Fig. 6. That specific difference at the end of the distinguisher and the 3 analysis rounds were found using Mixed Integer Linear Programming (MILP). Specifically, we have followed the guidelines in [7] to create an MILP model that describes SPARX-64/128 and solved it using the publicly available MILP optimizer Gurobi [8]. The detailed procedure of the attack is described as follows.

Data Collection. We first choose 2^m structures of plaintexts where in each structure the left 32 bits of the plaintexts take a fixed value and the right 32 bits take all the 2^{32} possible values. Each structure includes about $\binom{2^{32}}{2} \approx 2^{63}$ pairs of plaintexts, therefore we have $2^m \times 2^{63} = 2^{m+63}$ pairs of plaintexts in total. We encrypt these pairs and keep the ones whose ciphertext difference matches the difference shown in Fig. 6. The probability of such ciphertext difference is about 2^{-64}, therefore the expected number of remaining pairs after this phase is about $2^{m+63-64} = 2^{m-1}$.

Key Recovery. To verify if the pairs generated during the data collection phase follow our 12-round impossible differential, we need to guess $RK_{(0,15)}, RK_{(1,15)}, RK_{(0,14)}, RK_{(1,14)},$ and $RK_{(0,13)}$. However, as pointed out above, $RK_{(0,15)}, RK_{(1,15)}$ are related to $RK_{(0,14)}, RK_{(1,14)}$. This means that these round keys take 2^{96} values only. The details of this phase are as follows.

Step 1. For all the ciphertext pairs obtained in the data collection phase, we guess the 64-bit round keys $RK_{(0,15)}$ and $RK_{(1,15)}$, decrypt round 15 and check if the difference matches the one shown in Fig. 6. If it is not

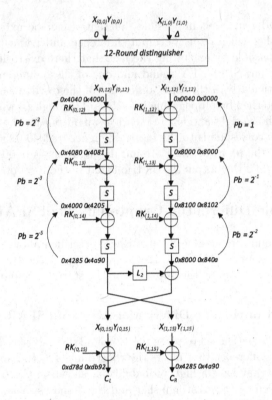

Fig. 6. 15-round impossible differential attack on SPARX-64/128

the case, the pair is discarded. The probability of this event is 2^{-7} and thus after this step the expected number of remaining pairs is about $2^{m-1-7} = 2^{m-8}$.

Step 2. We deduce $RK_{(0,14)}$ and $RK_{(1,14)}$ from the guessed $RK_{(0,15)}$ and $RK_{(1,15)}$, decrypt round 14 and check if the difference is the expected one according to Fig. 6. If it is not the case, the pair is discarded. The probability of this event is 2^{-4} and therefore the expected number of pairs surviving this step is about $2^{m-8-4} = 2^{m-12}$.

Step 3. We guess the 32-bit $RK_{(0,13)}$ and partially decrypt the left branch of round 13 and check if the difference meets the impossible differential difference. Once it is correct, we delete the 32-bit round key guesses of $RK_{(0,13)}$ since such a differential is impossible; each round key guess that proposes such a difference is a wrong key. After analyzing all the 2^{m-12} remaining pairs, we output the 96-bit round keys guess of $RK_{(0,15)}$, $RK_{(1,15)}$, and $RK_{(0,13)}$ as a candidate. The probability that the pairs pass this step is about 2^{-2}, therefore the time complexity of this step is the number of key guesses × 2 messages in each pair × the probability that the key guess is excluded after sequentially testing it against all the surviving pairs.

Table 1. Key recovery process of the attack on 15-round SPARX-64/128

Attack step	Guessed keys	# Surviving pairs	Time complexity
1	$RK_{(0,15)}$ $RK_{(1,15)}$	$2^{m-1-7} = 2^{m-8}$	$2^{64} \times 2 \times 2^{m-1} \times 1/15 \approx 2^{m+60.1}$
2	†	$2^{m-8-4} = 2^{m-12}$	$2^{64} \times 2 \times 2^{m-8} \times 1/15 \approx 2^{m+53.1}$
3	$RK_{(0,13)}$	–	$2^{96} \times 2 \times [1 + (1-2^{-2}) + (1-2^{-2})^2$ $+ \cdots + (1-2^{-2})^{2^{m-12}}] \times 1/(2 \times 15)$

†: No additional key guesses needed, i.e., the round keys are deduced from the previously guessed ones.

The steps of the key recovery phase are described in Table 1, whereas the second column gives the round keys to be guessed in the corresponding round for each attack step. The third column presents the number of surviving pairs after each step, and the fourth column is the time complexity of each step measured in 15-round encryption.

Attack Complexity. To balance the attack complexity between the different phases, we take $m = 19$. This means that after analyzing all the remaining pairs, there will be about $2^{96} \times (1 - 2^{-2})^{2^{m-12}} = 2^{96} \times (0.75)^{128} \approx 2^{42.9}$ remaining candidates for the 96-bit round keys. Then, we guess the 32-bit $RK_{(1,12)}$ which along with the surviving candidates allows us to recover the master key K via the key schedule. Afterwards, we test each one of these master key candidates using 2 plaintext/ciphertext pairs to find the correct master key. The time complexity of this exhaustive search step is $2 \times 2^{32} \times 2^{42.9} = 2^{75.9}$. Therefore the time complexity is dominated by step 3 of the attack and estimated to be $2^{96} \times 2 \times (1/2^{-2}) \times (1/30) \approx 2^{94.1}$. The data complexity of the attack is $2^{19+32} = 2^{51}$ chosen plaintexts. The memory complexity of the attack is dominated by the memory that is required to store the keys to be excluded, i.e., $2^{42.9} \times 96/64 \approx 2^{43.5}$ 64-bit blocks.

4.2   16-Round Impossible Differential Attack on SPARX-64/128

Although each round of SPARX-64/128 uses a 64-bit round key, there exists 3 specific rounds that contain only 2^{96} bits of key information as exemplified by the ones exploited in the previous attack. Nonetheless, any 4 rounds contain at least 128 bits of key information. Therefore, our 16-round attack on SPARX-64/128 has to start from round 2 and in this case, we use the 13-round impossible differential and prepend 3 rounds on its top as shown in Fig. 7. Again, we have used the Gurobi optimizer to find these 3 rounds after creating the MILP model that describes them.

In this attack, we do not use data structures as they do not generate enough pairs to launch the attack. Instead, we use known plaintexts and generate the pairs we need probabilistically. Hence, if we have $2^{61.5}$ known plaintexts, these can generate $\binom{2^{61.5}}{2} \approx 2^{122}$ pairs. Out of these pairs, we would have $2^{122-64} = 2^{58}$

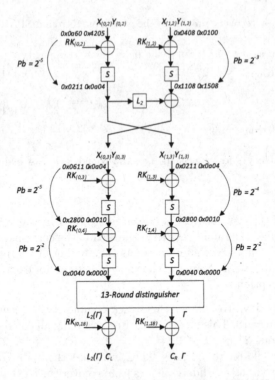

Fig. 7. 16-round impossible differential attack on SPARX-64/128

pairs that satisfy the plaintext difference shown in Fig. 7. Then, as the difference at the end of the distinguisher is the difference in the ciphertext, we have to filter the ciphertexts such as the right branch is a nonzero difference Γ and the left branch difference is $L_2(\Gamma)$ which means that we have $2^{58-32} = 2^{26}$ proper pairs.

In the key recovery phase which we perform on these 2^{26} pairs, the 3 round keys take 2^{96} values only and they are guessed on steps to reduce the time complexity of the attack as listed in Table 2. It is to be noted that, according to the key schedule, $RK_{(0,3)}$, $RK_{(1,3)}$ are deduced from the guessed $RK_{(0,2)}$, $RK_{(1,2)}$ and that $RK_{(1,4)}$ is deduced from $RK_{(0,3)}$.

After analyzing all the remaining pairs, there will be about $2^{96} \times (1-2^{-2})^{2^7} = 2^{96} \times (0.75)^{128} \approx 2^{42.9}$ remaining candidates for the 96-bit round keys. Then, we guess the remaining 32 bits of the master key and test each one of these master key candidates using 2 plaintext/ciphertext pairs to find the correct one. The time complexity of this exhaustive search step is $2 \times 2^{32} \times 2^{42.9} = 2^{75.9}$. Therefore the time complexity is dominated by step 4 of the attack (see Table 2) and estimated to be $2^{96} \times 2 \times (1/2^{-2}) \times (1/32) = 2^{94}$. The data complexity of the attack is $2^{61.5}$ known plaintexts. In this case, the memory complexity of the attack is dominated by the hash table [11] that is used to store the plaintexts while generating the required pairs, i.e., $2^{61.5}$ 64-bit blocks.

Table 2. Key recovery process of the attack on 16-round SPARX-64/128

Attack step	Guessed keys	# Surviving pairs	Time complexity
1	$RK_{(0,2)}$	$2^{26-8} = 2^{18}$	$2^{64} \times 2 \times 2^{26} \times 1/16 = 2^{87}$
	$RK_{(1,2)}$		
2	†	$2^{18-9} = 2^9$	$2^{64} \times 2 \times 2^{18} \times 1/16 = 2^{79}$
3	†	$2^{9-2} = 2^7$	$2^{64} \times 2 \times 2^9 \times 1/(2 \times 16) = 2^{69}$
4	$RK_{(0,4)}$	–	$2^{96} \times 2 \times [1 + (1 - 2^{-2}) + (1 - 2^{-2})^2$ $+ \cdots + (1 - 2^{-2})^{2^7}] \times 1/(2 \times 16)$

†: No additional key guesses needed, i.e., the round keys are deduced from the previously guessed ones.

5   Conclusion

In this paper, we have analyzed SPARX-64/128 against the impossible differential attack. We have presented 12 and 13-round impossible differential distinguishers that are used to attack 15 and 16-round SPARX-64/128 with the post-whitening key, respectively. The (data complexity in chosen/known plaintexts, time complexity in 15/16-round encryptions, memory complexity in 64-bit blocks) of these attacks are $(2^{51}, 2^{94.1}, 2^{43.5})$ and $(2^{61}, 2^{94}, 2^{61.5})$, respectively.

References

1. Beaulieu, R., Shors, D., Smith, J., Treatman-Clark, S., Weeks, B., Wingers, L.: The SIMON and SPECK families of lightweight block ciphers. Cryptology ePrint Archive, Report 2013/404 (2013). http://eprint.iacr.org/2013/404
2. Beaulieu, R., Shors, D., Smith, J., Treatman-Clark, S., Weeks, B., Wingers, L.: SIMON and SPECK: block ciphers for the internet of things. Cryptology ePrint Archive, Report 2015/585 (2015). http://eprint.iacr.org/2015/585
3. Biham, E., Biryukov, A., Shamir, A.: Cryptanalysis of Skipjack reduced to 31 rounds using impossible differentials. In: Stern, J. (ed.) EUROCRYPT 1999. LNCS, vol. 1592, pp. 12–23. Springer, Heidelberg (1999). doi:10.1007/3-540-48910-X_2
4. Daemen, J., Rijmen, V.: The wide trail design strategy. In: Honary, B. (ed.) Cryptography and Coding 2001. LNCS, vol. 2260, pp. 222–238. Springer, Heidelberg (2001). doi:10.1007/3-540-45325-3_20
5. Dinu, D., Perrin, L., Udovenko, A., Velichkov, V., Großschädl, J., Biryukov, A.: Design strategies for ARX with provable bounds: SPARX and LAX (Full Version). Cryptology ePrint Archive, Report 2016/984 (2016). http://eprint.iacr.org/2016/984
6. Dinu, D., Perrin, L., Udovenko, A., Velichkov, V., Großschädl, J., Biryukov, A.: Design strategies for ARX with provable bounds: SPARX and LAX. In: Cheon, J.H., Takagi, T. (eds.) ASIACRYPT 2016. LNCS, vol. 10031, pp. 484–513. Springer, Heidelberg (2016). doi:10.1007/978-3-662-53887-6_18
7. Fu, K., Wang, M., Guo, Y., Sun, S., Hu, L.: MILP-based automatic search algorithms for differential and linear trails for Speck. In: Peyrin, T. (ed.) FSE 2016. LNCS, vol. 9783, pp. 268–288. Springer, Heidelberg (2016). doi:10.1007/978-3-662-52993-5_14

8. Gurobi Optimization Inc.: Gurobi Optimizer Reference Manual (2016). http://www.gurobi.com
9. Knudsen, L.: DEAL: A 128-bit block cipher (1998). NIST AES Proposal
10. Lu, J., Kim, J., Keller, N., Dunkelman, O.: Improving the efficiency of impossible differential cryptanalysis of reduced Camellia and MISTY1. In: Malkin, T. (ed.) CT-RSA 2008. LNCS, vol. 4964, pp. 370–386. Springer, Heidelberg (2008). doi:10.1007/978-3-540-79263-5_24
11. Mala, H., Dakhilalian, M., Rijmen, V., Modarres-Hashemi, M.: Improved impossible differential cryptanalysis of 7-round AES-128. In: Gong, G., Gupta, K.C. (eds.) INDOCRYPT 2010. LNCS, vol. 6498, pp. 282–291. Springer, Heidelberg (2010). doi:10.1007/978-3-642-17401-8_20

Applications

Private Conjunctive Query over Encrypted Data

Tushar Kanti Saha$^{1(\boxtimes)}$ and Takeshi Koshiba^2

^{1} Division of Mathematics, Electronics, and Informatics,
Graduate School of Science and Engineering, Saitama University, Saitama, Japan
`saha.t.k.512@ms.saitama-u.ac.jp`
^{2} Faculty of Education and Integrated Arts and Sciences,
Waseda University, Tokyo, Japan
`tkoshiba@waseda.jp`

Abstract. In this paper, we propose an efficient protocol to process a private conjunctive query over encrypted data in the cloud using the somewhat homomorphic encryption (SwHE) scheme with a batch technique. In 2016, Cheon, Kim, and Kim (CKK) [IEEE Trans. Inf. Forensics Security] showed conjunctive query processing over encrypted data using search-and-compute circuits and an SwHE scheme and mentioned that their scheme should be improved in performance. To improve the performance of processing a private conjunctive query, we also propose a new packing method to support an efficient batch computation for our protocol using a few multiplications. Our implementation shows that our protocol works more than 50 times as fast as the CKK protocol for conjunctive query processing. In addition, the security level of our protocol is better than the security level of the CKK protocol.

Keywords: Private · Conjunctive · Query processing · Encrypted · Data · Packing method · Homomorphic encryption

1 Introduction

The conjunctive query plays a significant role in accessing data of a big database. Recent researches showed the importance of private conjunctive query over the encrypted database [1–4]. In a database, when the predicate of a query contains many conditions connected by 'and/∧' operator is called conjunctive query. For example, suppose a health research institute wants to know the number of male tuberculosis (TB) patients admitted to a hospital in 2015. The corresponding SQL can be represented as *select count(id) from mediRecord where sex = 'M' and diseases = 'TB' and year = '2015'*. Finding the result of such query of a client from thousand of records is a massive computational task. Therefore, the database owner and users like to outsource this computation to some third parties. Here neither database owner nor the user likes to reveal their information to the third party.

On the contrary, after the development of cloud computing, outsource computation is being increased rapidly. At the same time, users want the security of their

M. Joye and A. Nitaj (Eds.): AFRICACRYPT 2017, LNCS 10239, pp. 149–164, 2017.
DOI: 10.1007/978-3-319-57339-7_9

data along with the computation. Besides, secure computation has drawn users' attention after Gentry's ground-breaking work of fully homomorphic encryption (FHE) in 2009 [5]. Generally, a homomorphic encryption scheme allows meaningful computations like addition and multiplication on encrypted data without decrypting. Moreover, FHE allows any number of multiplications and additions on encrypted data. Gentry constructed the FHE scheme by applying the bootstrapping technique in somewhat homomorphic encryption (SwHE). But the FHE scheme is still far behind from practical use because of its speed [6]. Moreover, Brakerski and Vaikuntanathan [8] proposed one more somewhat homomorphic scheme using the ring learning with errors (ring-LWE) concept of Lyubashevsky et al. [7]. In this paper, we use an SwHE scheme in [8] because it works faster than the FHE and supports many additions and a few multiplications.

In this context, conjunctive query processing can be secured using SwHE. In addition, the database owners and their users can outsource such massive database computation to a third party like the cloud'. In 2016, Cheon et al. [2] showed a private query processing technique over encrypted data using the somewhat homomorphic encryption scheme of Brakerski et al. [9]. Here they considered circuit-based computation of different types of queries and showed their performances. They required about half an hour for executing a conjunctive query over an encrypted database consisting of 10000 records which is impractical in everyday use. In this paper, we take the challenge of performance improvement given by Cheon et al. [2] only for conjunctive query processing. So we propose an efficient method to answer private conjunctive query processing using SwHE.

1.1 Review of Recent Works

Here we review some recent works of the private conjunctive query using different cryptographic schemes. In 2013, Boneh et al. [1] proposed a new method to address private conjunctive query in the database. But their method required set intersection operations which reveal more information to the users. In 2014, Pappas et al. [11] showed a technique called 'Blind Seer' to support a rich query set over private DBMS using Yao's garbled circuits and oblivious transfer in the semi-honest model. Then Fisch et al. [12] improved the security of Blind Seer by adding malicious-client security. But both of them required a non-constant round of communication for searching data using Yao's garbled circuits. Besides, Cheon et al. [2] showed a method of conjunctive query processing in 2016. Here they provide privacy of the values in the predicate. But their performance can be improved for some practical uses. They also used a machine with 192 GB RAM which is impractical in cloud computing. Furthermore, another method of evaluating private database query is proposed by Kim et al. [3]. Here they reduced the communication cost of accessing m records required by previous methods. But they did not specify the individual query performances. Moreover, none of these aforementioned methods addressed conjunctive query with small multiplication depth that can be achieved to minimize the total complexity of the method. But they did not specify the individual query performances.

1.2   Our Contribution

The conventional solution to the evaluation of a private conjunctive query with k conditions in the predicate requires executing k sub-queries. Then the final result can be calculated by taking the intersection of the results from the sub-queries. For example, consider the conjunctive query like *select count(id) from record where* $\alpha_i = v_i \wedge \alpha_{i+1} = v_{i+1} \wedge \cdots \wedge \alpha_k = v_k$ which contains k conditions in its predicate. So it can be computed by counting 'id' from the result the k sub-queries with intersection as $\bigcap_{i=1}^{k} Q(\alpha_i = v_i)$ where $Q(\alpha_i = v_i) = \{id|$ the attribute α_i of id takes v_i as the value$\}$. If we outsource such computation to the cloud, then more information regarding records are disclosed to the cloud. To solve the problem, we propose a different protocol using ring-LWE based SwHE for providing security to the values in the predicate of a query. Moreover, we also propose a packing method for SwHE to address the conjunctive query in a few multiplications to make the protocol working faster.

Notations. In this paper, $\mathbb{Z}$ denotes the ring of integers. For a prime number q, the ring of integer is denoted by $\mathbb{Z}_q$. In addition, $\mathbb{Z}^n$ defines an n-dimensional integer vector space. Besides, $\mathbb{Z}[x]$ denotes the ring of polynomials over integers. For a vector $\mathbf{A} = (a_0, a_1, \ldots, a_{n-1})$, the maximum norm of $\|\mathbf{A}\|_\infty$ is defined as $\max |a_i|$. Let $\langle \mathbf{A}, \mathbf{B} \rangle$ denote the inner product between two vectors $\mathbf{A}$ and $\mathbf{B}$. Moreover, the function $\text{Enc}(m, pk) = ct$ defines the encryption of message m using the public key pk to produce the ciphertext ct. The ciphertexts ct_{add} and ct_{mul} denote homomorphic addition and multiplication of ciphertexts $ct = \text{Enc}(m, pk)$ and $ct' = \text{Enc}(m', pk)$. Furthermore, $s \leftarrow \chi$ denotes that s is chosen from the Gaussian distribution χ. The distribution $D_{\mathbb{Z}^n, \delta}$ indicates the n-dimensional discrete Gaussian distribution for some standard deviation $\delta > 0$. Besides, p, η, and m represent the total number of blocks, block size, and the total number of records respectively where a block is a collection of records.

2   Security Tool

In this section, we review the asymmetric SwHE scheme in [10] which is a variant of the security scheme in [8]. In 2011, Brakerski and Vaikunthanathan [8] proved the correctness of this scheme.

2.1   Asymmetric SwHE Scheme

In 2015, Lauter et al. [10] showed a SwHE scheme which is a public key variant of BV's SwHE scheme [8]. For the SwHE scheme in [10], we need to consider some parameters as follows.

- $\phi(x)$: is a cyclotomic polynomial where $\phi(x) = x^n + 1$.
- n: an integer which represents the lattice dimension of the ring $R_q = \mathbb{Z}_q[x]/\phi(x)$. Here n also represents the degree of polynomials which is a power of 2 such as 1024 or 2048.

- q: modulus q is an odd prime such that $q \equiv 1 (\bmod\ 2n)$ defining the ring $R_q = R/qR = \mathbb{Z}_q[x]/\phi(x)$ which denotes a ciphertext space.
- t: an integer $t < q$, which defines the message space of the scheme as $R_t = \mathbb{Z}_t[x]/\phi(x)$, the ring of integer polynomials modulo $\phi(x)$ and t.
- δ: is a parameter which defines a discrete Gaussian error distribution $\chi = D_{\mathbb{Z}^n, \delta}$ with an n-dimensional integer vector $\mathbb{Z}^n$ and a standard deviation δ where $\delta = 4$–8.

Now we can discuss the key generation, encryption, homomorphism, and decryption properties of SwHE scheme in [10] as follows:

Key Generation. Generate a ring element $s \leftarrow \chi$ for our secret key $sk = s$ where $s \in R$. We then sample a uniformly random element $a_1 \in R_q$ and an error $e \leftarrow \chi$ where $e \in R$. Now we get the public key pair as $pk = (a_0, a_1)$ with $a_0 = a_1 s + te$.

Encryption. For a given message $m \in R_t$ and a public key $pk = (a_0, a_1)$, the encryption algorithm first samples $u, f, g \leftarrow \chi$ where $u, f,$ and g are in R then encryption can be defined by a ciphertext pair $(c_0, c_1) = ct$ as follows.

$$\text{Enc}(m, pk) = (c_0, c_1) = (a_0 u + tg + m, -(a_1 u + tf)) \tag{1}$$

Here, the plaintext $m \in R_t$ is also in R_q because $t < q$.

Homomorphic Operations. Generally, homomorphic operations like addition ($\boxplus$) and multiplication ($\boxtimes$) are between two ciphertexts $ct = (c_0, \ldots, c_\psi)$ and $ct' = (c'_0, \ldots, c'_\omega)$. So the homomorphic operations between two ciphertexts can be defined as follows.

$$\begin{cases} ct_{add} = ct \boxplus ct' = \left(c_0 + c'_0, \ldots, c_{max(\psi,\omega)} + c'_{max(\psi,\omega)}\right) \\ ct_{mul} = ct \boxtimes ct' = \displaystyle\sum_{i=0}^{\psi+\omega} \hat{c}_i z^i = \left(\sum_{i=0}^{\alpha} c_i z^i\right)\left(\sum_{j=0}^{\beta} c'_j z^j\right) \end{cases} \tag{2}$$

where z denotes a symbolic variable such that $\{ct, ct'\} \in R_q[z]$. In addition, we can also define the subtraction as similar to component-wise addition.

Decryption. For a fresh or homomorphically operated ciphertext $ct = (c_0, \ldots, c_\psi)$ and $t \in R_t$, general decryption can be defined as

$$\text{Dec}(ct, sk) = [\tilde{m}]_q \bmod t \tag{3}$$

where $\tilde{m} = \sum_{i=0}^{\alpha} c_i s^i$. For the secret key vector $\mathbf{s} = (1, s, s^2, \ldots)$, we can simply rewrite $\text{Dec}(ct, sk) = [ct, \mathbf{s}]_q \bmod t$. For example, a fresh ciphertext $ct = (c_0, c_1)$ generated by (1) then we have

$$\langle ct, \mathbf{s} \rangle = (a_0 u + tg + m) - s \cdot (a_1 u + tf) = m + t \cdot (ue + g - sf) \tag{4}$$

in the ring R_q since $a_0 - a_1 s = te$. If the value $m + t \cdot (ue + g - sf)$ does not wrap-around mod q (all errors $e, f, g, u \leftarrow \chi$ must be sufficiently small) then we have $[\langle ct, \mathbf{s} \rangle]_q = m + t \cdot (ue + g - sf)$ in the base ring R. Here, it is clear that we can recover plaintext m by mod t operation. In addition, for two ciphertexts ct_1 and ct_2, we clearly have the following by the homomorphic operation if no wrap-around happens in the encrypted results after homomorphic operations.

$$\begin{cases} \langle ct_1 \boxplus ct_2, \mathbf{s} \rangle = \langle ct_1, \mathbf{s} \rangle + \langle ct_2, \mathbf{s} \rangle \\ \langle ct_1 \boxtimes ct_2, \mathbf{s} \rangle = \langle ct_1, \mathbf{s} \rangle \cdot \langle ct_2, \mathbf{s} \rangle \end{cases} \tag{5}$$

2.2   Security of SwHE Scheme

We can show the security of the SwHE scheme by the polynomial ring-LWE assumption (ring-LWE$_{n,q,\chi}$) as done by Lauter et al. [10] for the given parameters (n, q, t, δ). Let the ring $R_q = \mathbb{Z}_q/\phi(x)$ where $\phi(x) = (x^n + 1)$ be the cyclotomic polynomial of degree n. Let $s \leftarrow \chi = D_{\mathbb{Z}^n, \delta}$ be a random ring element. The assumption holds for any polynomial number of samples of the form

$$(a_i, b_i = a_i \cdot s + e_i) \in (R_q)^2$$

where a_i is uniformly random in R_q and e_i is drawn from the error distribution χ. Here the a_i's are uniformly random in R_q and b_i's ($b_i = a_i \cdot s + e_i$) are also uniform in R_q. Therefore, it is hard to distinguish (a_i, b_i) from a uniformly random pair $(a_i, b_i) \in (R_q)^2$. Besides, Lyubashevsky et al. [7] showed that the ring-LWE assumption is reducible to the worst-case hardness of problems on ideal lattices that is believed to be secure against the quantum computer.

Remark 1. In 2016, Castryck et al. [16] described the provably weak instances of ring-LWE. But these kinds of weak instances do not affect this SwHE scheme.

2.3   Correctness of SwHE Scheme

The correctness of the SwHE scheme depends on how the decryption can recover the original result from the ciphertext after some homomorphic operations. We can write the decryption process as follows.

$$\begin{cases} \mathrm{Dec}(ct_{add}, sk) = \mathrm{Dec}((ct \boxplus ct'), sk) = m + m' \\ \mathrm{Dec}(ct_{mul}, sk) = \mathrm{Dec}((ct \boxtimes ct'), sk) = m \cdot m' \end{cases} \tag{6}$$

The above process is already described in Sect. 1.1 in [8]. Here, ciphertext ct and ct' comes from $m \in R_q$ and $m' \in R_q$ respectively after encryption. The encryption scheme in Sect. 2.1 is the presentation of SwHE and its holds if the following lemma holds as shown in [13].

Lemma 1 (Condition for successful decryption). *For a ciphertext ct, the decryption $Dec(ct, sk)$ recovers the correct result if $\langle ct, s \rangle \in R_q$ does not wrap around mod q, namely, if the condition $\|\langle ct, s \rangle\|_\infty < \frac{q}{2}$ is satisfied where $\|a\|_\infty = \max |a_i|$ for an element $a = \sum_{i=0}^{n-1} a_i x^i \in R_q$. Specifically, for a fresh ciphertext ct, the ∞-norm $\|\langle ct, s \rangle\|_\infty$ is given by $\|m + t(ue + g - sf)\|_\infty$. Moreover, for a homomorphically operated ciphertext, the ∞-norm can be computed by (5).*

3   Private Conjunctive Query Protocol

In this section, we describe our protocol using a real-world scenario. Consider a hospital (Charlie) has a little computation ability which is maintaining a database of their admitted patients. Moreover, a government health research institute (Alice) wants to know the number of male patients admitted to that hospital in the last year who suffered from tuberculosis (TB). Aforementioned is a conjunctive equality query request to Charlie from Alice. Here Alice poses little computation ability like key generation, encryption, and decryption. Besides, Charlie cannot disclose his patients' information to Alice. Therefore, they want to outsource the computation to a third party like the cloud (Bob) without disclosing the query and corresponding data to Bob. In this scenario, consider that Alice has a conjunctive query with k equality conditions (*select count(id) from record where* $\alpha_i = v_i$ *and* $\alpha_{i+1} = v_{i+1}$ *and* ... *and* $\alpha_k = v_k$ where $k > 1$) in the predicate. Here we consider only the security of values in the predicate. So the values of k attributes $\{\alpha_1, \ldots, \alpha_k\}$ appeared in the predicate of the query is represented as a set $V = \{v_1, \ldots, v_k\}$ where $v_i = (a_{i,0}, \ldots, a_{i,l_i-1})$ is considered as a binary vector of length l_i with $1 \le i \le k$. Furthermore, Charlie has m records $\{\mathcal{R}_1, \ldots, \mathcal{R}_m\}$ in his patientRecord table of the hospital database with $(k + \lambda)$ attributes. Here we require only the k attributes $\{\alpha_1, \ldots, \alpha_k\}$ and their values in each record required for our conjunctive query processing. Now each record is represented as $\mathcal{R}_\mu = \{w_{\mu,1} \ldots, w_{\mu,k}\}$ where each value $w_{\mu,i} = (b_{\mu,i,0}, \ldots, b_{\mu,i,l_i-1})$ is considered as a binary vector of the same length l_i with $1 \le \mu \le m$ and $1 \le i \le k$. To find out the conjunctive equality of two sets V and $\mathcal{R}_\mu$, now we form a binary vector from the set V as $\mathbf{A} = (v_i, \ldots, v_k)$ by concatenating each binary vector in the set V where $v_i = (a_{i,0}, \ldots, a_{i,l_i-1})$ with $1 \le i \le k$. Similarly, we form another vector from the set $\mathcal{R}_\mu$ as $\mathbf{B}_\mu = (w_{\mu,1}, \ldots, w_{\mu,k})$ where $w_{\mu,i} = (b_{\mu,i,0}, \ldots, b_{\mu,i,l_i-1})$ with $1 \le i \le k$. Here $|\mathbf{A}| = |\mathbf{B}_\mu| = \sum_{i=1}^{k} l_i = L$. We can compute the conjunctive equality between two sets V and $\mathcal{R}_\mu$ by the following Hamming distance computation.

$$\mathbb{H}_\mu = \sum_{i=1}^{k} \sum_{j=1}^{l_i-1} |a_{i,j} - b_{\mu,i,j}| = \sum_{i=1}^{k} \sum_{j=1}^{l_i-1} (a_{i,j} + b_{\mu,i,j} - 2a_{i,j}b_{\mu,i,j}) \tag{7}$$

Here, $\mathbb{H}_\mu$ defines the Hamming distance between two binary vectors $\mathbf{A}$ and $\mathbf{B}_\mu$. Moreover, if $\mathbb{H}_\mu$ in Eq. (7) is 0 for some positions μ then we can say that $\mathbf{A} = \mathbf{B}_\mu$; otherwise $\mathbf{A} \ne \mathbf{B}_\mu$. In this way, Alice securely verifies her conjunctive equality query with the help of Bob and counts the number of 0 for some positions μ to find out total number of records for the given query. If we compute the Hamming distance $\mathbb{H}_\mu$ for all m records individually then the computation will be slow. So we need a technique to increase the performance of our protocol.

3.1   Boosting Performance Using the Batch Technique

Batching is the process of executing a single instruction on multiple data. The performance of our protocol can be increased by using this batch technique within our lattice dimension n that we call batch private conjunctive query (BPCQ) protocol. Generally, a big database consists many tables where each table contains numerous records. For our conjunctive query processing with batch technique, if we compare all the values of a certain attribute of a particular table using a single computation then we will be required higher lattice dimension n which requires more memory to compute. This high requirement of memory may exceed usual capacity of a machine in the cloud. So we divide all records of a table into blocks. For our given records, the total number of records accessed within the lattice dimension n is $\eta = \lfloor n/L \rfloor$ where $L = \sum_{i=1}^{k} l_i$. Furthermore, we divide the total records m into p blocks as $p = \lceil m/\eta \rceil$. If we access m-records of the Table one after another then it requires m-rounds communication between Charlie and Bob in the cloud. On the contrary, BPCQ protocol able to access a block of records $\{\mathcal{R}_1, \ldots, \mathcal{R}_\eta\}$ at a time. Then it will reduce the number of communication between Charlie and Bob. Here the batching technique allows us to reduce the communication complexity from m to $\lceil m/\eta \rceil$. Now we can pack the η records of each block in a single polynomial to support batch computation.

3.2 Batch Private Conjunctive Query Protocol

Here we call the protocol for secure batch processing of our conjunctive query as batch private conjunctive query (BPCQ) protocol. To describe the BPCQ protocol, we consider the same scenario as discussed in Sect. 3. Here, Charlie has m records $\{\mathcal{R}_1, \ldots, \mathcal{R}_m\}$ in his patientRecord table. Let us consider a block containing η records as $\beta_\sigma = \{\mathcal{R}_{\sigma,1}, \ldots, \mathcal{R}_{\sigma,\eta}\}$ where $1 \le \sigma \le p$. For $1 \le d \le \eta$ and $1 \le i \le k$, each record is represented as $\mathcal{R}_{\sigma,d} = \{w_{\sigma,d,1}, \ldots, w_{\sigma,d,k}\}$ where $w_{\sigma,d,i} = (b_{\sigma,d,i,0}, \ldots, b_{\sigma,d,i,l_i-1})$. Consider the same query set V and vector $\mathbf{A} = (v_i, \ldots, v_k)$ where $v_i = (a_{i,0}, \ldots, a_{i,l_i-1})$ as in Sect. 3. Similarly, we form another batch binary vector from all records for each block β_σ as $\mathbf{B}_\sigma = (\mathbf{B}_{\sigma,1}, \ldots, \mathbf{B}_{\sigma,\eta})$ where $\mathbf{B}_{\sigma,d} = (w_{\sigma,d,1}, \ldots, w_{\sigma,d,k})$. Here $|\mathbf{A}| = \sum_{i=1}^{k} l_i = L$ and $|\mathbf{B}_\sigma| = \eta \cdot L$. Here we can compute the conjunctive equality between two sets $\mathbf{A}$ and $\mathbf{B}_\sigma$ by the multiple Hamming distance computation as

$$\mathbb{H}_{\sigma,d} = \sum_{i=1}^{k} \sum_{j=1}^{l_i-1} |a_{i,j} - b_{\sigma,d,i,j}| = \sum_{i=1}^{k} \sum_{j=1}^{l_i-1} (a_{i,j} + b_{\sigma,d,i,j} - 2a_{i,j}b_{\sigma,d,i,j}) \quad (8)$$

where $1 \le d \le \eta$ and $1 \le \sigma \le p$. Here, $\mathbb{H}_{\sigma,d}$ defines the Hamming distance between two binary vectors $\mathbf{A}$ and $\mathbf{B}_{\sigma,d}$ where $\mathbf{B}_{\sigma,d} = (w_{\sigma,d,1}, \ldots, w_{\sigma,d,k})$ is the d-th sub-vector of the block vector $\mathbf{B}_\sigma$. Moreover, if $\mathbb{H}_{\sigma,d}$ in Eq. (8) is 0 for some positions d in the block σ then we can say that $\mathbf{A} = \mathbf{B}_{\sigma,d}$; otherwise $\mathbf{A} \ne \mathbf{B}_{\sigma,d}$. In this way, Alice securely verifies her conjunctive equality query with the help of Bob and counts the number of 0 for some positions d in the σ-th block to find out total number of records for the given query. Now we narrate our BPCQ protocol by the following steps.

Inputs: $\mathbf{A} = (v_i, \ldots, v_k)$ where $v_i = (a_{i,0}, \ldots, a_{i,l_i-1})$ and $\mathbf{B}_\sigma = (\mathbf{B}_{\sigma,1}, \ldots, \mathbf{B}_{\sigma,\eta})$
where $\mathbf{B}_{\sigma,d} = (w_{\sigma,d,1}, \ldots, w_{\sigma,d,k})$ for $1 \leq \sigma \leq p$ and $1 \leq d \leq \eta$.
Output: $|\{(\sigma, d) | \mathbf{B}_{\sigma,d} = \mathbf{A}\}|$
BPCQ protocol:

1. Alice creates the public key and private key by herself and sends the public key to Charlie through a secure channel. She also sends conjunctive attributes information $\{\alpha_1, \ldots, \alpha_k\}$ to Charlie.
2. Then she forms a vector $\mathbf{A} = (v_i, \ldots, v_k)$ using the k values appeared in the predicate of her conjunctive query using a certain order and length where $v_i = (a_{i,0}, \ldots, a_{i,l_i-1})$. Then she encrypts $\mathbf{A}$ using her public key and sends it to Bob.
3. Charlie also forms another vector $\mathbf{B}_\sigma = (\mathbf{B}_{\sigma,1}, \ldots, \mathbf{B}_{\sigma,\eta})$ from η records from each block σ where $\mathbf{B}_{\sigma,d} = (w_{\sigma,d,1}, \ldots, w_{\sigma,d,k})$ and $1 \leq \sigma \leq p$. He uses Alice's public key to encrypt $\mathbf{B}_\sigma$ and sends the value to Bob.
4. Bob does the secure computation of the batch equality tests as in Eq. (8) and sends the encrypted result $ct(\mathbb{H}_{\sigma,d})$ to Alice to find out the number of 0 appears for some positions d in the σ-block.
5. For $1 \leq m \leq k$, Alice decrypts $ct(\mathbb{H}_{\sigma,d})$ using her secret key and counts the number of zero appears in the value $ct(\mathbb{H}_{\sigma,d})$ for $1 \leq \sigma \leq p$ and $1 \leq d \leq \eta$ and thus find out the total number of records.

Remark 2. Here our protocol is secure under the assumption that Bob is semi-honest (also known as honest-but-curious), i.e., he always obeys the protocol but tries to learn information from it.

3.3 Data Representation for Conjunctive Query Processing

Usually, data may have several dimensions depending on its presentation as shown in Fig. 1. Here one-dimensional data contains a single value of a certain attribute. For example, a data set containing sex of the patients of a hospital of a particular disease in 2014 is an one-dimensional data as shown in Fig. 1(a). Here one-dimensional data is not required for computing the conjunctive query. In contrast, values of multi-dimensional data depend on several attributes. Again, a data set representing sex of the patient of several diseases over several years is a multi-dimensional data as shown in Fig. 1(b). For computing our conjunctive query protocol, we need an efficient packing method to process the multi-dimensional data using an SwHE.

3.4 Packing Method of Data

The packing method of data means binding many bits in a single polynomial. Some packing methods has already been used in researches [10,13,14] to do

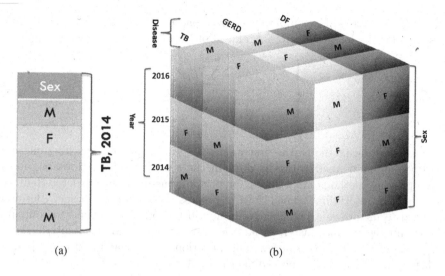

Fig. 1. (a) One-dimensional data and (b) Multi-dimensional data.

faster computation of their protocols using the SwHE scheme. Here we discuss the existing packing methods, our packing method, and inner product property of our packing method in the following sub-section.

Existing Packing Method. Since 2011, we have observed that several researchers [10,13,14] have proposed some packing methods for secure homomorphic computation in the cloud with.practical implementations. In 2011, Lauter et al. [10] used a packing method for encoding a long integer like 128 bits. They packed an integer in a single polynomial of n degree. They divided the n-bit integer into as an n-bit binary vector $\mathbf{A} = (a_0, a_1, \ldots, a_{n-1})$ and the polynomial was represented as by the equation as follows.

$$Poly(\mathbf{A}) = \sum_{i=0}^{n-1} a_i x^i$$

Here the vector $\mathbf{A}$ is encrypted as $ct_{pack}(\mathbf{A}) = Enc(Poly(\mathbf{A}), pk)$. Homomorphic operations may require several additions and multiplications. This packing method of n degree is useful for adding vectors like $\mathbf{A} = (a_0, a_1, \ldots, a_{n-1})$ and $\mathbf{B} = (b_0, b_1, \ldots, b_{n-1})$. But for multiplication, $\mathbf{A}$ and $\mathbf{B}$, it increases the polynomial degree greater than n. Since the degree is at most n, this packing method can do d multiplications if the polynomials are reduced to n/d degree polynomials.

In addition, an extension of the packing method in [10] was done by Yasuda et al. [13] to facilitate the secure Hamming distance and Euclidean distance computation using the inner product of two vectors $\mathbf{A} = (a_0, a_1, \ldots, a_{n-1})$ and $\mathbf{B} = (b_0, b_1, \ldots, b_{n-1})$ within the degree n. They proposed two different packing methods to serve wider computations over packed ciphertext. So, the packing methods for the integral vectors $\mathbf{A}$ and $\mathbf{B}$ of length n was represented by

$$
\begin{cases}
Poly_a(\mathbf{A}) = \sum_{i=0}^{n-1} a_i x^i \\
Poly_b(\mathbf{B}) = -\sum_{i=0}^{n-1} a_i x^{n-i}.
\end{cases}
\tag{9}
$$

In these packing methods, the first method is quite similar to packing method of [10], but main modification is done in the second method. However, all of these packing methods are suitable for packing one-dimensional data. Here we need to pack multi-dimensional data to process the conjunctive query. Besides, we need to introduce a batching technique in the packing method to compute the Hamming distance faster. In addition, we want to compute multiple Hamming distance in Eq. (8) in a few multiplications. So a new packing method is indispensable to pack the multi-dimensional data.

Our Packing Method. We require to pack multi-dimensional data as shown in Fig. 1(b) to process conjunctive query in a few multiplications. To support the batch computation, we need to compute the Hamming distance between the query vector and each record of the σ-th block using one instruction. The Eq. (8) contains multiple Hamming distance. To compute the multiple Hamming distance, let us form an integer vector $\mathbf{A} = (v_i, \ldots, v_k) \in R_t$ from the set V of length L where $v_i = (a_{i,0}, \ldots, a_{i,l_i-1})$. Then we form another integer block vector $\mathbf{B}_\sigma = (\mathbf{B}_{\sigma,1}, \ldots, \mathbf{B}_{\sigma,n}) \in R_t$ of length $\eta \cdot L$ by taking each record $\mathcal{R}_{\sigma,d}$ from the σ-th block where the length of each vector $\mathbf{B}_\sigma$ is $\eta \cdot L$ and $\mathbf{B}_{\sigma,d} = (w_{\sigma,d,1}, \ldots, w_{\sigma,d,k})$. Here the multiple Hamming distance means the distances between the vector $\mathbf{A}$ and each sub-vector in $\mathbf{B}_\sigma$. So we need to define another packing method than Eq. (9). Moreover, we know from [13] that the secure inner product $\langle \mathbf{A}, \mathbf{B}_\sigma \rangle$ helps to compute the Hamming distance between $\mathbf{A}$ and $\mathbf{B}_\sigma$. Here we pack these integer vectors by some polynomials with the highest degree$(x) = n$ in such a way so that inner product $\langle \mathbf{A}, \mathbf{B}_\sigma \rangle$ does not wrap-around a coefficient of x with any degrees. For the integer vectors $\mathbf{A}$ and $\mathbf{B}_\sigma$ with $n \geq \eta \cdot L$ and $1 \leq \sigma \leq p$, the packing method of [13] in the same ring $R = \mathbb{Z}[x]/(x^n + 1)$ can be rewritten as

$$
\begin{cases}
Poly_1(\mathbf{A}) = \sum_{i=1}^{k} \sum_{j=0}^{l_i-1} a_{i,j} x^{i \cdot j} \\
Poly_2(\mathbf{B}_\sigma) = \sum_{d=1}^{s} \sum_{i=1}^{k} \sum_{j=0}^{l_i-1} b_{\sigma,d,i,j} x^{d \cdot L - i \cdot j}.
\end{cases}
\tag{10}
$$

Here the multiplication of the above two polynomials helps the inner product computations which in turn helps the multiple Hamming distances computation between $\mathbf{A}$ and $\mathbf{B}_\sigma$. Here each Hamming distance can be found as a coefficient of x with different degrees.

Inner Product Property. Consider the above two vectors $\mathbf{A}$ and $\mathbf{B}_\sigma$ again. We already know that inner product of two vectors helps the Hamming distance computation. So the polynomial multiplications of $Poly_1(\mathbf{A})$ and $Poly_2(\mathbf{B})$ in the same base ring R can be represented as

$$\left(\sum_{i=1}^{k}\sum_{j=0}^{l_i-1} a_{i,j}x^{i\cdot j}\right) \times \left(\sum_{d=1}^{s}\sum_{i=1}^{k}\sum_{j=0}^{l_i-1} b_{\sigma,d,i,j}x^{d\cdot L-i\cdot j}\right)$$

$$= \sum_{d=1}^{s}\sum_{i=1}^{k}\sum_{j=0}^{l_i-1} a_{i,j}b_{\sigma,d,i,j}x^{d\cdot L}$$

$$= \sum_{d=1}^{s}\langle\mathbf{A},\mathbf{B}_{\sigma,d}\rangle x^{d\cdot L} + \text{ToHD} + \text{ToLD}. \tag{11}$$

Here, $\mathbf{A}$ is the vector of length L and $\mathbf{B}_{\sigma,d}$ is the d-th sub-vector of $\mathbf{B}_\sigma$ with the same length as $\mathbf{A}$ where $1 \leq \sigma \leq p$ and $1 \leq d \leq \eta$. Moreover, the ToHD (terms of higher degree) means $deg(x) > d \cdot L$ and the ToLD (terms of lower degrees) means $deg(x) < d \cdot L$. The result in Eq. (11) shows that one polynomial multiplication includes the multiple inner products of $\langle\mathbf{A},\mathbf{B}_{\sigma,d}\rangle$. In addition, the following proposition is needed to hold for computing the multiple inner products over packed ciphertexts. Here we define the packed ciphertext of $Poly_i(A) \in R$ with $i = \{1,2\}$ as

$$ct_i(A) = Enc(Poly_i(A), pk) \in (R_q)^2 \tag{12}$$

Proposition 1. *Let $\mathbf{A} = (v_i, \ldots, v_k) \in R_t$ be an integer vector where $v_i = (a_{i,0}, \ldots, a_{i,l_i-1})$ and $|\mathbf{A}| = L$. In addition, $\mathbf{B}_\sigma = (\mathbf{B}_{\sigma,1}, \ldots, \mathbf{B}_{\sigma,\eta}) \in R_t$ be another integer vector of length $\eta \cdot L$ where $\mathbf{B}_{\sigma,d} = (w_{\sigma,d,1}, \ldots, w_{\sigma,d,k})$. For $1 \leq d \leq \eta$, the vector $\mathbf{B}_\sigma$ includes η sub-vectors where the length of each sub-vector is L. If the ciphertext of $\mathbf{A}$ and $\mathbf{B}_\sigma$ can be represented as $ct_1(\mathbf{A})$ and $ct_2(\mathbf{B}_\sigma)$ respectively by Eq. (12) then under the condition of Lemma 1, decryption of homomorphic multiplication $ct_1(\mathbf{A}) \boxtimes ct_2(\mathbf{B}_\sigma) \in (R_q)^2$ will produce a polynomial of R_t with $x^{d\cdot L}$ including coefficient $\langle\mathbf{A},\mathbf{B}_{\sigma,d}\rangle = \sum_{d=1}^{s}\langle\mathbf{A},\mathbf{B}_{\sigma,d}\rangle x^{d\cdot L}$. Consequently, we can say that homomorphic multiplication of $ct_1(\mathbf{A})$ and $ct_2(\mathbf{B}_\sigma)$ simultaneously computes the multiple inner products for $1 \leq \sigma \leq p$ and $1 \leq d \leq \eta$.*

4 Secure Computation of Private Conjunctive Query

We discuss secure computation of the batch private conjunctive query (BPCQ) protocol (see Sect. 3 for details) in the following subsections.

4.1 Batch Private Conjunctive Query Protocol

We compute our batch private conjunctive query (BPCQ) protocol using the SwHE scheme in Sect. 2 and the packing method in the previous section.

In addition, according to Eq. (8), we need to find out the values of the Hamming distance $\mathbb{H}_{\sigma,d}$ for $1 \leq \sigma \leq p$ and $1 \leq d \leq \eta$. Let us consider the same integer vectors $\mathbf{A}$ and $\mathbf{B}_\sigma$ from which $\mathbb{H}_{\sigma,d}$ can be computed. Here, for $1 \leq \mu \leq m$, $\mathbb{H}_{\sigma,d}$ is computed by the Hamming distance between $\mathbf{A}$ and $\mathbf{B}_\sigma$ using the arithmetic computation as

$$\mathbb{H}_{\sigma,d} = \sum_{i=1}^{k} \sum_{j=1}^{l_i-1} (a_{i,j} + b_{\mu,i,j} - 2a_{i,j}b_{\mu,i,j}). \tag{13}$$

Computation over Packed Ciphertext. For the two integer vectors $\mathbf{A}$ and $\mathbf{B}_\sigma$ mentioned above, the Hamming distance $\mathbb{H}_{\sigma,d}$ in Eq. (13) can be computed by the packing method in Eq. (10) and inner product property in Eq. (11). Moreover, the packed ciphertext of the vectors $\mathbf{A}$ and $\mathbf{B}_\sigma$ is computed by the Eq. (12). Here $\mathbb{H}_{\sigma,d}$ is computed as $ct(\mathbb{H}_{\sigma,d})$ from the Proposition 1 and the packed ciphertext vector $ct_1(\mathbf{A}) \in R_q$ and $ct_2(\mathbf{B}_\sigma) \in R_q$. Due to the using of packed ciphertexts, we require only three homomorphic multiplications and two homomorphic additions to compute $ct(\mathbb{H}_{\sigma,d})$ as follows.

$$ct_1(\mathbf{A}) \boxtimes ct_2(\mathbf{V}_B) \boxplus ct_2(\mathbf{B}_\sigma) \boxtimes ct_1(\mathbf{V}_L) \boxplus (-2ct_1(\mathbf{A}) \boxtimes ct_2(\mathbf{B}_\sigma)) \tag{14}$$

Here $\mathbf{V}_B$ denotes an integer vector $(1,\dots,1)$ of length $\eta \cdot L$ and $\mathbf{V}_L$ denotes another integer vector $(1,\dots,1)$ of length L. The encrypted polynomial $ct(\mathbb{H}_{\sigma,d})$ includes the Hamming distances between the sub-vectors of $\mathbf{A}$ and sub-vectors of $\mathbf{B}_\sigma$. Here we need the Hamming distance $\mathbb{H}_{\sigma,d}$ in Eq. (13). Bob sends $ct(\mathbb{H}_{\sigma,d})$ to Alice for decryption. According to our Proposition 1 and BPCQ protocol, Alice decrypts $ct(\mathbb{H}_{\sigma,d})$ in the ring R_q using her secret key and extracts $\mathbb{H}_{\sigma,d}$ as a coefficient of $x^{d \cdot L}$ from the plaintext of $ct(\mathbb{H}_{\sigma,d})$. Then Alice counts the number of zero appears in the value of $\mathbb{H}_{\sigma,d}$ for some $1 \leq \sigma \leq p$ and $1 \leq d \leq \eta$ and thus finds out the total number of records satisfying the conjunctive conditions in the given query.

4.2 Solving Additional Information Leakage Problem

Here Alice can know some additional information from the computation of $ct(\mathbb{H}_{\sigma,d})$ than she requires due to sending encrypted polynomial $ct(\mathbb{H}_{\sigma,d})$ to her. But Alice needs to know only those coefficients which has degree $x^{d \cdot L}$. We solve this problem by adding a random polynomial at the cloud (Bob) ends where every coefficient is random except $x^{d \cdot L}$ and the coefficient of $x^{d \cdot L}$ is zero. Bob adds a random polynomial r to $ct(\mathbb{H}_{\sigma,d})$ for masking extra information since Alice needs to check only the coefficient of $x^{d \cdot L}$ from the large polynomial $ct(\mathbb{H}_{\sigma,d})$ produced by Bob. The random polynomial in the ring R can be represented by the following equations as

$$r = r_0 + \sum_{d=0}^{n/L} \sum_{i=1}^{L-1} r_{d \cdot L+i} x^{d \cdot L+i}.$$

Here $ct(\mathbb{H}_{\sigma,d})$ consists of three ciphertext components as $ct(\mathbb{H}_{\sigma,d}) = (c_0, c_1, c_2)$. So Bob adds r to the ciphertext as $ct(\mathbb{H}'_{\sigma,d}) = ct(\mathbb{H}_{\sigma,d}) \boxplus r = (c_0 \boxplus r, c_1, c_2)$. Here the resulting ciphertext $ct(\mathbb{H}'_{\sigma,d})$ contains all required information as a coefficient of $x^{d \cdot L}$ and hide all other coefficients using the randomization. In this way, $ct(\mathbb{H}'_{\sigma,d})$ does not leak any information to Alice except the coefficient of $x^{d \cdot L}$.

5   Performance Analysis

In this section, we experimented our BPCQ protocol and compared its performances with conjunctive query results in [2]. Here our protocol scenario is different from the scenario of Cheon et al. [2] protocol while both requires conjunctive query processing. Here we describe the theoretical performance comparison, used parameters of our experiments and their security levels. Here, we also discuss the practical performance of our protocol using ring-LWE SwHE.

5.1   Theoretical Evaluation

Here we evaluate the multiplication depth of equality computation for Cheon et al. [2] and our protocol. We use the Hamming distance computation to compute the equality of data which is realized by Eq. (13). Here the encrypted computation of this Hamming distance required only three polynomial multiplications as in Eq. (14). The method of Cheon et al. required a multiplication depth of $\log \mu$ for their equality circuit comparing two μ-bit integers whereas our method required only $\log 3$ due to using our packing method as shown in Table 1. In addition, our BPCQ protocol requires a communication complexity of $\mathcal{O}(m \cdot L \log q)$.

Table 1. Theoretical performance comparison of our protocol

Operation	Data size	Depth of multiplication	
		Cheon et al. [2]	Our method
Equality	μ-bit	$\log \mu$	$\log 3$

5.2   Experimental Settings

For our experiment, we considered the same database settings as referred in [2] for Charlie in our protocol with a different protocol scenario. We also considered the same conjunctive query with equality as a comparison operator. We show the experimental setting of our used parameters in Table 2. Here the number of equality conditions k appeared in the query are 2 and 4. Furthermore, we took the data size 15–16-bits and block size $\eta = 100$. We also considered appropriate values for the parameters (n, q, t, δ) of our security scheme as discussed in Sect. 2 for successful decryption and achieving a certain security level. Here, we need the lattice dimension to be greater than $n \geq \eta \cdot L$ for our BPCQ protocol as mentioned in Sect. 3.4. For this reason, we choose the lattice dimension

Table 2. Experimental parameter settings in BPCQ protocol

Index	Parameters (n, t, q, δ)	Data size	Block size (η)
1	(3000, 2048, 61-bits, 8)	15 bits	
2	(6000, 2048, 63-bits, 8)	15 bits	
3	(3200, 2048, 61-bits, 8)	16 bits	100
4	(6400, 2048, 63-bits, 8)	16 bits	
5	(3200, 2048, 61-bits, 8)	16 bits	
6	(6400, 2048, 63-bits, 8)	16 bits	

n=3000–6400. Furthermore, we set $t = 2048$ for our plaintext space R_t. According to the work in [10], we choose $\delta = 8$ and value of q must be greater than $16n^2t^2\delta^4 = 2^4 \cdot 2^{22} \cdot 2^{22} \cdot 2^{12} = 2^{60}$ for the ciphertext space R_q. Therefore, we fix our parameters as $(n, q, t, \delta) = (3000\text{–}6400, 61\text{–}63\text{-bits}, 2048, 8)$. Following the same computation procedure of Yasuda et al. [15], our parameters setting provides at least 257-bit security level to protect our method from some distinguishing attacks. In addition, NIST [17] defines different security levels for many security algorithms and their corresponding validity periods. Furthermore, they notified that a minimum strength of 112-bit level security has a security lifetime up to 2030. They also disclosed that a security algorithm with a minimum strength of 128-bit level security has a security lifetime beyond 2030.

5.3 Experimental Evaluation

Here we show the practical performance of our protocol in Table 3. Here, we wrote the required code of our protocol in C programming language with Pari library (version 2.7.5) [18]. Then we ran on a single machine configured with one 3.6 GHz Intel core-i7 CPU and 8 GB RAM in Linux environment. Here we also compared our private conjunctive query protocol performance with that of Cheon et al. [2] for three different data set of 100, 1000, and 10000 records. Here we use indexing as in the first column of Tables 2 and 3 to show the relation of

Table 3. Performance of the BPCQ protocol

Index	m	(k)	Total time (seconds)		Security level	
			Cheon et al. [2]	Our method	Cheon et al. [2]	Our method
1	100	2	5.12	0.109	80	257
2		4	10.24	0.249		648
3	1000	2	51.63	1.281		281
4		4	101.86	1.828		698
5	10000	2	913.180	8.734		281
6		4	1788.19	19.124		698

parameter settings and corresponding performances. However, our system used less RAM than that of Cheon et al. [2]. From the Table 3, we can say that our protocol worked more than 50 times as fast as the CKK protocol [2]. Here our conjunctive query processing worked faster because of low multiplicative depth of equality circuits and batch computation. As shown in Table 3, we achieved 257–698-bit security level for our protocol whereas Cheon et al. achieved only 80-bit security level.

6   Conclusions

In this paper, we proposed and implemented an efficient private conjunctive query protocol over encrypted data which performs more than 50 times as fast as the existing protocol using SwHE. While we compared our technique with the method of Cheon et al. [2] in a different settings from theirs. However, both of the protocols needed the conjunctive query processing. Here we have observed from our computation time that if we would use the same settings then it will not cost as much time as required by them. Moreover, we showed that our equality protocol required less multiplicative depth than that of Cheon et al. due to using our packing method. Besides, we believe that our packing method is extendable to many new queries processing. In future work, we will consider all types of queries (e.g. disjunctive and threshold queries) that are used for private database query processing using our packing method along with SwHE.

Acknowledgment. This research is supported by KAKENHI Grant Numbers JP26540002, JP-24106008, and JP16H0175.

References

1. Boneh, D., Gentry, C., Halevi, S., Wang, F., Wu, D.J.: Private database queries using somewhat homomorphic encryption. In: Jacobson, M., Locasto, M., Mohassel, P., Safavi-Naini, R. (eds.) ACNS 2013. LNCS, vol. 7954, pp. 102–118. Springer, Heidelberg (2013). doi:10.1007/978-3-642-38980-1_7
2. Cheon, J.H., Kim, M., Kim, M.: Optimized search-and-compute circuits and their application to query evaluation on encrypted data. IEEE Trans. Inf. Forensics Security **11**(1), 188–199 (2016)
3. Kim, M., Lee, H.T., Ling, S., Wang, H.: On the efficiency of FHE-based private queries. In: IACR Cryptology ePrint Archive 2015: 1176 (2015)
4. Kim, M., Lee, H.T., Ling, H., Ren, S.Q., Tan, B.H.M., Wang, H.: Better security for queries on encrypted databases. In: IACR Cryptology ePrint Archive 2016: 470 (2016)
5. Gentry, C.: Fully homomorphic encryption using ideal lattices. In: Symposium on Theory of Computing - STOC 2009, pp. 169–178. ACM, New York (2009)
6. Hu, Y.: Improving the efficiency of homomorphic encryption schemes. Ph.D. diss., Worcester Polytechnic Institute, Massachusetts (2013)
7. Lyubashevsky, V., Peikert, C., Regev, O.: On ideal lattices and learning with errors over rings. In: Gilbert, H. (ed.) EUROCRYPT 2010. LNCS, vol. 6110, pp. 1–23. Springer, Heidelberg (2010). doi:10.1007/978-3-642-13190-5_1

8. Brakerski, Z., Vaikuntanathan, V.: Fully homomorphic encryption from ring-LWE and security for key dependent messages. In: Rogaway, P. (ed.) CRYPTO 2011. LNCS, vol. 6841, pp. 505–524. Springer, Heidelberg (2011). doi:10.1007/978-3-642-22792-9_29

9. Brakerski, Z., Gentry, C., Vaikuntanathan, V.: (Leveled) Fully homomorphic encryption without bootstrapping. In: Proceedings of the 3rd Innovations in Theoretical Computer Science Conference, pp. 309–325. ACM (2012)

10. Lauter, K., Naehrig, M., Vaikuntanathan, V.: Can homomorphic encryption be practical? In: ACM Workshop on Cloud Computing Security Workshop, CCSW 2011, pp. 113–124. ACM, New York (2011)

11. Pappas, V., Vo, B., Krell, F., Choi, S., Kolesnikov, V., Keromytis, A., Malkin, T.: Blind Seer: a scalable private DBMS. In: 35th IEEE Symposium on Security and Privacy 2014, pp. 359–374. IEEE Computer Society Press (2014)

12. Fisch, B.A., Vo, B., Krell, F., Kumarasubramanian, A., Kolesnikov, V., Malkin, T., Bellovin, S.M.: Malicious-client security in Blind Seer: a scalable private DBMS. In: 36th IEEE Symposium on Security and Privacy, pp. 395–410. IEEE Computer Society Press (2015)

13. Yasuda, M., Shimoyama, T., Kogure, J., Yokoyama, K., Koshiba, T.: Practical packing method in somewhat homomorphic encryption. In: Garcia-Alfaro, J., Lioudakis, G., Cuppens-Boulahia, N., Foley, S., Fitzgerald, W.M. (eds.) DPM/SETOP -2013. LNCS, vol. 8247, pp. 34–50. Springer, Heidelberg (2014). doi:10.1007/978-3-642-54568-9_3

14. Yasuda, M., Shimoyama, T., Kogure, J., Yokoyama, K., Koshiba, T.: Secure statistical analysis using RLWE-based homomorphic encryption. In: Foo, E., Stebila, D. (eds.) ACISP 2015. LNCS, vol. 9144, pp. 471–487. Springer, Cham (2015). doi:10.1007/978-3-319-19962-7_27

15. Yasuda, M., Shimoyama, T., Kogure, J., Yokoyama, K., Koshiba, T.: Secure pattern matching using somewhat homomorphic encryption. In: ACM Workshop on Cloud Computing Security Workshop, CCSW 2013, pp. 65–76. ACM, New York (2013)

16. Castryck, W., Iliashenko, I., Vercauteren, F.: Provably weak instances of ring-LWE revisited. In: Fischlin, M., Coron, J.-S. (eds.) EUROCRYPT 2016. LNCS, vol. 9665, pp. 147–167. Springer, Heidelberg (2016). doi:10.1007/978-3-662-49890-3_6

17. Barker, E.: Recommendation for key management. In: NIST Special Publication 800–57 Part 1 Revision 4. NIST (2016)

18. The PARI Group, PARI/GP version 2.7.5, Bordeaux (2014). http://pari.math.u-bordeaux.fr/

Efficient Oblivious Transfer
from Lossy Threshold Homomorphic Encryption

Isheeta Nargis[✉]

University of Calgary, Calgary, Canada
`inargis@ucalgary.ca`

Abstract. In this article, a new oblivious transfer (OT) protocol, secure in the presence of erasure-free one-sided active adaptive adversaries is presented. The new bit OT protocol achieves better communication complexity than the existing bit OT protocol in this setting. The new bit OT protocol requires fewer number of public key encryption operations than the existing bit OT protocol in this setting. As a building block, a new two-party lossy threshold homomorphic public key cryptosystem is designed. It is secure in the same adversary model. It is of independent interest.

Keywords: Oblivious transfer · Active adversary · One-sided adaptive adversary · Threshold encryption · Lossy encryption · Public key encryption · Homomorphic encryption

1 Introduction

Oblivious transfer (OT) is one of the most critical problems in cryptography since many applications can be designed based on the existence of a secure OT protocol. In *one-sided active adaptive adversary model* for two-party computation, it is assumed that the adversary is active, adaptive and it can corrupt at most one party [13]. This is a relaxation from the standard adaptive adversary model for two-party computation, where the adversary can corrupt both parties. This relaxed model is used to achieve more efficient protocols. Let n denote the security parameter. Garay et al. [12] designed the most efficient OT protocol secure against active adaptive adversaries. For string OT of size q bits, their protocol requires $O(q)$ public key encryption (PKE) operations in the worst case. Here, q is a polynomial of n. Hazay and Patra [13] designed an OT protocol for one-sided active adaptive adversary model. For string OT of size q bits, their protocol requires constant number of PKE operations in the expected case. So, relaxing the notion of security has resulted in a protocol requiring significantly smaller number of PKE operations, in the expected case. In the *erasure-free adaptive adversary model*, it is assumed that the adversary can see all history of a party when it corrupts that party.

Hazay and Patra [13] designed an OT protocol for one-sided active adaptive adversary model. The OT protocol of [13] requires $O(n^2)$ communication

© Springer International Publishing AG 2017
M. Joye and A. Nitaj (Eds.): AFRICACRYPT 2017, LNCS 10239, pp. 165–183, 2017.
DOI: 10.1007/978-3-319-57339-7_10

complexity for bit OT. One research goal is to improve the communication complexity in this setting.

Contribution of this Article. In this article, a new OT protocol secure against erasure-free one-sided active adaptive adversaries is presented. The worst case analysis is used as the measure of efficiency in this article. The new bit OT protocol needs $O(n)$ communication complexity, which is a significant improvement over the $O(n^2)$ communication complexity of the bit OT protocol of [13]. The bit OT protocol of [13] requires $O(n)$ PKE operations in the worst case, and the new bit OT protocol needs a constant number of PKE operations in the worst case. The OT protocol of [13] is secure in the universally composable (UC) framework. The new OT protocol is secure according to the simulation-based security definition of Canetti [5], which satisfies sequential composition.

As a building block, a new two-party lossy threshold homomorphic PKE scheme is designed. This encryption scheme is of independent interest. It can be used in other two-party protocols.

Techniques. Aumann and Lindell [1] designed an OT protocol secure against covert static adversaries. The new OT protocol is designed by converting their OT protocol. It is secure against erasure-free one-sided active adaptive adversaries. The new OT protocol achieves a much stronger notion of security than the OT protocol of [1] in two senses. Firstly, the active adversary model is a stronger security model than the covert adversary model [1]. Secondly, the adaptive adversary model is more secure than the static adversary model [5]. The OT protocol of [1] is modified in two ways. The protocol of [1] uses a traditional homomorphic PKE scheme and the new OT protocol uses a two-party lossy threshold homomorphic PKE scheme. For verification, the protocol of [1] uses cut-and-choose technique and the new OT protocol uses adaptive zero-knowledge arguments.

2   Background

Notation. Let n denote the security parameter. Let $\mathbb{Z}_q = \{0, 1, \ldots, q-1\}$ where q is a prime. Let $\mathbb{Z}_q^* = \{1, 2, \ldots, q-1\}$. For all elements a and $b \neq 1$ in group $\mathbb{G}$, the discrete logarithm of a in base b is denoted by $\log_b(a)$. For a set R, let $r \xleftarrow{\$} R$ denote that r is selected uniformly at random from R. Let A be a probabilistic polynomial-time algorithm. Let $coins(A)$ denote the distribution of the internal randomness of A. $y \leftarrow A(x)$ means that y is computed by running A on input x and randomness r where $r \xleftarrow{\$} coins(A)$. Let $E_{pk}(m, r)$ denote the result of encryption of plaintext m using encryption key pk and randomness r. Let $D_{sk}(c)$ denote the result of decryption of ciphertext c using decryption key sk. Let $Com_\mu(a, r)$ denote the commitment of secret a using commitment key μ and randomness r.

The DDH Assumption. The decisional Diffie-Hellmann (DDH) assumption for cyclic group $\mathbb{G}$ of order prime q says that, for random generator $g \in \mathbb{G}^*$

($\mathbb{G}^*$ denotes the generators of $\mathbb{G}$), the tuples (g, g^a, g^b, g^{ab}) and (g, g^a, g^b, g^c) are computationally indistinguishable, where $a, b, c \xleftarrow{\$} \mathbb{Z}_q$.

Trapdoor Commitment Scheme. A *trapdoor commitment scheme* is a commitment scheme such that a trapdoor is generated during the key generation. With the trapdoor, one can efficiently compute a randomness to open a given commitment to any value of choice. Without the trapdoor, the binding property of the commitment scheme holds. Pedersen [18] designed a trapdoor commitment scheme based on the DDH assumption. In Pedersen's commitment scheme, the commitment key is $\mu = g^\delta$, and δ is the trapdoor.

Adaptive Zero-Knowledge Arguments. For definition of zero-knowledge argument, see [15]. An adaptive zero-knowledge argument is a zero-knowledge argument secure against adaptive adversaries. For definition of non-erasure Σ-protocol, see [8,17].

Additive Homomorphic PKE Scheme. In an *additive homomorphic PKE scheme*, one can efficiently compute an encryption c of $(m_1 + m_2)$ from ciphertexts c_1 and c_2 encrypting plaintexts m_1 and m_2, respectively. This is called *homomorphic addition* and denoted by $c = c_1 +_h m_2$. In an additive homomorphic PKE scheme, one can also efficiently compute an encryption c_2 of $(m_1 \times m_2)$ from an encryption c_1 of m_1 and the plaintext m_2. This is called *homomorphic multiplication by constant*, and denoted by $c_2 = m_2 \times_h c_1$.

Randomizable PKE Scheme. In a *randomizable PKE scheme*, there exists a probabilistic polynomial-time algorithm *Blind*, which, on input public key pk and an encryption c of plaintext m, produces another encryption c_1 of plaintext m such that c_1 is distributed identically to $E_{pk}(m,r)$ where $r \xleftarrow{\$} Coins(E)$.

Lossy Encryption Scheme

Definition 1 *(Lossy PKE Scheme [3]). A **lossy PKE scheme** is a tuple (G, E, D) of probabilistic polynomial time algorithms such that keys generated by $G(1^k, 1)$ and $G(1^k, 0)$ are called injective keys and lossy keys, respectively. The algorithms must satisfy the following properties.*

1. **Correctness on Injective Keys:** *For all plaintexts m,*

$$Pr\left[(pk, sk) \leftarrow G(1^k, 1); r \xleftarrow{\$} coins(E) : D_{sk}\big(E_{pk}(m, r)\big) = m\right] \overset{s}{=} 1.$$

2. **Indistinguishability of Keys:** *The lossy public keys are computationally indistinguishable from the injective public keys. If $proj : (pk, sk) \rightarrow pk$ is the projection map, then $\{proj(KG(1^k, 1))\} \overset{c}{=} \{proj(KG(1^k, 0))\}$.*
3. **Lossiness on Lossy Keys:** *If $(pk_\ell, sk_\ell) \leftarrow G(1^k, 0)$, then, for all m_0, m_1, the distributions $E_{pk_\ell}(m_0, R)$ and $E_{pk_\ell}(m_1, R)$ are statistically indistinguishable.*

4. **Openability:** If $(pk_\ell, sk_\ell) \leftarrow G(1^k, 0)$ and $r_0 \overset{\$}{\leftarrow} coins(E)$, then, for all m_0, m_1, with overwhelming probability, there exists $r_1 \in coins(E)$ such that $E_{pk_\ell}(m_0, r_0) = E_{pk_\ell}(m_1, r_1)$. That is, there exists a (possibly inefficient) algorithm Opener that can open a lossy ciphertext to any arbitrary plaintext with all but negligible probability.

The semantic security of a lossy encryption scheme is implied by definition [3].

Security Model. The security of the new protocols are proved following the simulation based security definition by Canetti [5].

3   Definition of Two-Party Lossy Threshold PKE Scheme

A definition of two-party lossy threshold PKE scheme secure against one-sided active adaptive adversaries is presented below.

Definition 2 (Lossy Threshold PKE Scheme Secure against Erasure-Free One-Sided Active Adaptive Adversaries). A **lossy threshold PKE scheme secure against erasure-free one-sided active adaptive adversaries** for the set of parties $P = \{P_1, P_2\}$, and security parameter n, is a 4-tuple (K, KG, E, Π_{DEC}) having the following properties.

Key Space: The key space K is a family of finite sets (pk, sk_1, sk_2). pk is the public key and sk_i is the secret key share of P_i. Let M_{pk} denote the message space for public key pk.

Key Generation: There exists a probabilistic polynomial-time key generation algorithm KG, which, on input $(1^n, mode)$, generates public output pk and a list $\{vk, vk_1, vk_2\}$ of verification keys, and secret output sk_i for P_i, where $(pk, sk_1, sk_2) \in K$. By setting mode to zero and one, key in lossy mode and injective mode can be generated, respectively. vk is called the verification key, vk_i is called the verification key of P_i.

Encryption: There exists a probabilistic polynomial-time encryption algorithm E, which, on input pk, $m \in M_{pk}$, $r \overset{\$}{\leftarrow} coins(E)$, outputs an encryption $c = E_{pk}(m, r)$ of m.

Decryption: There exists a two-party decryption protocol Π_{DEC} secure against erasure-free one-sided active adaptive adversaries. On common public input (c, pk, vk, vk_1, vk_2), and secret input sk_i for each $P_i, i \in \{1, 2\}$, where sk_i is the secret key share of P_i for the public key pk (as generated by KG), and c is an encrypted message, Π_{DEC} returns a message m, or the symbol $\perp$ denoting a decryption failure, as a common public output.

Lossy Encryption Properties: The encryption scheme is a lossy encryption scheme according to Definition 1.

Threshold Semantic Security: Consider the following game G for an erasure-free one-sided active adaptive adversary $\mathcal{A}$.

G1. $\mathcal{A}$ may corrupt at most one party. If $\mathcal{A}$ corrupts P_i, then $\mathcal{A}$ learns the history of P_i.

G2. *The challenger executes algorithm KG. The challenger broadcasts the public key and the verification keys. For each $i \in \{1, 2\}$, the challenger sends sk_i to P_i. If there is a corrupted party P_i, then $\mathcal{A}$ learns sk_i.*

G3. *$\mathcal{A}$ adaptively makes the following types of queries.*

   1. Corruption query

      $\mathcal{A}$ may corrupt a party, if no party was corrupted before. If $\mathcal{A}$ corrupts P_i, then $\mathcal{A}$ learns sk_i and the history of P_i.

   2. Decryption query

      $\mathcal{A}$ selects a message $m \in M_{pk}$, and sends it to the challenger. The challenger sends $\mathcal{A}$ the decryption shares and the validity proofs of P_1 and P_2, for an encryption of m.

      $\mathcal{A}$ repeats step G3 as many times as it wishes.

G4. *$\mathcal{A}$ selects two message m_0 and m_1 from M_{pk}, and sends them to the challenger. The challenger randomly selects a bit b, and sends an encryption c of m_b, to $\mathcal{A}$.*

G5. *$\mathcal{A}$ repeats step G3 as many times as it wishes. $\mathcal{A}$ cannot request message m_0 or m_1 in step G3(2).*

G6. *$\mathcal{A}$ outputs a guess bit b_1.*

*A threshold encryption scheme is said to be **semantically secure against erasure-free one-sided active adaptive adversaries** if, for any probabilistic polynomial-time erasure-free one-sided active adaptive adversary, $b = b_1$ with probability only negligibly greater than $\frac{1}{2}$.*

The verification keys are used for validity proofs in Π_{DEC}. During Π_{DEC}, each party P_i uses *validity proof* such that P_i can convince the remaining party that P_i performed its calculation in Π_{DEC} correctly, without disclosing its secret. Note that $\mathcal{A}$ can only request for ciphertexts for which it knows the plaintext. It is not like the chosen ciphertext attack (CCA) where the adversary can ask for decryption shares for any chosen ciphertext. Step G3(2) is used in game G to denote that, despite learning all the decryption shares and validity proofs for several chosen plaintexts, the adversary still does not gain any advantage in guessing the plaintext from the ciphertext.

Let $\mathcal{F}_{KG}$ be the ideal functionality for the key generation. In a two-party lossy threshold encryption scheme, there may exist a two-party distributed key generation (DKG) protocol that computes $\mathcal{F}_{KG}$ securely against erasure-free one-sided active adaptive adversaries.

4   A New Two-Party Lossy Threshold Homomorphic Encryption Scheme

In this section, a new two-party lossy threshold homomorphic public key encryption scheme $ELTA2E = (K, KG, E, \Pi_{DEC})$ is presented. The name $ELTA2E$ denotes an encryption scheme that is lossy, threshold, secure against adaptive adversaries, for two parties and based on the ElGamal encryption scheme.

$ELTA2E$ is based on the DDH assumption. All protocols of $ELTA2E$ work in the CRS model.[1]

Bellare and Yilek [4] designed a non-threshold lossy encryption scheme with efficient *Opener* algorithm, based on the DDH assumption. Let $EncLossy$ denote their encryption scheme. $ELTA2E$ is created by adding the threshold properties to $EncLossy$.

One possible group $\mathbb{G}$ for $ELTA2E$ is as follows. *Safe primes* are primes of the form $p = 2q + 1$ where q is also a prime. On input n, using known methods to generate safe primes, an n-bit safe prime p is generated, with generator g_0 of $\mathbb{Z}_p^*$. There is exactly one subgroup $\mathbb{G}$ of $\mathbb{Z}_p^*$ of order q. Let g be the generator of $\mathbb{G}$. $g = g_0^{\frac{p-1}{q}} = (g_0)^2$. (p, q, g) is the description of group $\mathbb{G}$. Unless otherwise specified, all computations are performed in group $\mathbb{G}$. Pedersen commitment scheme [18] is used as the trapdoor commitment scheme in $ELTA2E$.

Key Generation. Let the input be $(1^n, mode)$. Select $\alpha_1, \alpha_2 \xleftarrow{\$} \mathbb{Z}_q$. Set $\alpha = (\alpha_1 + \alpha_2) \bmod q, h = g^\alpha, h_1 = g^{\alpha_1}, h_2 = g^{\alpha_2}$. Select $\gamma \xleftarrow{\$} \mathbb{Z}_q$. Set $j = g^\gamma$. If $mode = 1$, then set $\ell = g^{\gamma\alpha}$. If $mode = 0$, then select $\rho \xleftarrow{\$} \mathbb{Z}_q \setminus \{\alpha\}$, and set $\ell = g^{\gamma\rho}$. The public key is $pk = (q, g, j, h, \ell)$. The secret key shares are $sk_1 = \alpha_1, sk_2 = \alpha_2$. The verification keys are $vk = g, vk_1 = h_1, vk_2 = h_2$.

Encryption. The encryption algorithm E works as follows. Let the plaintext be $m \in \{0, 1\}$. Select randomness $r = (s, t) \xleftarrow{\$} \mathbb{Z}_q \times \mathbb{Z}_q$. Compute $y = g^s j^t$, and $z = h^s \ell^t g^m$. Return the ciphertext $c = (y, z)$.

Protocol for Threshold Decryption. The threshold decryption protocol Π_{DEC} works as follows. P_1 sends $ds_1 = y^{(sk_1)}$. Adaptive zero-knowledge argument for equality of discrete logarithm is used as the validity proof in Π_{DEC}. P_1 proves that $\log_y(ds_1) = \log_{vk}(vk_1)$. If P_1 fails, then P_2 aborts. P_2 sends $ds_2 = y^{(sk_2)}$. P_2 proves that $\log_y(ds_2) = \log_{vk}(vk_2)$. If P_2 fails, then P_1 aborts. P_1 and P_2 compute $w = \frac{z}{ds_1 \cdot ds_2}$. From w, P_1 and P_2 compute m where $m \in \{0, 1\}$, and $g^m = w$ in $\mathbb{G}$. If there is no such value m, then P_1 and P_2 output $\perp$, denoting decryption failure.

It is also possible to perform *private threshold decryption* to just one party P_k. In that case, P_{2-k} sends $ds_{2-k} = y^{(sk_{2-k})}$, and proves as above. P_k computes ds_k, then computes the output similarly.

Distributed Key Generation Protocol. $ELTA2E$ has a DKG protocol Π_{DKG}. The protocol Π_{DKG} is presented below.

Protocol Π_{DKG}.

CRS: $\mu \in \mathbb{Z}_p$.
Group description: (p, q, g).
Input: $(1^n, mode)$.

---

[1] In the common reference string (CRS) model, it is assumed that all parties have access to a common string that is selected from some specified distribution.

1. P_1 selects $\alpha_1, \gamma_1, \beta_1, \theta_1 \xleftarrow{\$} \mathbb{Z}_q$. P_1 sets $sk_1 = \alpha_1$. P_1 computes $h_1 = g^{\alpha_1}, j_1 = g^{\gamma_1}$. P_1 computes commitments $b_1 = Com_\mu(h_1, \beta_1), c_1 = Com_\mu(j_1, \theta_1)$. P_1 sends (b_1, c_1).

2. P_1 proves the knowledge of committed secret for commitments b_1 and c_1. If P_1 fails in any proof, then P_2 aborts.

3. P_2 selects $\alpha_2, \gamma_2 \xleftarrow{\$} \mathbb{Z}_q$. P_2 sets $sk_2 = \alpha_2$. P_2 computes $h_2 = g^{\alpha_2}, j_2 = g^{\gamma_2}$. P_2 sends (h_2, j_2).

4. P_2 proves knowledge of discrete logarithm of h_2 and j_2. If P_2 fails in any proof, then P_1 aborts.

5. P_1 sends the openings (h_1, β_1) and (j_1, θ_1) of its commitments.

6. P_2 verifies that $b_1 = Com_\mu(h_1, \beta_1)$, and $c_1 = Com_\mu(j_1, \theta_1)$. If any of these two equalities does not hold, then P_2 aborts.

7. P_1 and P_2 set $vk = g, vk_1 = h_1, vk_2 = h_2, h = h_1 h_2, j = j_1 j_2$.

8. If $mode = 0$, then P_1 selects $\tau_1 \xleftarrow{\$} \mathbb{Z}_q \setminus \{\alpha_1\}$, and sets $\ell_1 = j^{\tau_1}$. If $mode = 1$, then P_1 sets $\ell_1 = j^{\alpha_1}$. P_1 sends ℓ_1.

9. If $mode = 1$, then P_1 proves that $\log_j(\ell_1) \neq \log_{vk}(vk_1)$. If $mode = 1$, then P_1 proves that $\log_j(\ell_1) = \log_{vk}(vk_1)$. If P_1 fails, then P_2 aborts.

10. P_2 sends $\ell_2 = j^{\alpha_2}$.

11. P_2 proves that $\log_j(\ell_2) = \log_{vk}(vk_2)$. If P_2 fails, then P_1 aborts.

12. P_1 and P_2 set $\ell = \ell_1 \ell_2, pk = (q, g, j, h, \ell)$.

13. P_1 outputs $(pk, sk_1, (vk, vk_1, vk_2))$.

14. P_2 outputs $(pk, sk_2, (vk, vk_1, vk_2))$.

The proofs in steps 2,4,9, and 11 are performed using adaptive zero-knowledge arguments. The CRS μ is used as the commitment key for Pedersen commitment scheme. The CRS μ also acts as the CRS for the zero-knowledge arguments. The reason for using commitments in Π_{DKG} is to ensure that no party can affect the distribution of the generated key.

Lemma 1. *ELTA2E is additive homomorphic.*

Proof. **Homomorphic Addition.** Let $c_1 = (g^{s_1} j^{t_1}, h^{s_1} \ell^{t_1} g^{m_1})$, and $c_2 = (g^{s_2} j^{t_2}, h^{s_2} \ell^{t_2} g^{m_2})$ be two ciphertexts encrypting plaintexts m_1 and m_2, respectively. $c = c_1 +_h c_2 = (g^{s_1} j^{t_1} \cdot g^{s_2} j^{t_2}, h^{s_1} \ell^{t_1} g^{m_1} \cdot h^{s_2} \ell^{t_2} g^{m_2}) = (g^{s_1+s_2} j^{t_1+t_2}, h^{s_1+s_2} \ell^{t_1+t_2} g^{m_1+m_2})$.

Homomorphic Multiplication by Constant. Let $c_1 = (y_1, z_1) = (g^{s_1} j^{t_1}, h^{s_1} \ell^{t_1} g^{m_1})$ be a ciphertext encrypting plaintext m_1. Let m_2 be a known plaintext. $c_2 = c_1 \times_h m_2 = ((g^{s_1} j^{t_1})^{m_2}, (h^{s_1} \ell^{t_1} g^{m_1})^{m_2}) = (g^{s_1 m_2} j^{t_1 m_2}, h^{s_1 m_2} \ell^{t_1 m_2} g^{m_1 m_2})$.

ELTA2E **is Randomizable.** Let $c = (y, z) = (g^s j^t, h^s \ell^t g^m)$ be a ciphertext encrypting plaintext m. The *Blind* function on input $(pk, c) = ((q, g, j, h, -\ell), (y, z))$ works as follows. Select $s_1, t_1 \xleftarrow{\$} \mathbb{Z}_q \times \mathbb{Z}_q$. Set $y_1 = y \cdot g^{s_1} j^{t_1}, z_1 = z \cdot h^{s_1} \ell^{t_1}$. Return $c_1 = (y_1, z_1)$.

5   Security of the DKG Protocol Π_{DKG}

Canetti et al. [6] introduced the *single inconsistent party* (SIP) technique. At the start of simulation, the identity of the single inconsistent party (SIP) is generated uniformly at random. The view of any party except the SIP in the simulation is computationally indistinguishable from its view in the real world. The view of the adversary is independent from the choice of SIP. This technique is used in the security proofs of $ELTA2E$. Let $\mathcal{A}$ be a one-sided active adaptive adversary and $\mathcal{Z}$ be the environment. Let $\mathcal{S}_{DKG}$ be the simulator for Π_{DKG} for adversary $\mathcal{A}$ and environment $\mathcal{Z}$. At start, $\mathcal{S}_{DKG}$ selects $I \xleftarrow{\$} \{P_1, P_2\}$ where I denotes the identity of the SIP. If $\mathcal{A}$ corrupts I, then $\mathcal{S}_{DKG}$ rewinds to the start of simulation, generates a new $I \xleftarrow{\$} \{P_1, P_2\}$, and proceeds again. $\mathcal{A}$ corrupts at most one party, so the probability of a randomly selected party I being corrupted is at most $\frac{1}{2}$. The expected number of rewinds of $\mathcal{S}_{DKG}$ is at most two, and the simulation can be performed in expected polynomial time. To bound the running time of $\mathcal{S}_{DKG}$ to strictly polynomial time, simulation can continue running up to n^{ℓ_1} steps where ℓ_1 is a constant. If $\mathcal{S}_{DKG}$ does not halt within n^{ℓ_1} steps, then $\mathcal{S}_{DKG}$ fails. The probability of failure of $\mathcal{S}_{DKG}$ is negligible.

Theorem 1. *Provided that the DDH assumption holds, and trapdoor commitment scheme and adaptive zero-knowledge arguments exist, protocol Π_{DKG} computes $\mathcal{F}_{KG}$ securely against erasure-free one-sided active adaptive adversaries.*

Proof (Sketch). The main idea of the proof is given here. The full proof is available in the full version [16]. At start, $\mathcal{S}_{DKG}$ selects $\delta \xleftarrow{\$} \mathbb{Z}_q$, and sets the CRS to $\mu = g^\delta$. Then, $\mathcal{S}_{DKG}$ knows the trapdoor δ of the commitment key μ. The simulator $\mathcal{S}_{DKG}$ for the case where P_1 is the SIP, works as follows. If $\mathcal{A}$ corrupts P_2 after any step, then $\mathcal{S}_{DKG}$ corrupts P_2 in the ideal world. If P_2 fails in some proof, then P_1 aborts in the real world. In the ideal world, $\mathcal{S}_{DKG}$ sends $abort_{P_2}$ to the trusted party and halts. The trusted party sends $abort_{P_2}$ to P_1, and P_1 halts. If P_2 does not fail in any proof, then the following things happen. In step 1, $\mathcal{S}_{DKG}$ selects $\alpha_2, \gamma_2 \xleftarrow{\$} \mathbb{Z}_q$, computes $h_2 = g^{\alpha_1}, j_2 = g^{\gamma_2}, h_1 = \frac{h}{h_2}, j_1 = \frac{j}{j_2}$. $\mathcal{S}_{DKG}$ selects $\beta_1, \theta_1 \xleftarrow{\$} \mathbb{Z}_q$ and computes $b_1 = Com_\mu(h_1, \beta_1), c_1 = Com_\mu(j_1, \theta_1)$. By the hiding property of the commitment scheme, the distribution of (b_1, c_1) in two worlds are identical. $\mathcal{S}_{DKG}$ honestly performs step 2. In step 3, if P_2 is honest, then $\mathcal{S}_{DKG}$ uses h_2, j_2 computed in step 1. In step 4 and 11, if P_2 is corrupted, $\mathcal{S}_{DKG}$ acts as an honest verifier. If P_2 passes the proofs, then $\mathcal{S}_{DKG}$ extracts α_2 and γ_2 using the knowledge extractor of the zero-knowledge argument, in step 4. If P_2 is honest, then $\mathcal{S}_{DKG}$ acts as an honest prover in step 4 and 11. In step 5, $\mathcal{S}_{DKG}$ computes $\hat{h}_1 = \frac{h}{h_2}, \hat{j}_1 = \frac{j}{j_2}$. Using the trapdoor δ of the commitment key μ, $\mathcal{S}_{DKG}$ computes $\hat{\beta}_1, \hat{\theta}_1$ such that $b_1 = Com_\mu(\hat{h}_1, \hat{\beta}_1)$, and $c_1 = Com_\mu(\hat{j}_1, \hat{\theta}_1)$. $\mathcal{S}_{DKG}$ uses $(\hat{h}_1, \hat{\beta}_1), (\hat{j}_1, \hat{\theta}_1)$ as the message from P_1. If $\mathcal{A}$ corrupts P_2 before step 3, then, corrupted P_2 sends h_2 and j_2 in step 3. The value of h and j are fixed since they are part of the input of $\mathcal{S}_{DKG}$. $\mathcal{A}$ sees that the openings of the commitments are consistent, and $h = \hat{h}_1 h_2$ and $j = \hat{j}_1 j_2$, as required. $(\hat{h}_1, \hat{j}_1)$ is identically distributed to (h_1, j_1). If P_2 is honest up to step 3, then $\hat{h}_1 = \frac{h}{h_2} = h_1, \hat{\beta}_1 = \beta_1, \hat{j}_1 = j_1, \hat{\theta}_1 = \theta_1$. In step 6, P_1 passes the verification tests in the ideal world. In step 7, $\mathcal{S}_{DKG}$ sets $vk = g$, $vk_1 = \hat{h}_1, vk_2 = h_2$, and uses (h, j) of the input. As argued earlier, these values in two worlds are identically distributed. In step 8, $\mathcal{S}_{DKG}$ computes $\hat{\ell}_1 = \frac{\ell}{j^{(\alpha_2)}} > \ell$ is fixed in both worlds. The distribution of α_2 in two worlds are identical. Then, $\hat{\ell}_1$ and ℓ_1 are identically distributed. In step 9, $\mathcal{S}_{DKG}$ generates a proof transcript using trapdoor δ. By definition of zero-knowledge argument, the proof transcript in two worlds are computationally indistinguishable. If P_2 is honest, then $\mathcal{S}_{DKG}$ honestly performs steps 10 and 11. In step 12, $\mathcal{S}_{DKG}$ uses (ℓ, pk) of input. In step 13, the output of honest P_1 is $(pk, vk, vk_1, vk_2, \alpha_1)$. Then, the output of the honest P_1 in two worlds are identically distributed. In step 14, if P_2 is honest, then $\mathcal{S}_{DKG}$ sets $(pk, vk, vk_1, vk_2, \alpha_2)$ as the output of P_2. The distribution of sk_2 in two worlds are identical. The simulator for the case where P_2 is the SIP is similar, so it is not given separately.

6 Security of Encryption Scheme $ELTA2E$

Lemma 2. *If the decisional Diffie-Hellman assumption holds, then $ELTA2E$ is a lossy encryption scheme. $ELTA2E$ has an efficient (polynomial-time) Opener algorithm.*

Proof. **Correctness of Decryption in the Injective Mode.** In the injective mode, $pk = (q, g, j, h, \ell) = (q, g, g^\gamma, g^\alpha, g^{\alpha \cdot \gamma})$. Then, $w = \frac{z}{ds_1 \cdot ds_2} = \frac{z}{y^{sk_1} \cdot y^{sk_2}} = \frac{z}{y^{\alpha_1 + \alpha_2}} = \frac{z}{y^\alpha} = \frac{h^s \ell^t g^m}{(g^s j^t)^\alpha} = \frac{(g^\alpha)^s (g^{\gamma\alpha})^t g^m}{\left(g^s (g^\gamma)^t\right)^\alpha} = \frac{g^{\alpha s + \alpha\gamma t + m}}{g^{\alpha s + \alpha\gamma t}} = g^m$.

Indistinguishability of Keys. In the injective mode, $pk = (q, g, j, h, \ell) = (q, g, g^\gamma, g^\alpha, g^{\gamma\alpha})$. In the lossy mode, $pk = (q, g, j, h, \ell) = (q, g, g^\gamma, g^\alpha, g^{\gamma\rho})$. By the DDH assumption, the public key in injective mode is computationally indistinguishable from the public key in lossy mode.

Lossiness on Lossy Keys. Let $pk = (q, g, j, h, \ell) = (q, g, g^\gamma, g^\alpha, g^{\gamma\rho})$ be a lossy public key. Encryption of a message m with randomness (s, t) is $c = (y, z) = (g^{s+\gamma t}, g^{\alpha s + \gamma\rho t} \cdot g^m)$. Since $\rho \neq \alpha$, $(s+\gamma t)$ and $(\alpha s + \gamma\rho t)$ are linearly independent combinations of s and t. Then, y and z are uniformly random group elements.

Efficient $Opener$ Algorithm. Let $pk = (q, g, j, h, \ell) = (q, g, g^\gamma, g^\alpha, g^{\gamma\rho})$ be a lossy public key. Let the corresponding secret key be $sk = (\gamma, \rho, \alpha)$. Let $c = (y, z)$ be an encryption of plaintext m with randomness $r = (s, t)$. Then, $c = (y, z) = (g^{s+\gamma t}, g^{\alpha s + \gamma\rho t} \cdot g^m)$. Let m_1 be the plaintext with which the ciphertext c has to be opened. On input $(pk, sk, (y, z), m, (s, t), m_1)$, the algorithm $Opener$ has to find randomness $r_1 = (s_1, t_1) \in \mathbb{Z}_q \times \mathbb{Z}_q$ such that $s + \gamma t = s_1 + \gamma t_1$, and $\alpha s + \gamma\rho t + m = \alpha s_1 + \gamma\rho t_1 + m_1$. These two equations are two linear equations on the variables (s_1, t_1). The $Opener$ algorithm solves these two equations to find s_1 and t_1 in polynomial time.

Lemma 3. *Provided that the decisional Diffie-Hellman assumption holds, and trapdoor commitment scheme and adaptive zero-knowledge arguments exist, the encryption scheme ELTA2E achieves threshold semantic security in the presence of erasure-free one-sided active adaptive adversaries.*

Proof (Sketch). The threshold semantic security is proved by reduction, following the idea in [11]. The lossy encryption properties of $EncLossy$ are proved in [4]. Since any lossy PKE scheme is semantically secure [3], $EncLossy$ is semantically secure. Assume that there exists a probabilistic polynomial-time one-sided active adaptive adversary $\mathcal{A}_1$ that can break the semantic security of the two-party lossy threshold encryption scheme $ELTA2E$. It is described how to construct a probabilistic polynomial-time one-sided active adaptive adversary $\mathcal{A}_2$, using $\mathcal{A}_1$, that can break the semantic security of the non-threshold lossy encryption scheme $EncLossy$. As $EncLossy$ is semantically secure, a contradiction is reached. To convert $\mathcal{A}_1$ to $\mathcal{A}_2$, it is necessary to simulate the extra information that are not available in the non-threshold lossy cryptosystem. The simulator is designed using the SIP technique. The inputs of the simulator are the public key $pk = (q, g, j, h, \ell)$, the mode parameter $mode$, and the identity I of the single inconsistent party. In step G1, if $\mathcal{A}_1$ corrupts a party P_i, then $\mathcal{A}_2$ corrupts P_i. $\mathcal{A}_2$ receives the history of P_i from $\mathcal{Z}$. In step G2, if P_1 is the SIP, $\mathcal{A}_2$ simulates as follows. $\mathcal{A}_2$ selects $\alpha_1, \alpha_2 \overset{\$}{\leftarrow} \mathbb{Z}_q$. $\mathcal{A}_2$ sets $sk_1 = \alpha_1, sk_2 = \alpha_2, vk = g, vk_2 = g^{\alpha_2}, vk_1 = \frac{h}{vk_2}$. $\mathcal{A}_2$ sends $((pk, vk, vk_1, vk_2, sk_1), (pk, vk, vk_1, vk_2, sk_2))$ to $\mathcal{A}_1$ in step G2. The distribution of sk_1, sk_2 are identical in two worlds. $\mathcal{A}_2$ sets $vk_1 = \frac{h}{vk_2}$. The value of h is fixed and the distribution of vk_2 in two worlds are identical. Therefore, the distribution of vk_1 in two worlds are identical. Here $h = vk_1 \cdot vk_2$, so it is consistent. If the adversary corrupts P_2, then it sees that $vk_2 = g^{sk_2}$ so everything is consistent for P_2. When P_2 is the SIP, the simulation is similar and not given separately. In step G3(1), if $\mathcal{A}_1$ corrupts a party P_i, then $\mathcal{A}_2$ corrupts P_i. $\mathcal{A}_2$ receives sk_i and the history of P_i from $\mathcal{Z}$. In step G3(2), $\mathcal{A}_1$ selects $m \in M_{pk}$ and sends m to $\mathcal{A}_2$. $\mathcal{A}_2$ computes $c_m = (y_m, z_m) = (g^s j^t, h^s \ell^t g^m)$. c_m is a valid encryption of m. If P_1 is the SIP, then $\mathcal{A}_2$ simulates the steps of protocol Π_{DEC} as follows. In step 1, $\mathcal{A}_2$ computes $ds_1 = (y_m)^{sk_1}$ where sk_1 is the secret key share of P_1 computed by $\mathcal{A}_2$ in step G2. As argued in step G2, the distribution of sk_1 in two worlds are identical. Then, the distribution of ds_1 in two worlds are identical. In step 2, $\mathcal{A}_2$ acts as an honest prover. In step 3, if P_2 is honest, then $\mathcal{A}_2$ computes $ds_2 = (y_m)^{sk_2}$ where sk_2 is the secret key share of P_2 computed by $\mathcal{A}_2$ in step G2. Proof argument is similar to step 1. In step 4, if P_2 is honest, then $\mathcal{A}_2$ acts as an honest prover. If P_2 is corrupted, then $\mathcal{A}_2$ acts as an honest verifier. If P_2 fails, then $\mathcal{A}_2$ sends $abort_{P_2}$ to the trusted party and halt. Then, the trusted party sends $abort_{P_2}$ to P_1 and honest P_1 halts. Honest P_1 aborts in the real world. In step 5, $\mathcal{A}_2$ computes $w = g^m$. The value of w is identical in two worlds. In step 6, $\mathcal{A}_2$ uses m. The simulation of step G3(2) when P_2 is the SIP is similar. So, it is not given separately. In step G4, $\mathcal{A}_1$ chooses two plaintexts $m_0, m_1 \in M_{pk}$ and sends them to $\mathcal{A}_2$. $\mathcal{A}_2$ sends (m_0, m_1) to the challenger of $EncLossy$. Then, the challenger of $EncLossy$ selects $b \overset{\$}{\leftarrow} \{0, 1\}$, computes an encryption c of m_b and returns c to $\mathcal{A}_2$. $\mathcal{A}_2$ sends c to $\mathcal{A}_1$. Step G5 is similar to step G3. In step G6, $\mathcal{A}_1$ returns a guess b_1. $\mathcal{A}_2$ returns b_1.

Theorem 2. *Provided that the DDH assumption holds, and trapdoor commitment scheme and adaptive zero-knowledge arguments exist, the encryption scheme $ELTA2E$ is a two-party lossy threshold encryption scheme secure against erasure-free one-sided active adaptive adversaries.*

Proof. By Lemma 2, $ELTA2E$ satisfies the lossy encryption properties. By Lemma 3, $ELTA2E$ satisfies the threshold semantic security requirement given in Definition 2. Then, $ELTA2E$ is a two-party lossy threshold encryption scheme secure against erasure-free one-sided active adaptive adversaries.

7 Oblivious Transfer Against One-Sided Active Adaptive Adversaries

In this section, a new protocol Π_{OTAA} for bit OT is presented. Let $\mathcal{F}_{OT}$ denote the ideal functionality for OT. Let $\mathcal{F}_{zk}$ denote the ideal functionality for adaptive zero-knowledge argument. The protocol Π_{OTAA} is presented below.

Protocol Π_{OTAA}.

CRS: $\mu \overset{\$}{\leftarrow} \mathbb{Z}_p$.
Input of $S : (x_0, x_1) \in \{0,1\}^2$.
Input of $R : \sigma \in \{0,1\}$.
Auxiliary Input: (n, p, q, g) where n is the security parameter, and (p, q, g) is the representation of a group $\mathbb{G}$ for the encryption scheme $ELTA2E = (K, KG, E, \Pi_{DEC})$.

1. S and R jointly generate an injective key for $ELTA2E$, by executing $\mathcal{F}_{KG}$ with input $(1^n, 1)$. Here, S and R acts as P_1 and P_2, respectively. Both parties get the public key $pk = (q, g, j, h, \ell)$, and the verification keys (vk, vk_1, vk_2). S gets its secret key share sk_1 and R gets its secret key share sk_2.
2. R selects $s_0, t_0, s_1, t_1 \overset{\$}{\leftarrow} \mathbb{Z}_q$. R computes $c_0 = E_{pk}(1 - \sigma, (s_0, t_0))$, and $c_1 = E_{pk}(\sigma, (s_1, t_1))$. R sends (c_0, c_1).
3. R proves that one of (c_0, c_1) is an encryption of zero, without disclosing which one. If R fails, then S aborts.
4. For each $i \in \{0,1\}$, S and R perform the following steps.
   (a) S computes $d_i = x_i \times_h c_i$. S computes $v_i = Blind(pk, d_i)$. S sends v_i.
   (b) S proves correctness of homomorphic multiplication by constant. If S fails, then R aborts.
5. For each $i \in \{0,1\}$, S and R jointly perform private decryption of v_i to R, as follows.
   (a) Let $v_i = (y_i, z_i)$. S sends $ds_{1,i} = (y_i)^{(sk_1)}$.
   (b) S proves that $\log_{(y_i)}(ds_{1,i}) = \log_{(vk)}(vk_1)$. If S fails, then R aborts.
   (c) R performs the following steps.

  i. R computes $ds_{2,i} = (y_i)^{(sk_2)}$.
  ii. R computes $\theta_i = \frac{z_i}{ds_{1,i} \cdot ds_{2,i}}$.
  iii. From θ_i, R computes w_i where $w_i \in \{0,1\}$ and $g^{w_i} = \theta_i$ in $\mathbb{G}$.
6. R outputs w_σ.

Π_{OTAA} works in the CRS model in the $(\mathcal{F}_{zk}, \mathcal{F}_{KG})$-hybrid world. One possibility to generate the auxiliary inputs p, q, g is as follows. S generates the description (p, q, g) of the group $\mathbb{G}$ for $ELTA2E$, using Bach's algorithm [2]. S sends (p, q, g) to R. R checks its validity. If the description is invalid, then S and R repeat the same process. The proofs in steps 3, 4(b) and 5(b) are performed by adaptive zero-knowledge arguments. The CRS μ acts as the CRS for functionality $\mathcal{F}_{KG}$ and all the zero-knowledge arguments. In step 3, R proves that one of c_0 and c_1 encrypts zero, without disclosing which one. If R could set both ciphertexts c_0 and c_1 to encryptions of one, then R could learn both x_0 and x_1 at step 5. This proof is incorporated to prevent this type of cheating by R. In step 4, S computes $d_i = x_i \times_h c_i, v_i = Blind(d_i)$. S sends v_i. R knows the ciphertext c_i. The $Blind$ function is included so that new randomness is added to the result d_i. Then, R cannot learn the constant x_i after seeing v_i.

Correctness of Protocol Π_{OTAA}. If S and R both follow the protocol, then the following events occur. S and R generate an injective key for $ELTA2E$. R honestly computes c_0 and c_1. c_σ encrypts one, and $c_{1-\sigma}$ encrypts zero. R passes the proof in step 3. S honestly performs step 4, and passes the proofs. v_σ encrypts x_σ and $v_{1-\sigma}$ encrypts zero. In step 5, S and R honestly perform two private decryption processes. By the "correctness on injective keys" property of $ELTA2E$, $w_\sigma = x_\sigma$ and $w_{1-\sigma} = 0$. Therefore, R learns x_σ.

Extension to String OT. In a string OT, S has a pair of bit strings of length q as input: $(\overline{x_0}, \overline{x_1}) = (\{x_0^1, x_0^2, \ldots, x_0^q\}, \{x_1^1, x_1^2, \ldots, x_1^q\})$. Here q is a polynomial of n. R has input $\sigma \in \{0,1\}$. The output of R is $\overline{x_\sigma} = \{x_\sigma^1, x_\sigma^2, \ldots, x_\sigma^q\}$ and S does not get any output. The bit OT protocol Π_{OTAA} is extended to a string OT protocol as follows. In step 4, for each $i \in \{0,1\}, j \in \{1,2,\ldots,q\}, S$ computes $v_i^j = x_i^j \times_h c_i$. In step 5, for each $i \in \{0,1\}; j \in \{1,2,\ldots,q\}$, S and R jointly perform private decryption of v_i^j to R, so R obtains result w_i^j. R outputs $\{w_\sigma^1, w_\sigma^2, \ldots, w_\sigma^q\}$.

8   Security of Protocol Π_{OTAA}

The following theorem describes the security of protocol Π_{OTAA}.

Theorem 3. *Assume that the DDH assumption holds and there exists a trapdoor commitment scheme and adaptive zero-knowledge arguments. Assume that there exists a two-party lossy threshold public key cryptosystem which is secure against erasure-free one-sided active adaptive adversaries, is additive homomorphic, randomizable, and has an efficient (polynomial-time) Opener algorithm. Then, protocol Π_{OTAA} is a protocol for oblivious transfer secure under sequential composition, in the presence of erasure-free one-sided active adaptive adversaries.*

Proof. Let $\mathcal{A}$ be an erasure-free one-sided active adaptive adversary and $\mathcal{Z}$ be the environment. Let $\mathcal{S}_{OT}$ be the simulator for protocol Π_{OTAA} for adversary $\mathcal{A}$ and environment $\mathcal{Z}$. The security is proved using the SIP technique. The full security proof is available in the full version [16]. The main intuition behind the security is described for two cases below. In both cases, at start, $\mathcal{S}_{OT}$ selects $\delta \xleftarrow{\$} \mathbb{Z}_q$, and sets the CRS to $\mu = g^{\delta}$. $\mathcal{S}_{OT}$ stores δ as the trapdoor of the commitment key μ.

Case 1: Security for the case where S is the SIP

In step 1, $\mathcal{S}_{OT}$ generates a lossy key pair of $ELTA2E$ as follows. $\mathcal{S}_{OT}$ selects $\alpha_1, \alpha_2 \xleftarrow{\$} \mathbb{Z}_q$. $\mathcal{S}_{OT}$ sets $\alpha = (\alpha_1 + \alpha_2) \bmod q, h = g^{\alpha}$. $\mathcal{S}_{OT}$ selects $\gamma \xleftarrow{\$} \mathbb{Z}_q$. $\mathcal{S}_{OT}$ sets $j = g^{\gamma}$. $\mathcal{S}_{OT}$ selects $\rho \xleftarrow{\$} \mathbb{Z}_q \setminus \{\alpha\}$, and sets $\ell = g^{\gamma\rho}$. $\mathcal{S}_{OT}$ sets $pk = (q, g, j, h, \ell)$. $\mathcal{S}_{OT}$ stores the corresponding secret key $sk = (\alpha, \gamma, \rho)$. $\mathcal{S}_{OT}$ generates the lossy key pair in a similar way to the way the key generation algorithm KG of $ELTA2E$ generates a lossy key pair. That means, the distribution of the key pair (pk, sk) is identically distributed to a lossy key pair generated by algorithm KG. The reason for generating the components of the keys, without using algorithm KG is as follows. When $\mathcal{S}_{OT}$ generates the values, it can obtain the values of α, γ and ρ. The secret key $sk = (\alpha, \gamma, \rho)$ is necessary to use the efficient *Opener* algorithm of $ELTA2E$. If protocol Π_{DKG} is used to implement step 1, then $\mathcal{S}_{OT}$ uses the simulator $\mathcal{S}_{DKG}$ of protocol Π_{DKG} on input $(pk, 0, P_1)$. That means $\mathcal{S}_{OT}$ invokes simulator $\mathcal{S}_{DKG}$ on input public key pk, mode parameter set to zero to denote lossy mode, and the identity I of the SIP set to P_1. By Theorem 1, the message that $\mathcal{S}_{DKG}$ generates in the hybrid world is computationally indistinguishable from the message that $\mathcal{A}$ views during the execution of Π_{DKG} in the real world. In the real world, an injective key pair is used. Since $\mathcal{A}$ corrupts at most one party, $\mathcal{A}$ cannot learn the secret key shares of both parties. So, $\mathcal{A}$ cannot learn the secret key. By the "indistinguishability of keys" property of $ELTA2E$, the public key in the hybrid world is computationally indistinguishable from the public key in the real world. If R is honest, then $\mathcal{S}_{OT}$ computes c_0, c_1 based on $\sigma = 0$ in step 2. By threshold semantic security of $ELTA2E$, the distribution of c_0, c_1 in two worlds are computationally indistinguishable. If $\mathcal{A}$ corrupts R after step 2, then $\mathcal{A}$ cannot replace the input σ as the value of σ is already fixed by the message supplied up to step 2. $\mathcal{S}_{OT}$ corrupts R in the hybrid world and receives its input σ from $\mathcal{Z}$. $\mathcal{S}_{OT}$ sends σ to the trusted party of $\mathcal{F}_{OT}$, and receives back its output x_{σ}. If R is corrupted, then $\mathcal{S}_{OT}$ acts as an honest verifier in step 3. If R fails, then $\mathcal{S}_{OT}$ sends $abort_R$ to the trusted party and halts. The trusted party sends $abort_R$ to S and S halts. In this case, honest S aborts in the real world. If R passes, then $\mathcal{S}_{OT}$ extracts the plaintexts of c_0 and c_1 by using the knowledge extractor of the zero-knowledge arguments. From these plaintexts, $\mathcal{S}_{OT}$ learns the possibly modified input σ_1 of corrupted R. $\mathcal{S}_{OT}$ sends σ_1 to the trusted party of $\mathcal{F}_{OT}$, and receives back its output x_{σ_1}. $\mathcal{S}_{OT}$ sets $\sigma = \sigma_1$ and the output of R to x_{σ_1}. In the real world, the generated key pair is injective, so $\mathcal{A}$ cannot open a ciphertext encrypting one to be a ciphertext encrypting zero. In the hybrid world, $\mathcal{S}_{OT}$ generates a lossy key pair. Since $\mathcal{A}$ corrupts at most one party, $\mathcal{A}$ cannot learn the secret key. Without the knowledge of the secret key, $\mathcal{A}$ cannot use the efficient *Opener* algorithm as the efficient *Opener* algorithm requires the secret key as one of its inputs. That means in the hybrid world, $\mathcal{A}$ cannot open a ciphertext encrypting one to be a ciphertext encrypting zero in polynomial time. That means the result of the zero-knowledge argument will be identical in both worlds. If $\sigma = 0$, then $\mathcal{S}_{OT}$ performs no additional updates in step 3, since $\mathcal{S}_{OT}$ calculated c_0, c_1 based on $\sigma = 0$. If $\sigma = 1$, then, in step 3, $\mathcal{S}_{OT}$ computes randomness $\hat{s}_0, \hat{t}_0, \hat{s}_1, \hat{t}_1$ using the efficient *Opener* algorithm, such that $c_0 = E_{pk}(0, (\hat{s}_0, \hat{t}_0))$ and $c_1 = E_{pk}(1, (\hat{s}_1, \hat{t}_1))$. $\mathcal{S}_{OT}$ supplies $\hat{s}_0, \hat{t}_0, \hat{s}_1, \hat{t}_1$ as the randomness for step 2. Since $\mathcal{S}_{OT}$ knows the secret key of the lossy key pair, algorithm *Opener* produces output in polynomial time. By the "openability" property of $ELTA2E$, the generated randomness is consistent. In step 4(a), $\mathcal{S}_{OT}$ selects $\hat{x}_i \xleftarrow{\$} \{0, 1\}$, computes $d_i = \hat{x}_i \times_h c_i, \hat{v}_i = Blind(pk, d_i)$. By threshold semantic security of $ELTA2E$, the distribution of v_i in two worlds are computationally indistinguishable. Correctness of decryption does not hold for a lossy key for $ELTA2E$. So, $\mathcal{S}_{OT}$ sets $w_\sigma = x_\sigma, w_{1-\sigma} = 0$. $\mathcal{S}_{OT}$ computes $\theta_i = g^{w_i}, ds_{2,i} = (vy_i)^{sk_2}, \widehat{ds}_{1,i} = \frac{vz_i}{\theta_i \cdot ds_{1,i}}$. In the real world, $\mathcal{A}$ receives $ds_{1,i} = (y_i)^{sk_1}$. Since S is honest, so $\mathcal{A}$ does not know sk_1. By the DDH assumption, the distribution of $ds_{1,i}$ in two worlds are computationally indistinguishable. The proofs of step 3 and step 4(b) do not work for a lossy key for $ELTA2E$. If R is honest, then $\mathcal{S}_{OT}$ generates a proof transcript for steps 3, 4(b), and 5(b) using the trapdoor δ. By definition of zero-knowledge argument, the proof transcripts in two worlds are computationally indistinguishable. If R is honest, then $\mathcal{S}_{OT}$ honestly performs step 5(c). If R is corrupted, then, in the hybrid world, $\mathcal{A}$ obtains w_i. In the real world, $\mathcal{A}$ obtains w_i due to the "correctness on injective keys" property of $ELTA2E$. If R is corrupted, then $\mathcal{A}$ will obtain the same value x_σ in step 6 in the hybrid world that it obtains in the real world. In an OT protocol, S has no output. So trivially, the output of the honest party S is identical (an empty string) in both worlds. If $\mathcal{A}$ corrupts R after any substep of step 4 or 5, then $\mathcal{S}_{OT}$ performs the same steps if $\mathcal{A}$ corrupts R after step 3.

Case 2: Security for the case where R is the SIP

In step 1, $\mathcal{S}_{OT}$ performs similar to step 1 in case 1. If Π_{DKG} is used to generate the key, then, $\mathcal{S}_{OT}$ uses the simulator $\mathcal{S}_{DKG}$ of protocol Π_{DKG} on input $(pk, 0, P_2)$. $\mathcal{S}_{OT}$ computes c_0, c_1 based on $\sigma = 0$ in step 2. Proof argument is similar to step 2 of case 1. In step 4, if S is honest, then $\mathcal{S}_{OT}$ selects $\hat{x}_i \xleftarrow{\$} \{0, 1\}$, computes $\hat{d}_i = \hat{x}_i \times_h c_i, \hat{v}_i = Blind(\hat{d}_i)$. Proof argument is similar to step 4 of case 1. If S is corrupted, then $\mathcal{S}_{OT}$ acts as an honest verifier in steps 4(b) and 5(b). If S fails in any proof, then $\mathcal{S}_{OT}$ sends $abort_S$ to the trusted party and halts. In this case, honest R aborts in the real world. If S passes, then $\mathcal{S}_{OT}$ extracts the possibly replaced input $\tilde{x}_i$ by using the knowledge extractor of the zero-knowledge argument. If S is honest, then $\mathcal{S}_{OT}$ generates proof transcripts for steps 4(b) and 5(b) using trapdoor δ. By definition of zero-knowledge argument, the proof transcript in two worlds are computationally indistinguishable. If $\mathcal{A}$ corrupts S after step 4, then $\mathcal{S}_{OT}$ corrupts S in the hybrid world, and receives its input (x_0, x_1) from $\mathcal{Z}$. In this case,

$\mathcal{A}$ cannot replace the input (x_0, x_1) as the value of (x_0, x_1) is already fixed by the message sent up to step 4. $\mathcal{S}_{OT}$ sets $(\tilde{x}_0, \tilde{x}_1)$ to (x_0, x_1). If $\hat{x}_i \neq \tilde{x}_i$, then $\mathcal{S}_{OT}$ computes randomness for the ciphertexts transmitted so far and the value of $\tilde{x}_i$, using the efficient *Opener* algorithm. By the "openability" property of $ELTA2E$, the randomness generated is consistent. Since $\mathcal{S}_{OT}$ knows the secret key, the *Opener* algorithm produces output in polynomial-time. $\mathcal{S}_{OT}$ sets $w_\sigma = \tilde{x}_\sigma, w_{1-\sigma} = 0$. If S is honest, then $\mathcal{S}_{OT}$ computes $\theta_i = g^{w_i}, ds_{2,i} = (y_i)^{sk_2}, \widehat{ds}_{1,i} = \frac{z_i}{\theta_i \cdot ds_{2,i}}$. In step 6, $\mathcal{S}_{OT}$ sends $(\tilde{x}_0, \tilde{x}_1)$ to the trusted party of $\mathcal{F}_{OT}$. Let σ be the input of honest R. Then the trusted party sends the output $\tilde{x}_\sigma$ to R. In the real world, honest R outputs the value $\tilde{x}_\sigma$, by the "correctness on injective keys" property of $ELTA2E$. Then, the output of the honest party R is identical in two worlds. If $\mathcal{A}$ corrupts R after any substep of step 3, 4 or 5, then $\mathcal{S}_{OT}$ performs the same steps if $\mathcal{A}$ corrupts R after step 3(a).

9   Efficiency and Comparison with Related Work

Efficiency. In $ELTA2E$, a ciphertext $c \in \mathbb{G} \times \mathbb{G}$. $\mathbb{G}$ is a subgroup of $\mathbb{Z}_p^*$ and p is an n-bit prime. The size of ciphertext is $2n$. The size of Pedersen commitment [18] is n. It is possible to use protocol Π_{DKG} for implementing step 1 of Π_{OTAA}. The communication complexity of Π_{DKG} is $50n$. The total communication complexity of Π_{OTAA} (including the communication complexity of Π_{DKG}) is $101n \in O(n)$. In step 2, R performs two encryption operations of $ELTA2E$. In step 5, S performs two homomorphic multiplication by constant and two *Blind* function evaluations. One homomorphic multiplication by constant and one *Blind* function together is similar in computational complexity of one encryption operation of $ELTA2E$. So, the total number of PKE operation of Π_{OTAA} is 4, in the worst case. For string OT of length n, the communication complexity is $(38n^2 + 98n)$, and the number of PKE operations is $(2n + 2)$, in the worst case.

Comparison with Related Work. Hazay and Patra [13] designed an OT protocol for erasure-free one-sided active adaptive adversaries. Their protocol for bit OT requires $(288n^2 + 100n + 16) \in O(n^2)$ communication complexity. Protocol Π_{OTAA} needs $101n \in O(n)$ communication complexity. The worst case number of PKE operations of the protocol of [13] for bit OT is $(16n + 6) \in O(n)$. The worst case number of PKE operations of Π_{OTAA} is constant (only four).

For OT of strings of size n, the OT protocol of [13] requires $(288n^2 + 110n + 16)$ communication complexity and $(16n + 6)$ PKE operations in the worst case. For OT of strings of size n, protocol Π_{OTAA} requires $(38n^2 + 98n)$ communication complexity and $(2n + 2)$ PKE operations in the worst case. For string OT of size n, protocol Π_{OTAA} requires seven times less communication complexity and eight times less PKE operations than the OT protocol of [13].

10   Efficiency of the OT Protocol by Hazay and Patra [13]

In this section, the main factor of the complexity of the OT protocol by Hazay and Patra [13] is described. They have different efficiency for polynomial-size message space and exponential-size message space, with respect to the security parameter n. Here, the efficiency of bit OT, which falls in the category of polynomial-size message space, is described.

The OT protocol of [13] uses a non-committing encryption (NCE) scheme secure against one-sided active adaptive adversaries. They designed a protocol Π_{OSC} that for this purpose. Π_{OSC} uses the somewhat non-committing encryption (SNCE) of [12]. The SNCE protocol of [12] uses the non-committing encryption scheme (NCE) of [10]. There was another more recent NCE scheme [7] which is error-free but requires more communication complexity and computational complexity than the NCE of [10,21]. The NCE scheme of [10] uses a subroutine named *attempt*. In [Theorem 2 [10]], it is mentioned that the NCE scheme of [10] has to repeat $4n$ calls of *attempt* in order to ensure that the probability of failure of subroutine *attempt* remains negligible in n. That means, the worst case number of repeats of *attempt* is $4n$. Each call of *attempt* has communication cost $(12n + 1)$. The communication complexity of the NCE scheme of [10] is $O(n^2)$ for message size of one bit. Each call of *attempt* uses one encryption operation of a simulatable public key encryption scheme, so the number of PKE operations for *attempt* is 1. Then, the NCE scheme of [10] needs $4n$ PKE operations in the worst case. The communication complexity of the SNCE protocol of [12], with equivocality parameter $\ell = 2$, is $O(n^2)$. It uses the NCE protocol of [10] for sending an index $i \in \{1, \ldots, \ell\}$. As mentioned in [12], the expected number of PKE operations for this step is $O(\log \ell)$. In the worst case, this step requires $4n \in O(n)$ PKE operations. The OT protocol of [13] uses a zero-knowledge proof that uses a constant number of PKE operations. The communication complexity of the bit OT protocol of [13] is $O(n^2)$. The number of PKE operations of the bit OT of [13], in the worst case, is $O(n)$.

Hazay and Patra claims that their OT protocol needs a constant number of PKE operations [Theorem 2 [13]]. One possibility is that they were counting one encryption of the NCE scheme Π_{OSC} secure against one-sided adaptive adversaries (or one encryption of the dual-mode encryption scheme of [19]), each of them as a single PKE operation. But the encryption scheme Π_{OSC} uses other PKE schemes (the non-committing encryption scheme for the sender (NCES) of [3], the non-committing encryption scheme for the receiver (NCER) of [14] and the SNCE scheme of [12]) as its subroutines inside its implementation. The notion of *atomic* PKE scheme is necessary for the analysis of the number of PKE operations. An *atomic* PKE scheme denotes a PKE scheme that does not use any other PKE scheme as a subroutine in its implementation. To get the actual number of PKE scheme of a protocol, it should be counted that how many operations of atomic PKE scheme are invoked inside that protocol.

11 Adaptive Zero Knowledge Arguments

In this section, the adaptive zero-knowledge arguments used in the protocols are described. First, the non-erasure Σ-protocols for the corresponding relations are presented. Then, it is described how to convert them to adaptive zero-knowledge arguments.

Scnorr [20] suggested a non-erasure Σ-protocol for knowledge of discrete logarithm [8]. A non-erasure Σ-protocol for equality of discrete logarithm is given in [8]. Damgård [8] presented a non-erasure Σ-protocol for proving knowledge of committed secret for Pedersen commitment scheme.

If $mode = 0$, then, in step 10 of Π_{DKG}, P_1 has to prove that $\log_j(\ell_1) \neq \log_{vk}(vk_1)$. This can be called a proof for inequality of discrete logarithm. Let the common input be $(x_1, x_2, y_1, y_2) = ((y_1)^{w_1} \bmod p, (y_2)^{w_2} \bmod p, y_1, y_2)$. The prover P knows witness $w_1, w_2 \in \mathbb{Z}_q$ such that $w_1 \neq w_2$. A new non-erasure Σ-protocol for proving the inequality of discrete logarithm is designed. P chooses $r \xleftarrow{\$} \mathbb{Z}_q$. P computes $a_1 = (y_1)^r \bmod p, a_2 = (y_2)^r \bmod p$. P sends $a = (a_1, a_2)$. V chooses a challenge $e \xleftarrow{\$} \mathbb{Z}_q$ and sends it. P computes $z_1 = r + ew_1 \bmod q, z_2 = r + ew_2 \bmod q$. P sends (z_1, z_2). V accepts if and only if $(y_1)^{z_1} = a_1(x_1)^e \bmod p, (y_2)^{z_2} = a_1(x_2)^e \bmod p, (y_1)^{z_2} \neq a_1(x_1)^e \bmod p$, and $(y_2)^{z_1} \neq a_2(x_2)^e \bmod p$.

A new non-erasure Σ-protocol, for proving multiplication correct for $ELTA2E$, is designed. P computes $c_2 = m_2 \times_h c_1, c_3 = Blind(pk, c_2)$. Let (s_3, t_3) be the randomness used in the $Blind$ function. Let the common input be $x = (c_1, c_3) = ((b_1, d_1), (b_3, d_3))$. Then, $b_3 = (b_1)^{m_2} g^{s_3} j^{t_3}, d_3 = (d_1)^{m_2} h^{s_3} \ell^{t_3}$. P knows witness $(m_2, s_3, t_3) \in \mathbb{G} \times \mathbb{G} \times \mathbb{G}$. The Σ-protocol is as follows. P chooses $r_1, r_2, r_3 \xleftarrow{\$} \mathbb{Z}_q$, sets $a_1 = (b_1)^{r_1} g^{r_2} j^{r_3} \bmod p, a_2 = (d_1)^{r_1} h^{r_2} \ell^{r_3} \bmod p$. P sends $a = (a_1, a_2)$. V chooses a challenge $e \xleftarrow{\$} \mathbb{Z}_q$ and sends it. P sets $z_1 = r_1 + em_2 \bmod q, z_2 = r_2 + es_3 \bmod q, z_3 = r_3 + et_3 \bmod q$. P sends $z = (z_1, z_2, z_3)$. V accepts if and only if $(b_1)^{z_1} g^{z_1} j^{z_3} = a_1(b_3)^e \bmod p$, and $(d_1)^{z_1} h^{z_1} \ell^{z_3} = a_2(d_3)^e \bmod p$.

Proving that a given ciphertext $c = (x, y)$ is an encryption of zero is equivalent to prove that $\log_g(x) = \log_h(y)$. For proving that one of two given ciphertexts encrypts zero without disclosing which one, the OR-construction of Σ-protocols [8] is performed.

Converting Σ-Protocol to Adaptive Zero-Knowledge Argument. Damgård [9] described how to convert a Σ-protocol $\Pi_{\Sigma R}$ for a given relation R to a zero-knowledge proof Π_{AdZKA}^R for the same relation. This conversion works in the CRS model and needs a trapdoor commitment scheme. The CRS μ is used as the commitment key. P computes its first message a of $\Pi_{\Sigma R}$, selects $r_a \xleftarrow{\$} \mathbb{Z}_p$, computes $c = Com_\mu(a, r_a)$, and sends c. V selects $e \xleftarrow{\$} \{0, 1\}^t$, and sends e. P computes its second message z of $\Pi_{\Sigma R}$, and sends (a, z, r_a) to V. V checks that (a, e, z) is an accepting conversation for $\Pi_{\Sigma R}$, and also that $Com_\mu(a, r_a) = c$. The security proof of this type of zero-knowledge proof against adaptive adversaries is given in [[17] Chap. 5]. When a trapdoor commitment scheme is used in a zero-knowledge proof, it only achieves computational soundness. By definition, the resulting system is a zero-knowledge argument.

12 Future Work

One future research work is to design an efficient two-party computation protocol for one-sided active adaptive adversary model, using the new efficient oblivious transfer protocol. Another research direction is to design efficient oblivious transfer protocol for the fully adaptive adversary model, that is, when the adversary may corrupt both parties at some point.

References

1. Aumann, Y., Lindell, Y.: Security against covert adversaries: efficient protocols for realistic adversaries. J. Cryptol. **23**(2), 281–343 (2010)
2. Bach, E.: Analytic Methods in the Analysis and Design of Number-Theoretic Algorithms. Massachusetts Institute of Technology, Cambridge (1985)
3. Bellare, M., Hofheinz, D., Yilek, S.: Possibility and impossibility results for encryption and commitment secure under selective opening. In: Joux, A. (ed.) EUROCRYPT 2009. LNCS, vol. 5479, pp. 1–35. Springer, Heidelberg (2009). doi:10.1007/978-3-642-01001-9_1
4. Bellare, M., Yilek, S.: Encryption schemes secure under selective opening attack. Cryptology ePrint Archive, Report 2009/101 (2009). http://eprint.iacr.org/
5. Canetti, R.: Security and composition of multiparty cryptographic protocols. J. Cryptol. **13**(1), 143–202 (2000)
6. Canetti, R., Gennaro, R., Jarecki, S., Krawczyk, H., Rabin, T.: Adaptive security for threshold cryptosystems. In: Wiener, M. (ed.) CRYPTO 1999. LNCS, vol. 1666, pp. 98–116. Springer, Heidelberg (1999). doi:10.1007/3-540-48405-1_7
7. Choi, S.G., Dachman-Soled, D., Malkin, T., Wee, H.: Improved non-committing encryption with applications to adaptively secure protocols. In: Matsui, M. (ed.) ASIACRYPT 2009. LNCS, vol. 5912, pp. 287–302. Springer, Heidelberg (2009). doi:10.1007/978-3-642-10366-7_17
8. Damgård, I.: On Σ-protocols. www.cs.au.dk/~ivan/Sigma.pdf
9. Damgård, I.: Efficient concurrent zero-knowledge in the auxiliary string model. In: Preneel, B. (ed.) EUROCRYPT 2000. LNCS, vol. 1807, pp. 418–430. Springer, Heidelberg (2000). doi:10.1007/3-540-45539-6_30
10. Damgård, I., Nielsen, J.B.: Improved non-committing encryption schemes based on a general complexity assumption. In: Bellare, M. (ed.) CRYPTO 2000. LNCS, vol. 1880, pp. 432–450. Springer, Heidelberg (2000). doi:10.1007/3-540-44598-6_27
11. Fouque, P.-A., Poupard, G., Stern, J.: Sharing decryption in the context of voting or lotteries. In: Frankel, Y. (ed.) FC 2000. LNCS, vol. 1962, pp. 90–104. Springer, Heidelberg (2001). doi:10.1007/3-540-45472-1_7
12. Garay, J.A., Wichs, D., Zhou, H.-S.: Somewhat non-committing encryption and efficient adaptively secure oblivious transfer. In: Halevi, S. (ed.) CRYPTO 2009. LNCS, vol. 5677, pp. 505–523. Springer, Heidelberg (2009). doi:10.1007/978-3-642-03356-8_30

13. Hazay, C., Patra, A.: One-sided adaptively secure two-party computation. In: Lindell, Y. (ed.) TCC 2014. LNCS, vol. 8349, pp. 368–393. Springer, Heidelberg (2014). doi:10.1007/978-3-642-54242-8_16

14. Jarecki, S., Lysyanskaya, A.: Adaptively secure threshold cryptography: introducing concurrency, removing erasures. In: Preneel, B. (ed.) EUROCRYPT 2000. LNCS, vol. 1807, pp. 221–242. Springer, Heidelberg (2000). doi:10.1007/3-540-45539-6_16

15. Naor, M., Ostrovsky, R., Venkatesan, R., Yung, M.: Perfect zero-knowledge arguments for NP using any one-way permutation. J. Cryptol. **11**, 87–108 (1998)

16. Nargis, I.: Efficient oblivious transfer from lossy threshold homomorphic encryption. Cryptology ePrint Archive, Report 2017/235 (2017). http://eprint.iacr.org/2017/235

17. Nielsen, J.B.: On protocol security in the cryptographic model. Ph.D. thesis, University of Aarhus (2004)

18. Pedersen, T.P.: Non-interactive and information-theoretic secure verifiable secret sharing. In: Feigenbaum, J. (ed.) CRYPTO 1991. LNCS, vol. 576, pp. 129–140. Springer, Heidelberg (1992). doi:10.1007/3-540-46766-1_9

19. Peikert, C., Vaikuntanathan, V., Waters, B.: A framework for efficient and composable oblivious transfer. In: Wagner, D. (ed.) CRYPTO 2008. LNCS, vol. 5157, pp. 554–571. Springer, Heidelberg (2008). doi:10.1007/978-3-540-85174-5_31

20. Schnorr, C.P.: Efficient signature generation by smart cards. J. Cryptol. **4**(3), 161–174 (1991)

21. Zhu, H., Araragi, T., Nishide, T., Sakurai, K.: Adaptive and composable non-committing encryptions. In: Steinfeld, R., Hawkes, P. (eds.) ACISP 2010. LNCS, vol. 6168, pp. 135–144. Springer, Heidelberg (2010). doi:10.1007/978-3-642-14081-5_9

Privacy-Friendly Forecasting for the Smart Grid Using Homomorphic Encryption and the Group Method of Data Handling

Joppe W. Bos[1], Wouter Castryck[2,3(✉)],
Ilia Iliashenko[2], and Frederik Vercauteren[2,4]

[1] NXP Semiconductors, Leuven, Belgium
[2] imec-Cosic, Department of Electrical Engineering, KU Leuven, Leuven, Belgium
`wouter.castryck@kuleuven.be`
[3] Laboratoire Paul Painlevé, Université de Lille-1, Villeneuve-d'Ascq, France
[4] Open Security Research, Shenzhen, China

Abstract. While the smart grid has the potential to have a positive impact on the sustainability and efficiency of the electricity market, it also poses some serious challenges with respect to the privacy of the consumer. One of the traditional use-cases of this privacy sensitive data is the usage for forecast prediction. In this paper we show how to compute the forecast prediction such that the supplier does not learn any individual consumer usage information. This is achieved by using the Fan-Vercauteren somewhat homomorphic encryption scheme. Typical prediction algorithms are based on artificial neural networks that require the computation of an activation function which is complicated to compute homomorphically. We investigate a different approach and show that Ivakhnenko's group method of data handling is suitable for homomorphic computation.

Our results show this approach is practical: prediction for a small apartment complex of 10 households can be computed homomorphically in less than four seconds using a parallel implementation or in about half a minute using a sequential implementation. Expressed in terms of the mean absolute percentage error, the prediction accuracy is roughly 21%.

1 Introduction

One of the promising solutions to cope with current and future challenges of electricity supply is the *smart grid*. With the prospect of having a positive impact on the sustainability, reliability, flexibility, and efficiency many countries around the world are investing significantly in such smart grid solutions. The deployment of smart meters is already well underway. For example, in the United Kingdom

This work was supported by the European Commission under the ICT programme with contract H2020-ICT-2014-1 644209 HEAT, and through the European Research Council under the FP7/2007-2013 programme with ERC Grant Agreement 615722 MOTMELSUM.

© Springer International Publishing AG 2017
M. Joye and A. Nitaj (Eds.): AFRICACRYPT 2017, LNCS 10239, pp. 184–201, 2017.
DOI: 10.1007/978-3-319-57339-7_11

the large energy suppliers were operating over 400, 000 smart gas and electricity meters, representing 0.9% of all the domestic meters operated by the large suppliers in 2014 [9]. This development is expected to continue and intensify: the EU third energy package has as an objective to replace at least 80% of electricity meters with smart meters by 2020 [15]. This change will fundamentally re-engineer the (electricity) service industry.

The replacement of the classical meters with their smart variants has advantages for both the consumer and industry. Some of the key benefits include giving consumers the information to gain control over their energy consumption, lowering the cost for managing the supply of energy across industry, and producing detailed consumption information data from these smart meters which in turn enable a wide range of services [9]. It is expected that the meters have an update rate of every 15 min at least [14]. When generating such a large amount of consumer data a lot of privacy sensitive information is being disclosed. There are various initiatives (e.g. [32,37]) which stress and outline the importance of having solutions for the smart grid where privacy protecting mechanisms are already built-in by design.

This work is concerned with enhancing the privacy of the smart meter readings in the setting of *forecast prediction*: energy suppliers need to forecast in order to buy energy generation contracts that cover their clients. Moreover, to ensure network capacity the network operators require longer term forecasting [10,23,37]. This forecasting is typically done by taking as input the (aggregated) data from a number of households. Based on this consumption data, together with other variables such as the date and the current temperature and weather, a forecast is computed to predict the short, medium, or long term consumption. The energy providers or network operators only need to know the desired forecast information based on their (potentially proprietary) forecasting algorithm and model. There is no need to observe the individual consumer data.

We investigate the potential of *fully homomorphic encryption* (FHE) to realize this goal. The notion of FHE was introduced in the late 1970s [34] and a concrete instantiation was found in 2009 by Gentry [19]. FHE allows an untrusted party to carry out arbitrary computation on *encrypted data* without learning anything about the content of this data. Currently, the Fan-Vercauteren (FV) FHE scheme [16] is regarded as the best choice with respect to security and practical performance. See Sect. 4 for a more detailed description of the FV scheme. Additively homomorphic encryption schemes [31] and other tools have been proposed to enhance the privacy in the setting of computing detailed billing in the context of the smart grid [13,18,24,25,30,33]. However, these approaches cannot be directly used in the setting of prediction algorithms since these more complex algorithms need to compute both additions and multiplications.

One popular class of algorithms which are used for prediction are based on artificial neural networks. One of the main ingredients in these forecasting algorithms is the computation of the so-called activation function, in practice it is common to use a sigmoid function where the logistic function $t \mapsto 1/(1 + e^{-t})$ is a popular choice. However, computing such a sigmoid function homomorphically is far from practical. One possible way to proceed is to simply ignore the sigmoidality require-ment and to proceed with a truncated Taylor series approximating this function or, more generally, to use any non-linear polynomial function which is *simple*. This was investigated by Livni et al. [26] regardless of cryptographic applications. Recent work by Xie et al. [39] and Dowlin et al. [12] suggests to apply the same approach to homomorphically encrypted data. However, by computing artificial neural networks in this fashion it becomes just an organized manner of fitting a polynomial through the given data set. In this paper we investigate an older tool for realizing this goal. Namely, we show that Ivakhnenko's group method of data han-dling (GMDH) which was proposed back in 1970 [22] is a perfect match for being computed homomorphically. Moreover, a recent comparison analysis between dif-ferent forecasting methods [36] showed that GMDH produced significantly more accurate results compared to the other methods considered.

We show that GMDH can be implemented homomorphically using the recent fixed point approach from [6,11]. Using a five-layered network (one input layer, three hidden layers and an output node) we are able to homomorphically pre-dict the next half-hour energy consumption for an apartment complex of 10 households. Our software implementation results indicate that this requires less than four seconds using a parallel implementation or about half a minute using a sequential implementation while the prediction accuracy expressed using the mean absolute percentage error (MAPE, see Sect. 3 for a definition) is only 21%. This shows that privacy preserving forecasting using homomorphic encryption is indeed practical.

2   The Smart Grid and Privacy Concerns

The authors of [35] define the smart grid as *"an electricity network that can cost efficiently integrate the behavior and actions of all users connected to it – generators, consumers and those that do both – in order to ensure economically efficient, sustainable power system with low losses and high levels of quality and security of supply and safety"*. This paper is concerned with the cryptographic solutions to privacy concerns within the smart grid. Within this scope we assume that the meters are protected against various types of side-channel attacks such that no secret data can be retrieved from the device when it is operating (e.g. key extraction). Moreover, we assume that the smart metering device acts honestly in accordance with the implementation or protocol given to it. These assumptions avoid the usual security threats and leave us with the privacy related concerns which we aim to address.

In the early 1990s, Hart showed a non-intrusive approach where by monitor-ing the electric load one can observe the individual appliances turning on and off [20]. Hence, detailed smart meter readings, which are expected to be gen-erated at least every 15 min in the context of the smart grid (cf. [14]), can be used to derive various privacy sensitive information about a house-hold or even an apartment complex. In order to grasp where the main privacy challenges are in smart metering it is good to understand how and when the meter readings are used in practice by the various parties involved. As identified by the survey paper [23], which in turn has collected this information from the privacy impact assessment by NIST [37] and the enumeration of data uses by the consultation of the British Department of Energy and Climate Change [10], the key usages of smart meter readings include the usage for *load monitoring and forecasting* and *smart billing*.

There has been a significant amount of work related to privacy-preserving smart billing solutions for the smart grid. One line of research allows complex non-linear tariff policies where the bill is computed and sent along with a zero-knowledge proof to ensure that the computations are correct [30,33]. Another approach is based on privacy-friendly aggregation schemes (e.g. using additively homomorphic encryption schemes such as the Paillier scheme [31]) where one can compute a function on the ciphertexts which corresponds to adding the plaintexts [13,18,24,25]. Such approaches heavily rely on the fact that only aggregation of the results is required. As soon as more complex operations need to be computed (such as a large number of multiplications) one has to look for other solutions.

One example where more operations are performed is in the setting of load monitoring and forecasting. There are many different forecasting approaches (see e.g. the survey paper [21] on this topic and the references therein). One of the popular and well-studied techniques is using artificial neural networks (see e.g. [1,17]). In the next section we describe how such neural networks operate, analyze the challenges they pose when being evaluated in the encrypted domain, and discuss how this naturally leads to considering the group method of data handling as an alternative forecasting tool.

3 Neural Networks versus the Group Method of Data Handling

Over time, artificial neural networks (ANNs) have manifested themselves among the most popular and reliable prediction tools for various purposes, including load forecasting. For our preliminary discussion, it suffices to think of an ANN as a real-valued function $f : \mathbf{R}^{n_0} \to \mathbf{R}$ that arises as the composition of a number of 'neurons' $\nu_{ij} : \mathbf{R}^{n_{i-1}} \to \mathbf{R}$, organized in layers $i = 1, \ldots, r$, as depicted in Fig. 1. Each neuron is of the form

$$\nu_{ij} : \mathbf{R}^{n_{i-1}} \to \mathbf{R} : (x_1, x_2, \ldots, x_{n_{i-1}}) \mapsto g\left(\sum_{k=1}^{n_{i-1}} w_{ijk}x_k - b_{ij}\right)$$

for weights and biases $w_{ijk}, b_{ij} \in \mathbf{R}$ and some fixed sigmoidal activation function $g : \mathbf{R} \to \mathbf{R}$, such as the logistic function $t \mapsto 1/(1 + e^{-t})$. The global shape of the network is decided in advance, and the goal is to determine the weights w_{ijk} and the biases b_{ij} such that f approximates an unknown target function $\tilde{f} : \mathbf{R}^{n_0} \to \mathbf{R}$, in our case load prediction, as well as possible. This is done during a so-called supervised learning phase. One starts from a reasonable guess, after which the network's performance is assessed by feeding to it a number of input-output pairs of $\tilde{f}$, taken from a given data set, and measuring the error. During a process called backpropagation, which is based on the chain rule for derivation, the weights and biases are then modified repeatedly, in the hope of converging to values that minimize the error.

The backpropagation method requires the activation function g to have a nice and easy derivative, while at the same time it should be sigmoidal, i.e. its graph should have the typical step-like activation shape, allowing the ANNs to do what they were designed for: to simulate computation in (an area of) the human brain. Unfortunately, the class of such functions does not contain examples that are easy to evaluate homomorphically. A natural attempt would be to use a Taylor approximation to the logistic function or to one of its known alternatives, but such approximations become highly non-sigmoidal away from the origin.

One way out is simply to ignore the sigmoidality requirement and to proceed with this truncated Taylor series, or more generally to replace g by any simple non-linear polynomial function, the easiest choice being $t \mapsto t^2$. This has been investigated by Livni et al. [26] for reasons of computational efficiency, regardless of cryptographic applications. Recent work by Xie et al. [39] and Dowlin et al. [12] suggested to apply the same approach to homomorphically encrypted data. The resulting neural networks were named 'crypto-nets'.

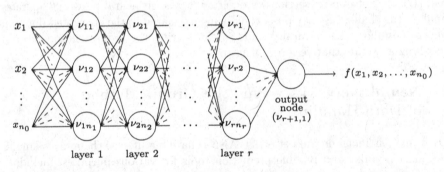

Fig. 1. Design of an Artificial Neural Network (ANN).

However in this way the ANN just becomes an organized way of fitting a polynomial through the given data set. There exist older and simpler prediction tools that do this. In this paper we study one of the oldest such tools, namely Ivakhnenko's group method of data handling (GMDH) from 1970 [22]. Besides being suited for applications using homomorphic encryption, one particular feature is that its performance in the context of load forecasting enjoys a large amount of existing literature, at times even with results that are superior to ANNs. Indeed, a comparison analysis between different forecasting methods from 2008 [36] showed that GMDH produced significantly more accurate results compared to the other methods considered.

The basic version of GMDH works as follows, although many variations are possible (and seem to deserve a further analysis). The goal is to approximate our target function $\tilde{f} : \mathbf{R}^{n_0} \to \mathbf{R}$ with a truncated Wiener series

$$a_0 + \sum_{i=1}^{n_0} a_i x_i + \sum_{i=1}^{n_0} \sum_{j=i}^{n_0} a_{ij} x_i x_j + \sum_{i=1}^{n_0} \sum_{j=i}^{n_0} \sum_{k=j}^{n_0} a_{ijk} x_i x_j x_k + \dots,$$

which is also called a Kolmogorov-Gabor polynomial. The idea is to approach this by a finite superposition of quadratic polynomials

$$\nu_{ij} : \mathbf{R}^2 \to \mathbf{R} : (x, y) \mapsto b_{ij0} + b_{ij1} x + b_{ij2} y + b_{ij3} xy + b_{ij4} x^2 + b_{ij5} y^2$$

along a diagram of the kind depicted in Fig. 2. One can think of this as some sort of ANN, and indeed the diagram is sometimes called a 'polynomial neural network'. As a first main difference, however, note that the wiring is incomplete: each neuron has two inputs only.

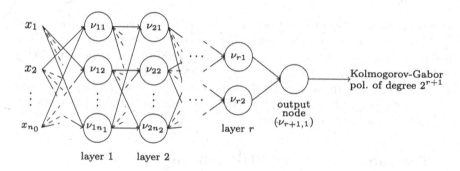

Fig. 2. Network-like illustration of the group method of data handling.

Also the learning phase is quite different from the one in conventional ANNs. Here the goal is to determine the coefficients b_{ijk} of the quadratic polynomials ν_{ij}, but also the concrete structure of the network, which is not fixed in advance. Indeed, one decides beforehand on the number of layers r and the number of neurons n_i in each layer, but the wiring between these is defined during the learning process. Recall that each node can have only two inputs, so the following constraint should be satisfied: $n_i \le \binom{n_{i-1}}{2}$. In order to prevent exponential growth of the number of neurons, the left hand side will in general be much smaller than the right hand side. As to *which* combinations end up being chosen, one first considers all possible combinations and then removes the $\binom{n_{i-1}}{2} - n_i$ worst neurons with respect to their error performance, in the sense explained below, while at the same time determining the coefficients b_{ijk} of the surviving neurons. One then proceeds with the next layer. In particular, there is no backpropagation. The node with the smallest error performance will be assigned as an output for the whole network; this may in fact be different from what was initially foreseen to become the output neuron. One sometimes applies the rule that if at some point all nodes in layer i perform worse than the best performing node in layer $i - 1$, then the algorithm stops, and the latter node is assigned as the output.

To assess the error performance of a neuron, while at the same time determining the coefficients of the corresponding quadratic polynomial, one uses a given data set of correct input-output pairs for $\tilde{f}$. Additionally, an error (or loss) function should be set up beforehand. Throughout this paper we use the Mean Squared Error (MSE) function

$$\text{MSE}((y_1^{\text{forecast}}, \ldots, y_n^{\text{forecast}}), (y_1^{\text{actual}}, \ldots, y_n^{\text{actual}})) = \frac{1}{n} \sum_{i=1}^{n} (y_i^{\text{forecast}} - y_i^{\text{actual}})^2,$$

but there are a couple of other standard choices, such as the Mean Absolute Error (MAE) and the Mean Absolute Percentage Error (MAPE):

$$\frac{1}{n} \sum_{i=1}^{n} |y_i^{\text{forecast}} - y_i^{\text{actual}}| \qquad \text{resp.} \qquad \frac{100}{n} \sum_{i=1}^{n} \left| \frac{y_i^{\text{forecast}} - y_i^{\text{actual}}}{y_i^{\text{actual}}} \right|.$$

For each neuron ν_{ij} the data set is randomly split into a learning set and a test set. This is done to avoid overfitting, where the network learns too much about the inherent noise always being present in real-world data. The learning set is used to determine the coefficients b_{ijk}, by choosing them such that the error is as small as possible. In the case of MSE this can be achieved by linearization of the quadratic polynomial and applying the least squares method. The test set is then used to assess the performance of the neuron.

4   The Fan-Vercauteren SHE Scheme

In this section we briefly describe a simplified version of the FV scheme [16], which we will present in its *somewhat* homomorphic encryption (SHE) form, meaning that it is suitable only for computations up to a given depth, thereby avoiding very expensive noise reduction operations (i.e. bootstrapping). It concerns a scale-invariant SHE scheme based on the hardness of the ring version of the learning with errors problem (RLWE) [27]. It works in the polynomial ring $R = \mathbf{Z}[X]/(f(X))$ with $f(X) = X^d + 1$ and $d = 2^n$. For an integer N we denote with R_N the reduction of R modulo N. Abusing notation, elements of R will often be identified with their unique representant in $\mathbf{Z}[X]$ of degree at most $d - 1$, and similarly elements of R_N are identified with their unique representant inside

$$\left\{ \alpha_{d-1}X^{d-1} + \alpha_{d-2}X^{d-2} + \ldots + \alpha_0 \,\middle|\, \alpha_i \in (-N/2, N/2] \text{ for all } i \right\},$$

but this should cause no confusion. For an element $a \in R$ or $a \in \mathbf{Z}[X]$ we write $[a]_N$ do denote its reduction inside the above set of representants.

The plaintext space in the FV scheme is given by the ring R_t for some small integer modulus $t > 1$, while a ciphertext is given by a pair of ring elements in R_q

where $q > 1$ is a much larger modulus. The key generation and the encryption operations in the FV scheme require sampling from two probability distributions defined on R, denoted χ_{key} and χ_{err}. The security of the scheme is determined by the degree d of f, the size of q, and by the probability distributions. Typically χ_{key} and χ_{err} are coefficient-wise discrete Gaussian distributions centered around 0 and having a small standard deviation, but in practice one often samples the coefficients of the key from a uniform distribution on a narrow set like $\{-1, 0, 1\}$. We remark that the errors are sampled coefficient-wise because R is a ring of 2-power cyclotomic integers: for more general rings one should proceed with the more complicated joint distribution described in [28]. The RLWE distribution on $R_q \times R_q$ is then constructed as follows: first choose a fixed element $s \leftarrow \chi_{\text{key}}$, and then generate samples of the form (a, b) with $a \leftarrow R_q$ uniformly random and $b = [-(as + e)]_q$ with $e \leftarrow \chi_{\text{err}}$. (The minus sign is not standard but makes a better fit with the discussion below.) The decision RLWE problem is then to distinguish between the RLWE distribution and the uniform distribution on $R_q \times R_q$. The search RLWE problem is to retrieve s from polynomially many samples. Both problems are believed to be very hard for an appropriate choice of parameters.

By construction, for a RLWE sample (a, b) we have that $e = -[as + b]_q$ and therefore that the right-hand side has small coefficients, with overwhelming probability. Furthermore note that the sample can be easily re-randomized without knowledge of s as follows: choose $u \leftarrow \chi_{\text{key}}$ and $e_1, e_2 \leftarrow \chi_{\text{err}}$ and form the new sample as $(ua + e_1, ub + e_2)$. In the encryption scheme below, the public key will consist of a single RLWE sample, which will be re-randomized during encryption. The new RLWE sample will then be used as an additive mask to encrypt a message $m \in R_t$. Before we present the FV scheme, we first describe some subroutines that are required in the algorithm:

- $\texttt{WordDecomp}_{w,q}(a)$: This function is used to decompose a ring element $a \in R_q$ in base w by splicing each coefficient of a. For $u = \lfloor \log_w(q) \rfloor$, it returns $a_i \in R$ with coefficients in $(-w/2, w/2]$, such that $a = \sum_{i=0}^{u} a_i w^i$.
- $\texttt{PowersOf}_{w,q}(a)$: This function scales an element $a \in R_q$ by the different powers of w. It is defined as $\texttt{PowersOf}_{w,q}(a) = (aw^i)_{i=0}^{u}$.

These two functions can be used to perform a polynomial multiplication in R_q through an inner product: $\langle \texttt{WordDecomp}_{w,q}(a), \texttt{PowersOf}_{w,q}(b) \rangle = a \cdot b$. This expression has advantage in reducing the noise during homomorphic multiplications, as the first vector contains small elements only.

The FV scheme consists of an encryption scheme augmented with additional functions $\texttt{Add}$, $\texttt{Mult}$, and $\texttt{ReLin}$ to compute homomorphically on encrypted data.

1. $\texttt{ParamsGen}(\lambda)$: For a given security parameter λ, choose a degree $d = 2^n$ and thus a polynomial $f(X) = X^d + 1$, moduli q and t and distributions χ_{err} and χ_{key}. Also choose the base w for $\texttt{WordDecomp}_{w,q}(\cdot)$. Return the system parameters $(d, q, t, \chi_{\text{err}}, \chi_{\text{key}}, w)$.
2. $\texttt{KeyGen}(d, q, t, \chi_{\text{err}}, \chi_{\text{key}}, w)$: Sample the secret key $s \leftarrow \chi_{\text{key}}$, sample $a \leftarrow R_q$ uniformly at random, and sample $e \leftarrow \chi_{\text{err}}$. Compute $b = [-(as + e)]_q$.

The public key is the pair $\mathbf{pk} = (b, a)$ and the secret key is $\mathrm{sk} = s$. The scheme uses another key $\mathbf{rlk}$ called *relinearization key* in the function $\mathtt{ReLin}$ below. Define $\ell = u+1 = \lfloor \log_w(q) \rfloor + 1$, sample a vector $\mathbf{a} \leftarrow R_q^\ell$ uniformly at random, sample $\mathbf{e} \leftarrow \chi_{\mathrm{err}}^\ell$, and let $\mathbf{rlk} = ([\mathtt{PowersOf}_{w,q}(s^2) - (\mathbf{e} + \mathbf{a} \cdot s)]_q, \mathbf{a}) \in R_q^\ell \times R_q^\ell$.

3. $\mathtt{Encrypt}(\mathbf{pk}, m)$: First encode the input message $m \in R_t$ into a polynomial $\Delta m \in R_q$ with $\Delta = \lfloor q/t \rfloor$. Next sample the error polynomials $e_1, e_2 \leftarrow \chi_{\mathrm{err}}$, sample $u \leftarrow \chi_{\mathrm{key}}$, and compute the two polynomials $c_0 = \Delta m + bu + e_1 \in R_q$ and $c_1 = au + e_2 \in R_q$. The ciphertext is the pair of polynomials $\mathbf{c} = (c_0, c_1)$.

4. $\mathtt{Decrypt}(\mathrm{sk}, \mathbf{c})$: First compute the polynomial $\tilde{m} = [c_0 + sc_1]_q$. Then recover the plaintext message m by a decoding the coefficients of $\tilde{m}$ by scaling down by Δ and rounding.

5. $\mathtt{Add}(\mathbf{c_1}, \mathbf{c_2})$: For two ciphertexts $\mathbf{c_1} = (c_{1,0}, c_{1,1})$ and $\mathbf{c_2} = (c_{2,0}, c_{2,1})$, return $\mathbf{c} = (c_{1,0} + c_{2,0}, c_{1,1} + c_{2,1}) \in R_q \times R_q$.

6. $\mathtt{Mult}(\mathbf{c_1}, \mathbf{c_2}, \mathbf{rlk})$: Compute $\tilde{\mathbf{c}}_{\mathrm{mult}} = (c_0, c_1, c_2)$ where $c_0 = \lfloor \frac{t}{q} \cdot c_{1,0} \cdot c_{2,0} \rceil$, $c_1 = \lfloor \frac{t}{q} \cdot (c_{1,0} \cdot c_{2,1} + c_{1,1} \cdot c_{2,0}) \rceil$, and $c_2 = \lfloor \frac{t}{q} \cdot c_{1,1} \cdot c_{2,1} \rceil$ and apply relinearization.

7. $\mathtt{ReLin}(\tilde{\mathbf{c}}_{\mathrm{mult}}, \mathbf{rlk})$: Write $\mathbf{rlk} = (\mathbf{b}, \mathbf{a})$ and $\tilde{\mathbf{c}}_{\mathrm{mult}} = (c_0, c_1, c_2)$, then compute a relinearized ciphertext as $\mathbf{c'} = (c_0', c_1')$ as $([c_0 + \langle \mathtt{WordDecomp}_{w,q}(c_2), \mathbf{b} \rangle]_q, [c_1 + \langle \mathtt{WordDecomp}_{w,q}(c_2), \mathbf{a} \rangle]_q)$.

Given an FV ciphertext $\mathbf{c} = (c_0, c_1)$, we can write $[c_0 + c_1 s]_q = \Delta m + e$, where e is called the noise inside the ciphertext. Every operation on ciphertexts causes the noise to increase. It is clear that when the noise gets too large, in particular if $\|e\|_\infty > \Delta/2$, correct decryption will fail, where $\|\cdot\|_\infty$ denotes the maximal absolute value of the coefficients.

From now on we assume that χ_{err} is a coefficient-wise discrete Gaussian with standard deviation σ and that χ_{key} samples the coefficients uniformly from $\{-1, 0, 1\}$. With overwhelming probability $B_{\mathrm{err}} = 6\sigma$ and $B_{\mathrm{key}} = 1$ are upper bounds on the absolute values of the coefficients of their respective samples. Therefore we can use $V = B_{\mathrm{err}}(1 + 2dB_{\mathrm{key}}) = B_{\mathrm{err}}(1 + 2d)$ as an upper bound on the noise of the input ciphertexts. When doing arithmetic the noise is affected in the following way. Firstly, adding ciphertexts $\mathbf{c_1}$ and $\mathbf{c_2}$ corresponds to adding the noises, potentially augmented by a carryover γ satisfying $\|\gamma\|_\infty < t$, as explained in [16]. Secondly, multiplying a ciphertext $\mathbf{c}$ by an unencrypted scalar $(\Delta\alpha, 0)$ for some $\alpha \in R_t$ corresponds to multiplying the noise by α, again with some carryover γ. For use below, fix an integer $\lambda \geq 1$ and assume that the coefficients of α are in $\{-1, 0, 1\}$ with at most λ of them being non-zero. Then in a similar way one sees that $\|\gamma\|_\infty < \lfloor \lambda/2 \rfloor \cdot t$. Thirdly, multiplying two ciphertexts $\mathbf{c_1}$ and $\mathbf{c_2}$ whose noise coefficients are bounded by E results in a ciphertext whose noise coefficients are at most

$$2 \cdot E \cdot t \cdot d \cdot (d+1) + 8 \cdot t^2 \cdot d^2 + \ell \cdot B_{\mathrm{err}} \cdot w \cdot d/2$$

in absolute value, by [16, Lemmas 2 and 3].

Now assume that we wish to evaluate a GMDH network $f : R_t^{n_0} \to R_t$ having r hidden layers in a fresh component-wise encryption of an n_0-tuple $(x_1, x_2, \ldots, x_{n_0}) \in R_t^{n_0}$. For the moment just think of this as a Kolmogorov-Gabor polynomial that we evaluate in the encrypted domain along a diagram of the kind depicted in Fig. 2; the purpose of this will become clear in the next section. The network parameters b_{ijk} are assumed to be small public scalars along the lines mentioned above: the coefficients are in $\{-1, 0, 1\}$ and at most λ of them are non-zero. Define $A_1 = 6 \cdot \lambda \cdot t \cdot d \cdot (d+1) + 2 \cdot \lambda$ and

$$A_2 = 3/2 \cdot \lambda \cdot \ell \cdot B_{\mathrm{err}} \cdot w \cdot d + 24 \cdot t^2 \cdot d^2 + 5 \cdot (\lfloor \lambda/2 \rfloor + 1) \cdot t.$$

One verifies that homomorphically evaluating a node $\nu_{1j} : R_t^{n_0} \to R_t^{n_1}$ from the first layer causes the noise coefficients to grow to at most $A_1 \cdot V + A_2$. Recursively applying this formula yields the upper bound $A_1^{r+1} \cdot V + (A_1^{r+1} - 1) \cdot A_2/(A_1 - 1)$ on the absolute values of the noise coefficients that are present in the output of the entire network f.

The parameters of the FV scheme are not only determined by the noise growth, but also by the security requirements. It is easy to see that when d and σ/q grow, amounting to larger polynomials and more noise in the ciphertexts, then RLWE becomes harder. A precise security analysis is beyond the scope of this paper, but to derive our security estimates we closely follow the work by Albrecht, Player and Scott [3] and the open source LWE-estimator implemented by Albrecht [2]. In particular, the LWE-estimator allows one to estimate the concrete hardness of the LWE problem given the dimension d, the modulus q and the standard deviation σ. Note that the actual tool takes as input the parameter $\alpha = \sqrt{2\pi}\sigma/q$, instead of σ directly.

For the design reasons explained in Sect. 6 we will take $r = 3$, $\lambda = 9$, while for compatibility reasons with the software library FV-NFLlib [7] we wish to take $w = 2^{32}$ and $\log_2 q$ an integral multiple of 62. Targetting a security level of 80 bits, we can address the restrictions coming from both the noise growth and the security considerations by using the parameter set $d = 4096$, $q \simeq 2^{186}$ and $\sigma = 102$ (corresponding to $\alpha = 256/q$). These parameters will be used throughout the remainder of the paper and allow for usage of all plaintext moduli $t \le 396$. Note that one ciphertext takes up 186 kB space.

5 Representing Fixed-Point Numbers in Plaintext Space

Our final goal is to evaluate a trained GMDH network in the encrypted domain using the FV scheme. As explained in the previous section, the plaintext space is of the form R_t, which is the reduction modulo a certain integer $t > 1$ of $R = \mathbf{Z}[X]/(X^d + 1)$, where $d = 2^n$ for some $n \in \mathbf{Z}_{>0}$. Therefore an important task is to encode the input values $x_1, x_2, \ldots, x_{n_0} \in \mathbf{R}$, as well as the coefficients $b_{ijk} \in \mathbf{R}$, as elements of R_t. This should be done in such a way that real additions and multiplications agree with the corresponding operations in the ring R_t, up to a certain depth of computation. Dowlin et al. [11] proposed two ways of addressing this issue, which were revisited in a recent paper by Costache et al. [6], who showed them to be essentially equivalent, and also provided lower bounds on t and d guaranteeing that the arithmetic in $\mathbf{R}$ is indeed compatible with that in R_t to the extent desired. We briefly recall their main conclusions, adapted to our setting.

On the real number side, we use fixed-point arithmetic. We assume that the x_i's and the b_{ijk}'s are given in balanced ternary expansion to some finite precision, that is, they are of the form

$$b_{\ell_1-1}b_{\ell_1-2}\ldots b_0 . b_{-1}b_{-2}\ldots b_{-\ell_2} \tag{1}$$

with $b_i \in \{-1,0,1\}$ for $i = -\ell_2,\ldots,\ell_1 - 1$. This should be read as

$$b_{\ell_1-1}3^{\ell_1-1} + b_{\ell_1-2}3^{\ell_1-2} + \ldots + b_0 3^0 + b_{-1}3^{-1} + b_{-2}3^{-2} + \ldots + b_{-\ell_2}3^{-\ell_2}.$$

As usual we say that (1) has ℓ_1 integral digits and ℓ_2 fractional digits; throughout we assume that $\ell_1 \geq 1$ and $\ell_2 \geq 0$. In order to encode (1) as an element of R_t one simply replaces the base 3 by X. This yields

$$b_{\ell_1-1}X^{\ell_1-1} + b_{\ell_1-2}X^{\ell_1-2} + \ldots + b_0 X^0 + b_{-1}X^{-1} + b_{-2}X^{-2} + \ldots + b_{-\ell_2}X^{-\ell_2}, \tag{2}$$

which one can rewrite as

$$b_{\ell_1-1}X^{\ell_1-1} + b_{\ell_1-2}X^{\ell_1-2} + \ldots + b_0 X^0 + b_{-1}X^{d-1} + b_{-2}X^{d-2} + \ldots + b_{-\ell_2}X^{d-\ell_2},$$

using the relation $X^d \equiv -1$.

To decode a given element of R_t one first considers its unique representant inside $\{ \alpha_{d-1}X^{d-1} + \alpha_{d-2}X^{d-2} + \ldots + \alpha_0 \mid \alpha_i \in (-t/2, t/2] \text{ for all } i \}$, after which one replaces all suitably high powers X^i by $-X^{i-d}$, and one evaluates the resulting Laurent polynomial at 3. The outcome is a rational number whose denominator is a power of 3, so it can be easily rewritten in balanced ternary expansion. For simplicity we think of 'suitably high' as $i > d/2$, although to improve the bound on d in Lemma 1 below, a more careful (but easy) estimation should be made, that takes into account the lengths of the integral and fractional parts of the fixed-point numbers involved.

Clearly, the ring operations in R_t are compatible with fixed-point arithmetic on the real number side as long as they do not involve 'wrapping around' modulo t and/or modulo $X^d + 1$. (In the latter case this means that neither the terms of high degree nor the terms of low degree are allowed to cross the separation point $X^{d/2}$.) Thus t and d should be taken large enough to ensure this, for which Costache et al. elaborated concrete lower bounds. We will not explicitly rely on these bounds, but rather apply the underlying ideas to obtain a more implicit statement. For all integers $\ell \geq 0, \lambda \geq 0, r \geq -1$ we define $d_{\ell,\lambda,r} := 2^{r+1}\ell + (2^{r+1} - 1)\lambda$. Moreover for all $\ell_1 \geq 1, \lambda_1 \geq 1, \ell_2 \geq 0, \lambda_2 \geq 0, r \geq -1$ we introduce a polynomial $D_{\ell_1,\lambda_1,\ell_2,\lambda_2,r}(X) \in \mathbf{Z}[X]$, which is recursively defined by putting $D_{\ell_1,\lambda_1,\ell_2,\lambda_2,-1}(X) = 1 + X + X^2 + \ldots + X^{\ell_1+\ell_2-1}$ and for $r \geq 0$ letting $D_{\ell_1,\lambda_1,\ell_2,\lambda_2,r}(X)$ be $X^{2d_{\ell_2,\lambda_2,r-1}} + 2X^{d_{\ell_2,\lambda_2,r-1}}D_{\ell_1,\lambda_1,\ell_2,\lambda_2,r-1}(X) + 3D_{\ell_1,\lambda_1,\ell_2,\lambda_2,r-1}(X)^2$ multiplied with $1 + X + X^2 + \ldots + X^{\lambda_1+\lambda_2-1}$. We then define $c_{\ell_1,\lambda_1,\ell_2,\lambda_2,r} = \|D_{\ell,\lambda,r}(X)\|_\infty$ where as before $\|\cdot\|_\infty$ denotes the maximal absolute value of the coefficients. Note that $\deg D_{\ell,\lambda,r}(X) = d_{\ell_1+\ell_2-1,\lambda_1+\lambda_2-1,r}$. This all looks a bit cumbersome but the idea underlying these definitions should become apparent from the proof below.

Lemma 1. *Suppose that the input values $x_1, x_2, \ldots, x_{n_0}$ resp. the coefficients b_{ijk} are given by balanced ternary expansions of at most ℓ_1 resp. λ_1 integral digits and ℓ_2 resp. λ_2 fractional digits. Let x_{out} be the evaluation of our GMDH network at the x_i's, obtained by using fixed-point arithmetic. Let $\phi(X) \in R_t$ be the evaluation of our GMDH network at the encodings of the x_i's (using the encodings of the b_{ijk}'s as coefficients), obtained by using the respective ring operations in R_t. If*

$$t \geq 2 \cdot c_{\ell_1, \lambda_1, \ell_2, \lambda_2, r} \quad \text{and} \quad d \geq 2 \cdot \max\{d_{\ell_1 + \ell_2 - 1, \lambda_1 + \lambda_2 - 1, r}, d_{\ell_2, \lambda_2, r} + 1\}$$

then $\phi(X)$ decodes to x_{out}.

Proof. Consider the evaluation of our GMDH network when carried out in $\mathbf{Z}[X, X^{-1}]$, using encodings of the form (2). We claim that the outcome is of the form $X^{-m}g(X)$ with $m \leq d_{\ell_2, \lambda_2, r}$ and $g(X) \in \mathbf{Z}[X]$ of degree at most $d_{\ell_1 + \ell_2 - 1, \lambda_1 + \lambda_2 - 1, r}$ and having coefficients bounded (in absolute value) by $c_{\ell_1, \lambda_1, \ell_2, \lambda_2, r}$. This claim clearly implies the lemma.

The key observation is that if one replaces all inputs by $X^{-\ell_2} + X^{-\ell_2 + 1} + X^{-\ell_2 + 2} + \ldots + X^{\ell_1 - 1}$ while replacing all encoded b_{ijk}'s by $X^{-\lambda_2} + X^{-\lambda_2 + 1} + X^{-\lambda_2 + 2} + \ldots + X^{\lambda_1 - 1}$ then these quantities can only increase, by the triangle inequality for the absolute value. By induction on r, it is easy to show that the corresponding evaluation is precisely $X^{-d_{\ell_2, \lambda_2, r}} \cdot D_{\ell_1, \lambda_1, \ell_2, \lambda_2, r}(X)$, from which the claim follows. ∎

These bounds are easy to compute in practice, using a computer algebra package. For example with $\ell_1 = 4$, $\ell_2 = 1$, $\lambda_1 = 1$, $\lambda_2 = 8$ and $r = 3$, we obtain the bounds

$$t \geq 93659577705415581454099599864654 \approx 2^{106.207} \quad \text{and} \quad d \geq 368. \quad (3)$$

This concrete choice of parameters will reoccur later in the paper.

One sees that the obtained bound on t is very large, which is problematic for a direct application of the FV scheme: remember from the previous section that we need $t \leq 396$. To address this issue we follow an idea mentioned in [4, Sect. 5.5], namely to decompose the plaintext space using the Chinese Remainder Theorem (CRT). That is, if one lets t be a large enough product of small mutually coprime numbers $t_1, t_2, \ldots, t_m$ then we have the well-known ring isomorphism

$$R_t \to R_{t_1} \times R_{t_2} \times \ldots \times R_{t_m} : g(X) \mapsto (g(X) \bmod t_1, \ldots, g(X) \bmod t_m).$$

Instead of evaluating our GMDH network directly in R_t, we can work in each of the R_{t_i}'s separately, simply by reducing things modulo t_i. The outcomes can then be combined very efficiently in order to end up in R_t again. As a consequence it suffices to carry out the FV scheme using the much smaller plaintext spaces R_{t_i}, although one needs to do it for each i separately. For the above example, the 13 mutually coprime numbers 269, 271, 277, 281, 283, 285, 286, 287, 289, 293, 307, 311 and 313 multiply together to $t = 9505948353308781246117151527620 \approx 2^{106.229}$, which indeed satisfies the bound from (3). Thus it suffices to work with $R_{269}, R_{271}, \ldots, R_{313}$.

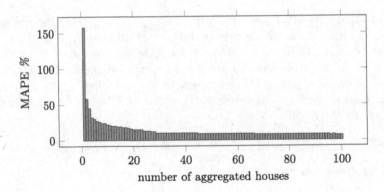

Fig. 3. The MAPE when forecasting the power consumption for the next half hour when using a varying number of aggregated households.

6   Prediction Approach for the Smart Grid

6.1   Prediction Model: Apartment Complexes

It is known that it is intrinsically difficult to make accurate short-term predictions based on data from one household when using an artificial neural network [38], and the same volatile behaviour is to be expected when following a GMDH approach. In order to confirm this, we designed and trained for each value of $n = 1, \ldots, 100$ a GMDH network that predicts the energy consumption during the next half hour for n aggregated households. This was done along the design criteria (and using the data set) described in Sect. 6.2 below. The observed prediction qualities, expressed in terms of the mean absolute percentage error (MAPE), are given in Fig. 3. One sees that the results for one household are particularly bad, showing a MAPE of over 158%. However, the results start to improve significantly when using aggregated measurements of 10 households: here the MAPE is slightly above 20%, while it drops to 7% for $n = 100$. These observations are well in line with the ones for ANNs [38]. Due to this volatile nature we decided to aim for *aggregated* prediction, albeit for a *low* number of households. More precisely, we chose $n = 10$, which matches small apartment complexes in rural areas.

The cryptographic setting we have in mind is that the individual meter readings are homomorphically encrypted by the smart meter or gateway, and then sent to a third party who will perform homomorphic computations. Our security assumption is that the third party is honest but curious: it runs the protocol and computations as specified (i.e. it evaluates the GMDH network), but it will try to learn as much as possible about its inputs and outputs. The third party has received the concrete parameters (such as the coefficients b_{ijk}) of a trained GMDH network from the final party who wants to know the consumption prediction (e.g. the electricity supplier or the network operator). After homomorphically aggregating the data per 10 households, the third party obtains the encrypted inputs $x_1, x_2, \ldots, x_{n_0}$ on which the GMDH network is evaluated homomorphically. The result is an encrypted forecast, which is then forwarded to the final party, who is able to decrypt using the private cryptographic key corresponding to the public key which is installed in the smart meters. The second security assumption we have to make is that the third party does not collude with the final party, since otherwise the third party could simply forward the encrypted individual meter readings.

6.2 Design of the Network

As explained in Sect. 3 the exact layout of our GMDH network is determined during a learning phase, for which we need access to some real smart meter data. We used the data that was collected through the Irish smart metering electricity customer behaviour trials [5] which ran in 2009 and 2010 with over 5,000 Irish homes and businesses participating. The data consists of electricity consumed during 30 min intervals (in kW). Per household there are $25,728$ electricity measurements for a total of 536 days. We use the measurements of the first year as training data and the remaining half year to validate and measure how good the network is performing.

An important balancing act is to find a network layout that minimizes the number of layers (and therefore the multiplicative depth of the prediction algorithm) while at the same time preserving a reasonable prediction accuracy, preferably comparable to [38]. Through some trial and error the simplest GMDH network we found to meet these requirements consists of $r = 3$ hidden layers with $n_1 = 8$, $n_2 = 4$ and $n_3 = 2$ nodes, respectively. As input layer a set of $n_0 = 51$ nodes is used, where 48 nodes represent the half hour measurements that were made during the previous 24 h. The remaining 3 inputs correspond to the temperature, the month, and the day of the week. The single output node $\nu_{4,1}$ then returns the predicted electricity consumption for the next half hour.

Let $\tilde{f} : \mathbf{R}^{51} \to \mathbf{R}$ denote the function that we want to approximate, for which a set of m input-output pairs $((x_{i1}, x_{i2}, \ldots, x_{in_0}), y_i^{\mathrm{actual}})_{i=1,\ldots,m}$, with $y_i^{\mathrm{actual}} = \tilde{f}(x_{i1}, x_{i2}, \ldots, x_{in_0})$, is given through the Irish data set. As explained in Sect. 3 these are used to inductively determine the coefficients b_{ijk}, while at the same time selecting the best performing nodes. Assuming that layer $i - 1$ was dealt with, for node ν_{ij} this is done by minimizing the quantity

$$\mathrm{MSE}\left(\left(f_{ij}(x_{11}, \ldots, x_{1n_0}), \ldots, f_{ij}(x_{m1}, \ldots, x_{mn_0})\right), (y_1^{\mathrm{actual}}, \ldots, y_m^{\mathrm{actual}})\right),$$

where $f_{ij} : \mathbf{R}^{51} \to \mathbf{R}$ denotes the function obtained from the network by temporarily considering ν_{ij} as an output node. The minimization can be done using standard linear regression. The useful feature of this approach is that one can apply L2-regularization and kill two birds with one stone. On the one hand regularization helps to avoid the *overfitting* problem, while on the other hand it allows to control the magnitude of the b_{ijk}'s. In this way one can achieve that ν_{ij} is a quadratic polynomial function with small coefficients and a reasonable MSE. We would like to point out that while we use MSE in the learning phase, the quality of the eventually resulting GMDH network is measured in terms of MAPE, in order to allow for a meaningful comparison with the forecasting results reported upon in the scientific literature.

As outlined in Sect. 5 we carry out fixed-point arithmetic using balanced ternary expansions, rather than binary expansions. To represent the input values $x_1, x_2, \ldots, x_{n_0}$ we use 1 fractional digit and, since the maximal data value is 27.265, at most 4 integral digits. The coefficients b_{ijk} are represented using 1 integral and 8 fractional digits. With these choices we attain basically the same average MAPE around 21% as in the floating point setting: a further increase of the precision does not give any significant improvement, although it gradually makes the fixed-point MSE converge to the floating-point one.

6.3   Benchmark Results

In order to assess the practical performance and verify the correctness of our selected parameters we implemented the privacy-preserving homomorphic forecasting approach as introduced in this paper. Our implementation (which will be made publicly avaiable soon) uses the FV-NFLlib software library [7] which implements the FV homomorphic encryption scheme which in turn uses the NFLlib software library (as described in [29] and released at [8]) for computing polynomial arithmetic. Our presented benchmark figures are obtained when running the implementation on an average laptop equipped with an Intel Core i5-3427U CPU (running at 1.80 GHz).

Let us recall and summarize the exact forecasting setting and the parameters we selected for the implementation. It is our goal to predict the energy consumption for the next half hour of an apartment complex of 10 households while not revealing any energy consumption information to the party computing on this data using the GMDH approach as outlined in Sect. 3. Inherent to this approach we expect a MAPE which is slightly over 20% (see Sect. 6.1). In order to work efficiently with real numbers we use the fixed-point representation with the parameters as outlined in Sect. 5, using the CRT approach for decomposing plaintext space. We use the FV scheme for the homomorphic computation with the parameters as presented in Sect. 4. Hence, we target a security level of 80 bits and use the ring $R_{2^{186}} = \mathbf{Z}_{2^{186}}[X]/(X^{2^{12}} + 1)$ along with a standard deviation of 102. This means a ciphertext size of 186 kB. Recall that the coefficients b_{ijk} are not being encrypted, which limits the noise growth when carrying out scalar multiplications.

As outlined in Sect. 6.2 the layout of our network consists of an input layer of 51 nodes, three hidden layers of 8, 4 and 2 nodes respectively and a single

Table 1. The time (in ms) to compute the various basic (homomorphic) operations for our selected parameters.

op	enc	dec	key gen	add	mul	scalar mul
ms	2.1	5.8	77	0.1	33	29

output node. Remember that when building a new layer the learning algorithm excludes nodes corresponding to node pairs from the previous layer. So not all nodes of the resulting GMDH network affect on the final output and thus can be ignored during evaluation. Each node performs 8 multiplications out of which 5 are by polynomial coefficients and 5 additions. Since there are at most 15 nodes being evaluated this means computing 120 multiplications (out of which 75 by polynomial coefficients) and 75 additions. Table 1 summarizes the performance cost (expressed in milliseconds) for the various basic building blocks used in our homomorphic prediction algorithm. As can be seen from this table, and this is confirmed by running the entire forecasting algorithm in practice, the average computation of the prediction over 100 aggregated datasets is around 2.5 s depending on the node wiring. However, as explained in Sect. 5, this process has to be repeated 13 times for the CRT approach. In practice, the entire forecasting can be computed in half a minute. Due to the embarrassingly parallel nature of the CRT approach, a parallel implementation can compute this in less than 4 s or 2.5 s on average.

7 Conclusions and Future Work

We have shown that Ivakhnenko's group method of data handling from the 1970s is very suitable for homomorphic computation. This seems to be a better method with respect to the applicability to implement prediction homomorphically compared to the related artificial neural network based approaches in this cryptographic setting. We have studied this prediction approach in the setting of enhancing the privacy of the consumer for forecasting in the smart grid. Our privacy-preserving implementation of this approach to homomorphically forecast for 10 households shows is that this can be computed in less than four seconds for parallel and in half a minute for a sequential implementation.

We would like to point out that this approach has applications beyond the scope of just the smart grid. Other areas which need reliable prediction algorithms but work with privacy sensitive data can directly benefit as well. Examples include computing on financial data or biometric data.

References

1. Ahmad, A., Hassan, M., Abdullah, M., Rahman, H., Hussin, F., Abdullah, H., Saidur, R.: A review on applications of ANN and SVM for building electrical energy consumption forecasting. Renew. Sustain. Energ. Rev. **33**, 102–109 (2014)
2. Albrecht, M.: Complexity estimates for solving LWE (2000–2004). https://bitbucket.org/malb/lwe-estimator/raw/HEAD/estimator.py
3. Albrecht, M.R., Player, R., Scott, S.: On the concrete hardness of learning with errors. J. Math. Cryptology **9**(3), 169–203 (2015)
4. Bos, J.W., Lauter, K., Loftus, J., Naehrig, M.: Improved security for a ring-based fully homomorphic encryption scheme. In: Stam, M. (ed.) IMACC 2013. LNCS, vol. 8308, pp. 45–64. Springer, Heidelberg (2013). doi:10.1007/978-3-642-45239-0_4

5. Commission for Energy Regulation: Electricity smart metering customer behaviour trials (CBT) findings report. Technical Report CER11080a (2011). http://www.cer.ie/docs/000340/cer11080(a)(i).pdf
6. Costache, A., Smart, N.P., Vivek, S., Waller, A.: Fixed point arithmetic in SHE schemes. In: SAC 2016. LNCS. Springer (2016)
7. CryptoExperts: FV-NFLlib (2016). https://github.com/CryptoExperts/FV-NFLlib
8. CryptoExperts, INP ENSEEIHT, and Quarkslab: NFLlib (2016). https://github.com/quarkslab/NFLlib
9. Department of Energy & Climate Change: Smart metering implementation programme. Technical Report Third Annual Report on the Roll-out of Smart Meters (2014). https://www.gov.uk/government/uploads/system/uploads/attachment_data/file/384190/smip_smart_metering_annual_report_2014.pdf
10. Department of Energy, Climate Change: Smart metering implementation programme - data access, privacy. https://www.gov.uk/government/uploads/system/uploads/attachment_data/file/43043/4933-data-access-privacy-con-doc-smart-meter.pdf
11. Dowlin, N., Gilad-Bachrach, R., Laine, K., Lauter, K., Naehrig, M., Wernsing, J.: Manual for using homomorphic encryption for bioinformatics. Technical report, Technical report MSR-TR-2015-87, Microsoft Research (2015)
12. Dowlin, N., Gilad-Bachrach, R., Laine, K., Lauter, K.E., Naehrig, M., Wernsing, J.: Cryptonets: applying neural networks to encrypted data with high throughput and accuracy. In: Balcan, M., Weinberger, K.Q. (eds.) International Conference on Machine Learning, vol. 48, pp. 201–210. JMLR.org (2016)
13. Erkin, Z., Tsudik, G.: Private computation of spatial and temporal power consumption with smart meters. In: Bao, F., Samarati, P., Zhou, J. (eds.) ACNS 2012. LNCS, vol. 7341, pp. 561–577. Springer, Heidelberg (2012). doi:10.1007/978-3-642-31284-7_33
14. European Commission: Commission recommendation of 9 on preparations for the roll-out of smart metering systems. Official Journal of the European Union (2012). http://eur-lex.europa.eu/legal-content/EN/ALL/?uri=CELEX:32012H0148
15. European Commission: Benchmarking smart metering deployment in the EU-27 with a focus on electricity. Technical Report 365, June 2014. http://eur-lex.europa.eu/legal-content/EN/TXT/PDF/?uri=CELEX:52014DC0356&from=EN
16. Fan, J., Vercauteren, F.: Somewhat practical fully homomorphic encryption. IACR Cryptology ePrint Archive 2012, 144 (2012)
17. Koo, B.G., Lee, S.W., Kim, W., Park, J.H.: Comparative study of short-term electric load forecasting. In: Conference on Intelligent Systems, Modelling and Simulation, pp. 463–467, January 2014
18. Garcia, F.D., Jacobs, B.: Privacy-friendly energy-metering via homomorphic encryption. In: Cuellar, J., Lopez, J., Barthe, G., Pretschner, A. (eds.) STM 2010. LNCS, vol. 6710, pp. 226–238. Springer, Heidelberg (2011). doi:10.1007/978-3-642-22444-7_15
19. Gentry, C.: Fully homomorphic encryption using ideal lattices. In: ACM Symposium on Theory of Computing - STOC 2009, pp. 169–178. ACM (2009)
20. Hart, G.W.: Nonintrusive appliance load monitoring. Proc. IEEE 80(12), 1870–1891 (1992)
21. Hernandez, L., Baladron, C., Aguiar, J.M., Carro, B., Sanchez-Esguevillas, A.J., Lloret, J., Massana, J.: A survey on electric power demand forecasting: future trends in smart grids, microgrids and smart buildings. IEEE Commun. Surv. Tutorials 16(3), 1460–1495 (2014)

22. Ivakhnenko, A.: Heuristic self-organization in problems of engineering cybernetics. Automatica **6**(2), 207–219 (1970)
23. Jawurek, M., Kerschbaum, F., Danezis, G.: Privacy technologies for smart grids - a survey of options. Technical Report MSR-TR-2012-119, November 2012. http://research.microsoft.com/apps/pubs/default.aspx?id=178055
24. Kursawe, K., Danezis, G., Kohlweiss, M.: Privacy-friendly aggregation for the smart-grid. In: Fischer-Hübner, S., Hopper, N. (eds.) PETS 2011. LNCS, vol. 6794, pp. 175–191. Springer, Heidelberg (2011). doi:10.1007/978-3-642-22263-4_10
25. Li, F., Luo, B., Liu, P.: Secure information aggregation for smart grids using homomorphic encryption. In: Smart Grid Communication, pp. 327–332. IEEE (2010)
26. Livni, R., Shalev-Shwartz, S., Shamir, O.: On the computational efficiency of training neural networks. In: Advances in Neural Information Processing Systems, pp. 855–863 (2014)
27. Lyubashevsky, V., Peikert, C., Regev, O.: On ideal lattices and learning with errors over rings. In: Gilbert, H. (ed.) EUROCRYPT 2010. LNCS, vol. 6110, pp. 1–23. Springer, Heidelberg (2010). doi:10.1007/978-3-642-13190-5_1
28. Lyubashevsky, V., Peikert, C., Regev, O.: On ideal lattices and learning with errors over rings. J. ACM **60**(6), 35 (2013). Article 43
29. Aguilar-Melchor, C., Barrier, J., Guelton, S., Guinet, A., Killijian, M.-O., Lepoint, T.: NFLLIB: NTT-based fast lattice library. In: Sako, K. (ed.) CT-RSA 2016. LNCS, vol. 9610, pp. 341–356. Springer, Cham (2016). doi:10.1007/978-3-319-29485-8_20
30. Molina-Markham, A., Shenoy, P.J., Fu, K., Cecchet, E., Irwin, D.E.: Private memoirs of a smart meter. In: Ruzzelli, A.G. (ed.) Workshop on Embedded Sensing Systems for Energy-Efficiency in Buildings, pp. 61–66. ACM (2010)
31. Paillier, P.: Public-key cryptosystems based on composite degree residuosity classes. In: Stern, J. (ed.) EUROCRYPT 1999. LNCS, vol. 1592, pp. 223–238. Springer, Heidelberg (1999). doi:10.1007/3-540-48910-X_16
32. Recommendation to the European Commission: Essential regulatory requirements and recommendations for data handling, data safety, and consumer protection. Technical Report version 1.0 (2011). https://ec.europa.eu/energy/sites/ener/files/documents/Recommendations
33. Rial, A., Danezis, G.: Privacy-preserving smart metering. In: Workshop on Privacy in the Electronic Society, WPES 2011, pp. 49–60. ACM (2011)
34. Rivest, R.L., Adleman, L., Dertouzos, M.L.: On data banks and privacy homomorphisms. Found. Secure Comput. **4**(11), 169–180 (1978)
35. Smart Grid Coordination Group: Smart grid information security, November 2012. http://ec.europa.eu/energy/sites/ener/files/documents/xpert_group.1_security.pdf
36. Srinivasan, D.: Energy demand prediction using GMDH networks. Neurocomputing **72**(1), 625–629 (2008)
37. The Smart Grid Interoperability Panel - Smart Grid Cybersecurity Committee: Guidelines for smart grid cybersecurity: volume 1 - smart grid cybersecurity strategy, architecture, and high-level requirements. Technical Report NISTIR 7628 Rev 1 (2014). http://nvlpubs.nist.gov/nistpubs/ir/2014/NIST.IR.7628r1.pdf
38. Veit, A., Goebel, C., Tidke, R., Doblander, C., Jacobsen, H.-A.: Household electricity demand forecasting: benchmarking state-of-the-art methods. In: Conference on future energy systems, pp. 233–234. ACM (2014)
39. Xie, P., Bilenko, M., Finley, T., Gilad-Bachrach, R., Lauter, K.E., Naehrig, M.: Crypto-nets: neural networks over encrypted data. CoRR, abs/1412.6181 (2014)

Number Theory

On Indifferentiable Hashing into the Jacobian of Hyperelliptic Curves of Genus 2

Michel Seck[1], Hortense Boudjou[2], Nafissatou Diarra[1]([⊠]),
and Ahmed Youssef Ould Cheikh Khlil[1]

[1] Department of Mathematics-Informatics,
Cheikh Anta Diop University, Dakar, Senegal
{michel.seck,nafissatou.diarra,ahmed.youssef}@ucad.edu.sn
[2] Maroua University, Maroua, Cameroon
hortense_boudjou@yahoo.fr

Abstract. Many authors have studied the problem of constructing indifferentiable and deterministic hash functions into elliptic and hyperelliptic curves with well-distributed encodings. In this work, we have designed three encodings suitable for indifferentiable hashing for the following hyperellitic curves of genus 2: $\mathbb{H}^1 : y^2 = F_1(x) = x^5 + ax^4 + cx^2 + dx$, $\mathbb{H}^2 : y^2 = F_2(x) = x^5 + bx^3 + dx + e$; $\mathbb{H}^3 : y^2 = F_3(x) = x^5 + ax^4 + e$. Since they are *well-distributed*, our encodings can be used to design indifferentiable and deterministic hash functions into the Jacobian of these hyperelliptic curves, using the technique developed by Farashahi *et al.* in 2013 (J. Math. Comput). Because of square rooting steps, these new encodings have the same asymptotic complexity as the work of Kammerer *et al.* at Pairing 2010, namely $\mathcal{O}(\log^{2+o(1)} q)$.

Keywords: Indifferentiable deterministic hashing · Injective encoding · Elliptic curve-based cryptography · Jacobian · Elligator · Random bit-string

1 Introduction

Elliptic curves based cryptography offers services that combine high speed, high security and small space consumption. It is thus increasingly desired by the designers of cryptographic protocols. Typically in the identity-based schemes, most often the identity of the user is associated to a point on the (hyper)elliptic curve, like in Franklin and Boneh's encryption scheme [2] for the particular case of supersingular curves. Also, when using a protocol that needs a hash function, we need to be able to encode on the group attached to the curve, namely the group of points of an elliptic curve or the Jacobian of an hyperelliptic curve. For the case of elliptic curves, there exist many deterministic encodings [1,5,7,9,12,17] and constructions of hash functions based on well-distributed encodings [3,6,8,18]. In 2007, Ulas [19] designed deterministic encodings for the family of hyperelliptic curves of the form $y^2 = x^n + ax + b$ or $y^2 = x^n + ax^2 + bx$,

© Springer International Publishing AG 2017
M. Joye and A. Nitaj (Eds.): AFRICACRYPT 2017, LNCS 10239, pp. 205–222, 2017.
DOI: 10.1007/978-3-319-57339-7_12

where $n \geq 5$. Ulas's encoding is based on Skalba equalities and requires to compute a square root in $\mathbb{F}_q$. A simplified version of Ulas's encoding was later proposed by Brier *et al.* in [3] in CRYPTO 2010, where they also explain how to construct hash functions indifferentiable from random oracles and based on deterministic encodings over elliptic curves (like Icart's function [12] or SWU's algorithm [19]).

In 2010, Kammerer *et al.* [13] proposed new encodings for some models of hyperelliptic curves, including $y^2 = (x^3 + 3ax + 2)^2 + 8bx^3$ over the field $\mathbb{F}_q$, with $q \equiv 2 \mod 3$. Farashahi *et al.* [8] have proposed another technique based on character sums which established the indifferentiability of hash functions based on the existence of deterministic well-distributed encodings into Jacobians of hyperelliptic curves, with application on the hyperelliptic curve proposed by Kammerer *et al.* [13]. In the case of odd hyperelliptic curves(which are curves defined by $y^2 = f(x)$ with $f(-x) = -f(x)$), Fouque and Tibouchi using the technique of Farashahi *et al.* [8], have proposed an explicit method to obtain elements of the Jacobian with injective well-distributed encodings on the associated hyperelliptic curves.

Contributions: It is well-known that cryptography based on hyperelliptic curves relies on hashing into the Jacobian of the underlying curve. And to do this, one needs to be able to compute a point of the curve in deterministic polynomial time. Our main contribution in this paper is the construction of deterministic encodings for three families of hyperelliptic curves that were not covered by the previous works. Moreover, we show that each of these encodings is $2 : 1$, invertible under some conditions and can be extended to all $\mathbb{F}_q$. Once this was done, and using the definitions of Farashahi *et al.* [8], we show how one can use these new *almost-injective* encodings to construct indifferentiable hash functions into the Jacobian of the hyperelliptic curves $\mathbb{H}_i$, $i = 1, 2, 3$.

Note that all these encodings can be computed in $\mathcal{O}(\log^{2+o(1)} q)$ operations over $\mathbb{F}_q$.

2   Preliminaries

Let q be an odd prime power.

1. The **quadratic character** is the function χ defined by $\chi : \mathbb{F}_q \to \mathbb{F}_q : u \mapsto \chi(u) = u^{(q-1)/2}$ and verifying: $\chi(u) = 1$ if u is a non-zero square; $\chi(u) = -1$ if u is a non-square; and $\chi(u) = 0$ if $u = 0$. The following properties are also verified: $\chi(uv) = \chi(u) \cdot \chi(v)$ for any $u, v \in \mathbb{F}_q$; $\chi(a^2) = 1$ for any $a \in \mathbb{F}_q^*$; and if $q \equiv 3 \mod 4$, $\chi(-1) = -1$, $\chi(\chi(u)) = \chi(u)$, for any $u \in \mathbb{F}_q$. If $q \equiv 1 \mod 4$, then $\chi(-1) = 1$.

2. Let $q \equiv 3 \mod 4$ and let A be the set of quadratic residues over the finite field $\mathbb{F}_q$. Define the square root function $\sqrt{\cdot}$ on A as follows:

$$\sqrt{\cdot} : A \to \mathbb{F}_q : a \mapsto \sqrt{a} = a^{(q+1)/4}$$

Then $\sqrt{a}$ is called the **principal square root** of a. Also note that if q is an odd prime, one can take $\sqrt{A} = \{0, 1, \ldots, (q-1)/2\}$.

3   *Almost-Injective* and Invertible Encodings into Three Families of Hyperellitic Curves

In this section, we propose 3 *almost-injective* encodings for hyperelliptic curves following the technique of Elligator [1] for elliptic curves. All our encodings are *almost-injective* in the following sense:

Let $\phi_i : \mathbb{F}_q \to \mathbb{H}^i$ for (i = 1, 2 or 3) be an encoding verifying:

- $\phi_i(r) = \phi_i(-r)$, $\forall r \in \mathbb{F}_q^*$;
- and $\#\phi_i^{-1}(\phi_i(r)) = 2$, $\forall r \in \mathbb{F}_q^*$, where $\#\phi_i^{-1}(\phi_i(r))$ denotes the number of elements of the set $\phi_i^{-1}(\phi_i(r))$.

So for a subset $S \subset \mathbb{F}_q$ verifying $S \bigcap(-S) = \{0\}$, the encoding $\phi_i : S \to \mathbb{H}^i$ is an injective one, that we call an *almost-injective* encoding.

To our knowledge, this paper is not covered by the previous works (summarized in the introduction) on how to encode into hyperelliptic curves. Note that in [8], the encoding works for odd hyperelliptic curves. In [3,13,19], the encodings are not almost-injective. All the encodings we propose here are almost-injective and can be used for non-odd hyperelliptic curves. Furthermore, in Sect. 4, we prove that our encodings are well-distributed; hence they can be used for indifferentiable hashing.

3.1   An *Almost-Injective* Encoding on $\mathbb{H}^1$

Let us consider the family of hyperelliptic curves $\mathbb{H}^1$ over $\mathbb{F}_q$ ($q = p^n$ where p is a prime, with $p = \text{char}(\mathbb{F}_q) \neq 2,5$), given by the equation:

$$\mathbb{H}^1 : y^2 = F_1(x) = x^5 + ax^4 + cx^2 + dx \text{ with } 8ad^2 + c^3 = 0 \text{ where } a,c,d \in \mathbb{F}_q^*.$$

Let u be a parameter such that $\chi(u) = -1$ and define the set

$$\mathcal{R}_1 = \left\{ r \in \mathbb{F}_q^*, \ ur^2c^2 + 4d^2 \neq 0; \right.$$

$$\left. -d - \frac{c^2}{2d}ur^2 + d(\frac{ur^2c^2}{4d^2} + 1)^3 \neq -\frac{16d^4(\frac{ur^2c^2}{4d^2} + 1)^4}{c^4} \right\}.$$

Input : The hyperelliptic curve $\mathbb{H}^1$ and r an element of $\mathcal{R}_1$.
Output: A point (x, y) on $\mathbb{H}^1$.
$v := -\dfrac{2d(\frac{ur^2c^2}{4d^2} + 1)}{c}$;
$\varepsilon := \chi(v^5 + av^4 + cv^2 + dv)$;
$x := \dfrac{v}{2}(\dfrac{vc + \varepsilon(vc + 4d)}{vc + 2d})$;
$y := -\varepsilon\sqrt{x^5 + ax^4 + cx^2 + dx}$;
return (x, y).

Algorithm 1. Hashing-1-Genus2-Encoding

Definition 1. *In the situation of Algorithm 1 the encoding function for $\mathbb{H}^1$ is the function $\phi_1 : \mathcal{R}_1 \to \mathbb{H}^1 : r \mapsto \phi_1(r) = (x, y)$. The map ϕ_1 is extending at 0 by $\phi_1(0) = (0, 0)$.*

Theorem 1. *Algorithm 1 computes a deterministic almost-injective encoding $\phi_1 : \mathcal{R}_1 \to \mathbb{H}^1 : r \mapsto \phi_1(r) = (x, y)$, in time $\mathcal{O}(\log^{2+o(1)} q)$, where $\mathcal{S}_1 = \mathbb{F}_q \setminus \mathcal{R}_1$ is a subset of $\mathbb{F}_q$ of at most 10 elements.*

Proof. 1. v exists and is nonzero since $cd \neq 0$ and $ur^2 c^2 + 4d^2 \neq 0$ by hypothesis.

2. $\varepsilon \neq 0$ if $F_1(v(r^2)) \neq 0$. Since $v = v(r^2) \neq 0$, using the definition of v and the condition $8ad + c^3 = 0$ we have:

$$\varepsilon \neq 0$$

$$\Leftrightarrow \left(-\frac{2d(\frac{ur^2 c^2}{4d^2}+1)}{c} \right)^4 + a \left(-\frac{2d(\frac{ur^2 c^2}{4d^2}+1)}{c} \right)^3 + c \left(-\frac{2d(\frac{ur^2 c^2}{4d^2}+1)}{c} \right) + d \neq 0$$

$$\Leftrightarrow \frac{16d^4(\frac{ur^2 c^2}{4d^2}+1)^4}{c^4} - \frac{8ad^3(\frac{ur^2 c^2}{4d^2}+1)^3}{c^3} - 2d(\frac{ur^2 c^2}{4d^2}+1) + d \neq 0$$

$$\Leftrightarrow \frac{16d^4(\frac{ur^2 c^2}{4d^2}+1)^4}{c^4} + d(\frac{ur^2 c^2}{4d^2}+1)^3 - \frac{c^2}{2d}ur^2 - d \neq 0$$

$$\Leftrightarrow -d - \frac{c^2}{2d}ur^2 + d(\frac{ur^2 c^2}{4d^2}+1)^3 \neq -\frac{16d^4(\frac{ur^2 c^2}{4d^2}+1)^4}{c^4},$$

which is true by definition of $\mathcal{R}_1$.

3. We have $x = \frac{v}{2}\left(\frac{vc + \varepsilon(vc + 4d)}{vc + 2d} \right)$: Note that $\varepsilon \neq 0$ by (2). Let us verify that x exists and is nonzero.

   - x exists if and only if $vc + 2d \neq 0 \Leftrightarrow v \neq -\frac{2d}{c} \Leftrightarrow -\frac{2d(\frac{ur^2 c^2}{4d^2}+1)}{c} \neq -\frac{2d}{c} \Leftrightarrow r \neq 0$ which is true by definition of $\mathcal{R}_1$.
   - Since $x = v$ (if $\varepsilon = 1$) or $x = \frac{\frac{-2d}{c}v}{v + \frac{2d}{c}}$ (if $\varepsilon = -1$) and $v \neq 0$ by (1), then $x \neq 0$.

4. We have $x = \frac{v}{2}\left(\frac{vc + \varepsilon(vc + 4d)}{vc + 2d} \right)$. Note that $\varepsilon \neq 0$ by (2).
   - If $\varepsilon = 1$ then $x = v$ and $F_1(v)$ is a nonzero square. Hence, $y = -\varepsilon\sqrt{x^5 + ax^4 + cx^2 + dx} = -\varepsilon\sqrt{v^5 + av^4 + cv^2 + dv}$ is well defined and is nonzero.
   - If $\varepsilon = \chi(v^5 + av^4 + cv^2 + dv) = -1$ then $x = \frac{\frac{-2d}{c}v}{v + \frac{2d}{c}}$ and

$$x^5 + ax^4 + cx^2 + dx = \left(\frac{\frac{-2d}{c}v}{v + \frac{2d}{c}} \right)^5 + a\left(\frac{\frac{-2d}{c}v}{v + \frac{2d}{c}} \right)^4 + c\left(\frac{\frac{-2d}{c}v}{v + \frac{2d}{c}} \right)^2 + d\left(\frac{\frac{-2d}{c}v}{v + \frac{2d}{c}} \right)$$

$$= \frac{(\frac{-2d}{c})^5}{(v + \frac{2d}{c})^5}\left(v^5(1 - \frac{ac}{2d} - \frac{c^4}{16d^3}) + v^4(-a - \frac{c^3}{4d^2}) + v^3(-\frac{3c^2}{2d} + \frac{3c^2}{2d}) + v^2(-c + 2c) + dv \right)$$

$$= \left(\frac{(-\frac{2d}{c})^5}{(v + \frac{2d}{c})^5} \right)(v^5 + av^4 + cv^2 + dv),$$

since $8ad^2 + c^3 = 0$.

Therefore $\chi(x^5 + ax^4 + cx^2 + dx) = \chi((-\frac{2d}{c})(v + \frac{2d}{c}))\chi(v^5 + av^4 + cv^2 + dv)) = \chi(u)\chi(r^2)\chi(v^5 + av^4 + cv^2 + dv) = (-1)(1)(-1) = 1$, since $r \neq 0$ in $\mathcal{R}_1$. Hence $x^5 + ax^4 + cx^2 + dx$ is a nonzero square, thus y is well defined and is nonzero. $\qquad\square$

Lemma 1. *In the situation of Theorem 1 and Definition 1, we have* $\phi_1(r) = \phi_1(-r)$, $\forall r \in \mathcal{R}_1$ *and* $\#(\phi_1^{-1}(\phi_1(r))) = 2$, $\forall r \in \mathcal{R}_1$.

Proof. – We have $\phi_1(r) = (x, y)$ with $x := \frac{v}{2}\left(\frac{vc+\varepsilon(vc+4d)}{vc+2d}\right)$ and $v := -\frac{2d(\frac{ur^2c^2}{4d^2}+1)}{c}$. Since $v = v(r^2)$ then $\phi_1(r) = \phi_1(-r)$.
- Let $r, r' \in \mathcal{R}_1$ such that $\phi_1(r) = \phi_1(r')$. As in Algorithm 1 for r (resp r') define v, ε, x, y (resp v', ε', x', y'). Since $x = x'$ and $y = y'$ then $-\varepsilon\sqrt{F_1(x)} = -\varepsilon'\sqrt{F_1(x)}$. Hence $\varepsilon = \varepsilon'$. From $x = x'$ we deduce that $v = v'$ (for $\varepsilon = \varepsilon' = 1$ and $\varepsilon = \varepsilon' = -1$) thus $r^2 = r'^2$. Which implies that $r' = \pm r$. $\qquad\square$

Input : The Hyperelliptic curve $\mathbb{H}^1$ and $(x, y) \in \mathbb{H}^1$.
Output: $\bar{r}$ such that $\phi_1(\bar{r}) = (x, y)$, $\bar{r} \in \mathcal{R}_1$ or $\perp$ (which means that "(x, y) is not in $\phi_1(\mathcal{R}_1)$").
if $-2cdu(x + 2d/c)$ *is a non-zero square* **then**
    if $y = \sqrt{f(x)}$ **then**
        $\bar{r} := \frac{2d}{c}\sqrt{-\frac{2d}{cu(x+\frac{2d}{c})}}$;
    else
        if $y = -\sqrt{f(x)}$ **then**
            $\bar{r} := \sqrt{-\frac{2d}{cu}(x + \frac{2d}{c})}$;
        end
    end
    return $\bar{r}$;
else
    return $\perp$.
end

Algorithm 2. Hashing-1-Genus2-Inverting

Theorem 2. *In the situation of Theorem 1 and Definition 1, Algorithm 2 defines the decoding function of ϕ_1. Furthermore, $\mathrm{Im}(\phi_1)$ is the set of $(x, y) \in \mathbb{H}^1$ verifying $-2cdu(x + 2d/c)$ is a nonzero square in $\mathbb{F}_q$.*

Proof. 1. Fix $r \in \mathcal{R}_1$, pose $\phi_1(r) = (x, y)$. As defined in $\mathcal{R}_1$, $r \neq 0$ then (x, y) is defined in Theorem 1.
- If $y \in \sqrt{\mathbb{F}_q^2}$ then $x = \frac{-\frac{2d}{c}v}{v+\frac{2d}{c}} \neq 0$ since $v \neq 0$, thus $y \neq 0$ and $-2cdu\big(x + \frac{2d}{c}\big) = -2cdu\left(\frac{-\frac{2d}{c}v}{v+\frac{2d}{c}} + \frac{2d}{c}\right) = -2cdu\left(\frac{-\frac{2d}{c}v}{v+\frac{2d}{c}}\right) - 4d^2u = 4d^2u\left(\frac{v}{v+\frac{2d}{c}} - 1\right) = \frac{16d^4}{c^2r^2}$ is a nonzero square.

- If $y \notin \sqrt{\mathbb{F}_q^2}$ then $x = v \neq 0$ thus $y \neq 0$ and $-\frac{2d}{cu}\left(x + \frac{2d}{c}\right) = -\frac{2d}{cu}\left(v + \frac{2d}{c}\right) = -\frac{2d}{cu}\left(-\frac{2d\left(\frac{ur^2c^2}{4d^2}+1\right)}{c} + \frac{2d}{c}\right) = \frac{4d^2}{c^2u}\left(\frac{ur^2c^2}{4d^2}\right) + \frac{4d^2}{c^2u} - \frac{4d^2}{c^2u} = r^2$. Thus $-2dcu\left(x + \frac{2d}{c}\right) = u^2r^2c^2$ is a nonzero square.

Assume that $(x, y) \in \mathbb{H}^1$ and that $-2cdu\left(x + \frac{2d}{c}\right)$ is a nonzero square. Let show that $(x, y) \in \phi_1(\mathcal{R}_1)$.

Since $-2cdu\left(x + \frac{2d}{c}\right)$ is a nonzero square, then $\bar{r}$ is defined and nonzero.

Define $\bar{v}, \bar{\varepsilon}, \bar{x}, \bar{y}$ as in the algorithm of Theorem 1.

If $y \in \sqrt{\mathbb{F}_q^2}$ then $\bar{r}^2 = -\frac{(2d)^3}{c^3u\left(x+\frac{2d}{c}\right)}$ so $u\bar{r}^2 = \frac{-\frac{2^3d^3}{c^3}}{\left(x+\frac{2d}{c}\right)}$ hence, we have $\bar{v} = \frac{-\frac{2d}{c}x}{\left(x+\frac{2d}{c}\right)}$. After some computations, using $8ad^2 + c^3 = 0$, we deduce that: $\bar{v}^5 + a\bar{v}^4 + c\bar{v}^2 + d\bar{v} = \left(\frac{-\frac{2d}{c}}{x+\frac{2d}{c}}\right)^5 (x^5 + ax^4 + cx^2 + dx)$.

Therefore $\chi(\bar{v}^5 + a\bar{v}^4 + c\bar{v}^2 + d\bar{v}) = \chi\left(\frac{-\frac{2d}{c}}{\left(x+\frac{2d}{c}\right)} \cdot (x^5 + ax^4 + cx^2 + dx)\right) = \chi\left(-2dc\left(x + \frac{2d}{c}\right)\right) \cdot \chi(x^5 + ax^4 + cx^2 + dx) = \chi(u)\chi(x^5 + ax^4 + cx^2 + dx) = -1 \cdot 1 = -1$, since $-2dcu\left(x+\frac{2d}{c}\right)$ and $x^5 + ax^4 + cx^2 + dx$ are nonzero square.

We deduce that $\bar{\varepsilon} = \chi(\bar{v}^5 + a\bar{v}^4 + c\bar{v}^2 + d\bar{v}) = -1$. Thus, we have $\bar{x} = \frac{-\frac{2d}{c}\bar{v}}{\left(\bar{v}+\frac{2d}{c}\right)} = x$ and $\bar{y} = -\bar{\varepsilon}\sqrt{F_1(x)} = \sqrt{F_1(x)} = y$.

Otherwise, if $y \notin \sqrt{\mathbb{F}_q^2}$, i.e., $y = -\sqrt{F_1(x)}$ then $\bar{r}^2 = -\frac{2d}{cu}\left(x + \frac{2d}{c}\right)$ so $u\bar{r}^2c^2 + 4d^2 = -2cdx$ so $\bar{v} = x$ and $F_1(\bar{v}) = F_1(x)$. Thus $\bar{\varepsilon} = 1$. Consequently $\bar{x} = \bar{v} = x$ and $\bar{y} = -\bar{\varepsilon}\sqrt{F_1(x)} = -\sqrt{F_1(x)} = y$.

Since for all v, we have $v^5 + av^4 + cv^2 + dv \neq 0$, then $-d - \frac{c^2}{2d}ur^2 + d\left(\frac{ur^2c^2}{4d^2}+1\right)^3 \neq -\frac{16d^4\left(\frac{ur^2c^2}{4d^2}+1\right)^4}{c^4}$, thus $\bar{r} \in \mathcal{R}_1$.

2. Follows from the above proof. $\square$

Remark 1 (Extension to $\mathbb{F}_q$).

- If $q \equiv 1 \mod 4$, and $\frac{F_1(x)}{x} = x^4 + ax^3 + cx + d$ don't have a root which is square, then ϕ_1 is defined in $\mathbb{F}_q$.
- If $q \equiv 3 \mod 4$, and $\frac{F_1(x)}{x} = x^4 + ax^3 + cx + d$ don't have a root which is square, then ϕ_1 is defined in $\mathbb{F}_q \setminus \left\{\pm\sqrt{\frac{-4d^2}{c^2u}}\right\}$. We can extend ϕ_1 at $\pm\sqrt{\frac{-4d^2}{c^2u}}$ as follows. Since $a = \frac{-c^3}{8d^2}$, choose d or c such that $\frac{-2d}{c}$ is a square, therefore $F_1\left(\frac{-2d}{c}\right)$ is a square, thus we can put $\phi_1\left(\pm\sqrt{\frac{-4d^2}{c^2u}}\right) = \left(\frac{-2d}{c}, \sqrt{F_1\left(\frac{-2d}{c}\right)}\right)$.

3.2   An *Almost-Injective* Encoding on $\mathbb{H}^2$

Let $\mathbb{F}_q$ be a finite field with $\text{char}(\mathbb{F}_q) = p \neq 2, 5$, $q = p^n$ is an odd prime power. We assume that $q \equiv 1 \mod 8$ or $q \equiv 7 \mod 8$, then 2 is a square.

Let $s \in \mathbb{F}_q^*$ such that $7s^2 + 20s - 100 = 0$:

- If $p \neq 7$ and "$q \equiv 1 \mod 8$ or $q \equiv 7 \mod 8$", we have $\Delta_s = 3200 = 2 \times 4^2 10^2$
  which is a square, then $s = \dfrac{-10 \pm 20\sqrt{2}}{7}$.
- If $p = 7$ and "$q \equiv 1 \mod 8$ or $q \equiv 7 \mod 8$", then $s = 5$.

Let $w \in \mathbb{F}_q^*$ be an arbitrary parameter.

Let $\mathbb{H}^2 : y^2 = F_2(x) = x^5 + bx^3 + dx + e$, with $b = sw^2$, $d = \frac{sw^4}{2}$, $e = \frac{s-10}{10}w^5$ be an hyperelliptic curve of genus 2 over $\mathbb{F}_q$ with the previous conditions on q.
Let u be a parameter such that $\chi(u) = -1$ and define the set

$$\mathcal{R}_2 = \left\{ r \in \mathbb{F}_q^*, [ur^2(-50 - 35s) - 1]^5 + s[ur^2(-50 - 35s) - 1]^3 \right.$$

$$\left. + \frac{s}{2}ur^2(-50 - 35s) \neq \frac{2s + 5}{5} \right\}$$

Input: The hyperellitic curve $\mathbb{H}^2$, an element $r \in \mathcal{R}_2$
Output: A point (x, y) on $\mathbb{H}^2$
$v := w[ur^2(-50 - 35s) - 1]$;
$\varepsilon := \chi(v^5 + v^3 + dv + e)$;
$x := \dfrac{1+\varepsilon}{2}v + \dfrac{1-\varepsilon}{2}\left(\dfrac{w(-v + w)}{v + w}\right)$;
$y := -\varepsilon\sqrt{x^5 + bx^3 + dx + e}$;
return (x, y).

Algorithm 3. Hashing-2-Genus2-Encoding

Definition 2. *In the situation of Algorithm 3, the encoding function for the hyperelliptic curve* $\mathbb{H}^2$ *is the function* $\phi_2 : \mathcal{R}_2 \to \mathbb{H}^2 : r \mapsto \phi_2(r) = (x, y)$.

Theorem 3. *Algorithm 3 computes a deterministic almost-injective encoding* $\phi_2 : \mathcal{R}_2 \to \mathbb{H}^2 : r \mapsto \phi_2(r) = (x, y)$, *in time* $\mathcal{O}(log^{2+\circ(1)}q)$, *where* $\mathcal{S}_2 = \mathbb{F}_q \setminus \mathcal{R}_2$ *is a subset of* $\mathbb{F}_q$ *of at most 10 elements.*

Proof. 1. Since $v = w[ur^2(-50 - 35s) - 1]$ we have $v^5 + bv^3 + dv + e = w^5[ur^2(-50-35s)-1]^5 + bw^3[ur^2(-50-35s)-1]^3 + dw[ur^2(-50-35s)-1]+e$.
Now, using $b = sw^2, d = \frac{sw^4}{2}, e = \frac{s-10}{10}w^5$, we deduce that $v^5 + bv^3 + dv + e \neq 0 \Leftrightarrow [ur^2(-50-35s)-1]^5 + s[ur^2(-50-35s)-1]^3 + \frac{s}{2}ur^2(-50-35s) - \frac{2s+5}{5} \neq 0$
which is true by definition of $\mathcal{R}_2$. Therefore $\varepsilon = \chi(v^5 + bv^3 + dv + e) \neq 0$.
2. x is well defined since $v + w = 0 \Leftrightarrow r = 0$ and $0 \notin \mathcal{R}_2$
3. Let us prove that $F_2(x)$ is nonzero square
   - $\varepsilon = 1$ i.e. $F_2(v)$ is a nonzero square and $x = v$ then $F_2(x)$ is a nonzero square.
   - $\varepsilon = -1$, we have $x = \left(\frac{w(-v+w)}{v+w}\right)$ and

$$F_2(x) = \frac{w^5(-v+w)^5 + bw^3(-v+w)^3(v+w)^2 + dw(-v+w)(v+w)^4 + e(v+w)^5}{(v+w)^5}.$$

Using $b = sw^2, d = \frac{sw^4}{2}$ and $e = \frac{s-10}{10}w^5$ in $F_2(x)$, after some computations, yields, the following:

$$F_2(x) = \frac{v^5(-7w^5\frac{s}{5}-2w^5)+v^3(2w^7s-20w^7)+v(w^9s-10w^9)+8w^{10}\frac{s}{5}}{(v+w)^5}.$$

Denote $\alpha_5 = (-7w^5\frac{s}{5} - 2w^5)$, $\alpha_3 = (2w^7s - 20w^7)$, $\alpha_1 = (w^9s - 10w^9)$ and $\alpha_0 = 8w^{10}\frac{s}{5}$ then $F_2(x) = \frac{\alpha_5}{(v+w)^5}\left[v^5 + \frac{\alpha_3}{\alpha_5}v^3 + \frac{\alpha_1}{\alpha_5}v + \frac{\alpha_0}{\alpha_5}\right]$.

We have:

$$\frac{\alpha_3}{\alpha_5} = \frac{2w^7s - 20w^7}{-7w^5\frac{s}{5} - 2w^5} = \frac{w^7(2s-20)}{\frac{w^5}{5}(-7s-10)} = \frac{w^2(10s-100)}{-7s-10}.$$

Now $7s^2+20s-100 = 0 \Rightarrow 10s - 100 = -7s^2-10$. So $\frac{\alpha_3}{\alpha_5} = \frac{w^2(-7s^2-10s)}{-7s-10} = sw^2 = b$.

$$\frac{\alpha_1}{\alpha_5} = \frac{w^9s - 10w^9}{-7w^5\frac{s}{5} - 2w^5} = \frac{w^9(5s-100)}{w^5(-7s-10)} = \frac{w^4\frac{10s-100}{2}}{-7s-10} \cdot \frac{w^4\frac{-7s^2-10s}{2}}{-7s-10}$$

$$= s\frac{w^4}{2} = d.$$

$$\frac{\alpha_0}{\alpha_5} = \frac{8w^{10}\frac{s}{5}}{-7w^5\frac{s}{5} - 2w^5} = \frac{8w^{10}s}{w^5(-7s-10)} = \frac{8sw^5}{-7s-10}.$$

Now $7s^2 + 20s - 100 = 0 \Longrightarrow -7s^2 - 20s + 100 = 0 \Longrightarrow -7s^2 - 80s + 60s + 100 = 0 \Longrightarrow -7s^2 + 60s + 100 = 80s \Longrightarrow (-7s - 10)(s - 10) = 80s = 10 \times 8s \Longrightarrow \frac{s-10}{10} = \frac{8s}{-7s-10}$. So we have $\frac{\alpha_0}{\alpha_5} = e$.

Thus, we have $\chi(F_2(x)) = \chi(\alpha_5)\chi(v+w)\chi(F_2(v)) = -\chi(\alpha_5(v+w))$ since $F_2(v)$ is a nonzero non-square.   •

But, we have $\alpha_5(v + w) = w^5\frac{-10-7s}{5}\left(wur^2(-50 - 35s)\right) = uw^6r^2(-10 - 7s)^2 = u\left(rw^3(-10 - 7s)\right)^2$. Since $r \neq 0$ in $\mathcal{R}_2$, then $\chi(\alpha_5(v + w)) = \chi(u) = -1$. Hence $\chi(F_2(x)) = 1$, therefore $F_2(x)$ is a nonzero square, we deduce that $y = -\varepsilon\sqrt{F_2(x)}$ is well-defined.   □

Lemma 2. *In the situation of Theorem 3 and Definition 2, we have* $\phi_2(r) = \phi_2(-r)$, $\forall r \in \mathcal{R}_2$ *and* $\#(\phi_2^{-1}(\phi_2((r))) = 2$, $\forall r \in \mathcal{R}_2$.

Proof. Similar to the proof of Lemma 1.   □

Input : The Hyperelliptic curve $\mathbb{H}^2$ and $(x, y) \in \mathbb{H}^2$.
Output: $\bar{r}$ s.t. $\psi(\bar{r}) = (x, y)$, $\bar{r} \in \mathcal{R}_2$ or $\perp$ (which means that "(x,y) is not in $\phi_2(\mathcal{R}_2)$").
if $uw(x + w)(-50 - 35s)$ *is a non-zero square* **then**
> **if** $y = \sqrt{f(x)}$ **then**
>> $\bar{r} := \sqrt{\frac{2w}{u(x+w)(-50-35s)}};$
>
> **else**
>> **if** $y = -\sqrt{f(x)}$ **then**
>>> $\bar{r} := \sqrt{\frac{x+w}{uw(-50-35s)}};$
>>
>> **end**
>
> **end**
> **return** $\bar{r};$

else
> **return** $\perp$.

end

Algorithm 4. Hashing-2-Genus2-Inverting

Theorem 4. *In the situation of Theorem 3 and Definition 2, Algorithm 4 defines the decoding function of ϕ_2. Furthermore, $\mathrm{Im}(\phi_2)$ is the set of $(x, y) \in H_1$ verifying $\chi(uw(x + w)(-50 - 35s)) = 1$.*

Proof. 1.   – Assume that $x, y \in \mathrm{Im}(\phi_2)$.

- If $\varepsilon = 1$ then $x = v$. Hence $x + w = 0 \Leftrightarrow v + w = 0 \Leftrightarrow r = 0$ but we know that $0 \notin \mathcal{R}_2$.

  Now, we have $uw(x + w)(-50 - 35s) = uw(v + w)(-50 - 35s) = uw[ur^2w(-50 - 35s)](-50 - 35s) = u^2w^2r^4(-50 - 35s)^2$.

- If $\varepsilon = -1$ then $x = \frac{w(-v+w)}{v+w} =$, Hence $x + w = 0 \Leftrightarrow \frac{-vw+w^2}{v+w} = -w \Leftrightarrow w = 0$ but we know that $w \in \mathbb{F}_q^*$.

  Now we have $uw(x+w)(-50-35s) = uw[\frac{w(-v+w)}{v+w} + w](-50-35s) = \frac{uw}{v+w}(2w^2)(-50 - 35s) = \frac{2w^2}{r^2}$, since $v + w = uwr^2(-50 - 35s)$ by Theorem 3, 2 is a square when $q \equiv 1 \mod 8$ or $q \equiv 7 \mod 8$.

We conclude that $uw(x + w)(-50 - 35s)$ is a nonzero square.

– Conversely assume that $uw(x + w)(-50 - 35s)$ is a nonzero square in $\mathbb{F}_q$. Let us prove that $(x, y) \in \mathrm{Im}(\phi_2)$. Put $\bar{r} = \sqrt{\frac{x+w}{uw(-50-35s)}}$ if $y \notin \sqrt{\mathbb{F}_q^2}$ and $\bar{r} = \sqrt{\frac{2w}{u(x+w)(-50-35s)}}$ if $y \in \sqrt{\mathbb{F}_q^2}$. By above assumptions $\frac{x+w}{uw(-50-35s)}$ and $\frac{2w}{u(x+w)(-50-35s)}$ are well-defined and are nonzero square, then $\bar{r}$ is always well-defined.

Now, we are going to prove that $\bar{r} \in \mathcal{R}_2$ and $(x, y) \in \mathrm{Im}(\phi_2)$. Define $\bar{v}, \bar{\varepsilon}, \bar{x}, \bar{y}$ as in Algorithm 3.

If $y \notin \sqrt{\mathbb{F}_q^2}$ then $\bar{r} = \sqrt{\frac{x+w}{uw(-50-35s)}} \implies \bar{v} = w[u\bar{r}^2(-50 - 35s) - 1] = w[\frac{x+w}{w} - 1] = x$. So we have $\bar{\varepsilon} = \chi(\bar{v}^5 + b\bar{v}^3 + d\bar{v} + e) = \chi(x^5 +$

$bx^3 + dx + e) = 1$. Hence $\bar{x} = \frac{1+\bar{\varepsilon}}{2}\bar{v} + \frac{1-\bar{\varepsilon}}{2}\left(\frac{\omega(-\bar{v}+\omega)}{\bar{v}+\omega}\right) = \bar{v} = x$ and

$\bar{y} = -\bar{\varepsilon}\sqrt{x^5 + bx^3 + dx + e} = -\sqrt{x^5 + bx^3 + dx + e} = y$.

Now if $y \in \sqrt{\mathbb{F}_q^2}$ then $\bar{r} = \sqrt{\frac{2w}{u(x+w)(-50-35s)}}$, since $\bar{v} = w[u\bar{r}^2(-50 -$

$35s) - 1]$ then $\bar{v} = w\frac{w-x}{x+w}$.

Now, after some computations, we have:

$$\bar{\varepsilon} = \chi(\bar{v}^5 + b\bar{v}^3 + d\bar{v} + e) = \chi\left(\left(\frac{-7\omega^5\frac{s}{5}-2\omega^5}{(x+\omega)^5}\right)(x^5 + bx^3 + dx + e)\right)$$

$$= \chi(5w(x+w)(-7s-10) = -1$$

because $uw(x + w)(-50 - 35s)$ is a nonzero square and $\chi(u) = -1$. Since $\bar{\varepsilon} = -1$ then $\bar{x} = \frac{\omega(-\bar{v}+\omega)}{\bar{v}+\omega} = x$ and $\bar{y} = -\bar{\varepsilon}\sqrt{x^5 + bx^3 + dx + e} = \sqrt{x^5 + bx^3 + dx + e} = y$. In both cases we have $\bar{x} = x$ and $\bar{y} = y$.

Since for all v, we have $v^5 + bv^3 + dv + e \neq 0$, then $[u\bar{r}^2(-50 - 35s) - 1]^5 + s[u\bar{r}^2(-50-35s) - 1]^3 + \frac{s}{2}[u\bar{r}^2(-50-35s) - 1] \neq \frac{2s+5}{5}$, thus $\bar{r} \in \mathcal{R}_2$.

2. Follows from the above proof.   □

Remark 2 (*Extension to* $\mathbb{F}_q$).

– We propose to extend ϕ_2 at 0 as follows. Recall that $b = sw^2$, $d = \frac{sw^4}{2}$, $e = \frac{s-10}{10}w^5$. Since $F_2(-w) = -w^5 - sw^5 - \frac{sw^5}{4} + \frac{s-10}{10}w^5 = \left(-1-s-\frac{s}{2}+\frac{s-10}{10}\right)w^5 = \frac{-14s-11}{10}w^5$. Now choose w such that $f(-w)$ is a nonzero square and put $\phi_2(0) = (-w, \sqrt{F_2(-w)})$.

– Let Z_2 be the set of roots of $F_2(x) = x^5 + bx^3 + cx + e$ which are square. If Z_2 is not empty, with the above choice, we put $F_2(\pm t) = (-w, -\sqrt{F_2(-w)}) \forall t/t^2 \in \mathcal{S}_2$.

3.3 An *Almost-Injective* Encoding on $\mathbb{H}^3$

Let $q = 5^n$ be a power of 5 and $u \in \mathbb{F}_q^*$ be a nonzero square. Let $w \in \mathbb{F}_q^*$ be an arbitrary parameter and $\mathbb{H}^3$ be an hyperelliptic curve of genus 2 given by $\mathbb{H}^3 : y^2 = F_3(x) = x^5 + ax^4 + e$ over $\mathbb{F}_q$ where $a \in \mathbb{F}_q^*$ and $e = 4aw^4$.

Let u be a parameter such that $\chi(u) = -1$ and define the set

$$\mathcal{R}_3 = \left\{r \in \mathbb{F}_q^*, w(1 - ur^2)^5 + a(1 - ur^2)^4 + 4a \neq 0\right\}.$$

Input: The hyperelliptic curve $\mathbb{H}^3$, an element $r \in \mathcal{R}_3$.
Output: A point (x, y) on $\mathbb{H}^3$
$v = v(r^2) := w(1 - ur^2)$;
$\varepsilon := \chi(v^5 + av^4 + 4aw^4)$;
$$x := \frac{1+\varepsilon}{2}v + \frac{1-\varepsilon}{2}\left(\frac{-wv}{v-w}\right);$$
$y := -\varepsilon\sqrt{x^5 + ax^4 + 4aw^4}$;
return (x, y).

Algorithm 5. Hashing-3-Genus2-Encoding

Definition 3. *In the situation of Algorithm 5, the encoding function for the hyperelliptic curve* $\mathbb{H}^3$ *is the function* $\phi_3 : \mathcal{R}_3 \rightarrow \mathbb{H}^3 : r \mapsto \phi_3(r) = (x, y)$.

Theorem 5. *Algorithm 5 computes a deterministic almost-injective encoding* $\phi_3 : \mathcal{R}_3 \rightarrow \mathbb{H}$ *where* $\mathcal{S}_3 = \mathbb{F}_q \setminus \mathcal{R}_3$ *is a subset of size at most 10.*

Proof. – x is well-defined since $v - w = 0 \Leftrightarrow v = w \Leftrightarrow w - wur^2 = w \Rightarrow w = 0$ or $r = 0$, which is impossible.
- Since $v = w(1 - ur^2)$, by definition of $\mathcal{R}_2$, we have $v^5 + av^4 + 4aw^4 = w^4 \left[w(1 - ur^2)^5 + a(1 - ur^2) + 4a \right] \neq 0$. So $\varepsilon \neq 0$.
- Now let us prove that $F_3(x)$ is a nonzero square.
  - If $\varepsilon = 1$, we have $x = v$ and $\chi(F_3(v)) = 1$. Consequently $F_3(v)$ is a nonzero non-square.
  - If $\varepsilon = -1$, then $F_3(v)$ is a nonzero non-square and $x = \frac{-wv}{v-w}$. So if we replace x by its value in $x^5 + ax^4 + 4aw^4$, we have after some computations $\chi(F_3(x)) = \chi\left(\frac{-w^5}{(v-w)^5} \right) \chi(F_3(v))$. Since $\chi(F_3(v)) = -1$, then $\chi(F_3(x)) = -\chi\left(\frac{-w}{v-w} \right) = -\chi\left(-w(w - uwr^2 - w) \right) = -\chi(uw^2r^2) = -\chi(u) = 1$. Thus $F_3(x)$ is a nonzero square. □

Lemma 3. *In the situation of Theorem 5 and Definition 3, we have* $\phi_3(r) = \phi_3(-r)$, $\forall r \in \mathcal{R}_3$ *and* $\#(\phi_3^{-1}(\phi_3(r))) = 2$, $\forall r \in \mathcal{R}_3$.

Proof. Similar to the proof of Lemma 4. □

Input   : The Hyperelliptic curve $\mathbb{H}^3$ and $(x, y) \in \mathbb{H}^3$.
Output: $\bar{r} s.t. \phi_3(\bar{r}) = (x, y)$, $\bar{r} \in \mathcal{R}_3$ or $\perp$ (which means that "(x,y) is not in $\phi_3(\mathcal{R}_3)$").
if $y = \sqrt{f(x)}$ then
  if $uw(w + x)$ is a nonzero square then
    | $\bar{r} = \sqrt{\dfrac{w}{u(w + x)}}$; return $\bar{r}$;
  else
    | return $\perp$
  end
else
  if $uw(w - x)$ is a nonzero square then
    | $\bar{r} = \sqrt{\dfrac{w - x}{uw}}$; return $\bar{r}$;
  else
    | return $\perp$
  end
end

Algorithm 6. Hashing-3-Genus2-Inverting

Theorem 6. *In the situation of Theorem 5 and Definition 3, Algorithm 6 defines the decoding function of ϕ_3. Furthermore, $\mathrm{Im}(\phi_3)$ is the set of $(x, y) \in H_3$ verifying $\chi(uw(w - x)) = 1$ if $y \notin \sqrt{\mathbb{F}_q^2}$ and $\chi(uw(w + x)) = 1$ $y \in \sqrt{\mathbb{F}_q^2}$.*

Proof. 1. (a) Assume that $(x, y) \in \mathrm{Im}(\phi_3)$, there exists $\bar{r}$ such that $\phi_3(\bar{r}) = (x, y)$ by Theorem 5.

- $y \notin \sqrt{\mathbb{F}_q} \Rightarrow \varepsilon = 1$ then $x = v$. Since $v = w(1 - ur^2)$ and $r \neq 0$, then $(w - x)uw = (w - v)uw = wur^2uw = uw^2r^2$ is a nonzero square.
- $y \in \sqrt{\mathbb{F}_q} \Rightarrow \varepsilon = -1$ then $x = \frac{-wv}{v-w}$. Since $v = w(1 - ur^2)$ and $r \neq 0$, then $uw(w + x) = \frac{1}{r^2}$ is a nonzero square.

(b) Conversely suppose that $uw(w - x)$ is a square and $w - x \neq 0$ if $y \notin \sqrt{\mathbb{F}_q}$, and $uw(w + x)$ is a square and $w + x \neq 0$ if $y \in \sqrt{\mathbb{F}_q}$. Now our goal is to show that $\bar{r}$ is defined and $\phi_4(\bar{r}) = (x, y)$. Define $\bar{v}, \bar{\varepsilon}, \bar{x}, \bar{y}$ as in Algorithm 4. They are two cases.

- The first case: $y \notin \sqrt{\mathbb{F}_q^2}$. In this case we have $\bar{r} = \sqrt{\frac{w-x}{uw}}$. Then $\bar{r}$ is well-defined since $uw(w - x)$ is a square and $\bar{r} \neq 0$ since $w - x \neq 0$ by hypothesis. So $\bar{v} = w - uw\bar{r}^2 = w - uw\left(\frac{w-x}{uw}\right) = w - w + x = x$; so $\bar{\varepsilon} = \chi(\bar{v}^5 + a\bar{v}^4 + 4aw^4) = \chi(\bar{x}^5 + a\bar{x}^4 + 4aw^4) = 1$, which implies that $\bar{x} = \frac{1+\bar{\varepsilon}}{2}\bar{v} + \frac{1-\bar{\varepsilon}}{2}\left(\frac{-w\bar{v}}{\bar{v}-w}\right) = \bar{v} = x$ and $\bar{y} = -\bar{\varepsilon}\sqrt{f(\bar{x})} = -\sqrt{F_3(x)} = y$.

- The second case is that $y \in \sqrt{\mathbb{F}_q^2}$. Then $\bar{r} = \sqrt{\frac{w}{u(w+x)}}$ is well-defined since $uw(w+x)$ is a square and $\bar{r} \neq 0$ since $w(w+x) \neq 0$ by hypothesis, so $\bar{v} = w - uw\bar{r}^2 = w - uw\left(\frac{w}{u(x+w)}\right) = \frac{wx}{w+x}$.

  We have $\bar{v}^5 + a\bar{v}^4 + 4aw^4 = \frac{(wx)^5 + a(wx)^4(w+x) + 4aw^4(w+x)^5}{(w+x)^5}$. Hence
  $$\bar{\varepsilon} = \chi(\bar{v}^5 + a\bar{v}^4 + 4aw^4) = \chi\left(\frac{w^5}{(x+w)^5}(x^5 + ax^4 + 4aw^4)\right) =$$
  $$\chi\left(\frac{w^5}{(x+w)^5}\right)\chi(F_3(x)) = \chi(w(w + x)) = \chi(u) = -1 \text{ because } uw(w + x)$$
  and $F_3(x)$ are nonzero squares. So we have $\bar{x} = \frac{1+\bar{\varepsilon}}{2}\bar{v} + \frac{1-\bar{\varepsilon}}{2}\left(\frac{-w\bar{v}}{\bar{v}-w}\right) =$
  $$\frac{-w\bar{v}}{\bar{v}-w} = \frac{\frac{-w^2x}{w+x}}{\frac{wx}{w+x}-w} = \frac{-w^2x}{-w^2} = x \text{ and } \bar{y} = -\bar{\varepsilon}\sqrt{F_3(\bar{x})} = \sqrt{F_3(x)} = y.$$
  Since for all v, we have $v^5 + av^4 + e \neq 0$, then $w(1 - ur^2)^5 + a(1 - ur^2)^4 + 4a \neq 0$, thus $\bar{r} \in \mathcal{R}_3$.

2. Follows from the above proof. $\qquad\qquad\qquad\qquad\qquad\qquad\qquad\qquad\square$

*Remark 3 **(Extension to $\mathbb{F}_{5^n}$)**.*

- As for ϕ_2, we propose to extend ϕ_3 at 0 as follows. Recall that $e = 4aw^4$. Since $F_3(w) = w^5$. Now choose $w = z^2$ then $F_3(w)$ is a nonzero square. We put $\phi_3(0) = (w, \sqrt{F_3(w)})$.
- Let Z_3 be the set of roots of $F_3(x) = x^5 + ax^4 + 4aw^4$ which are square. If Z_3 is not empty, with the above choice, we put $F_3(\pm t) = (w, -\sqrt{F_3(w)})$ $\forall t/t^2 \in Z_3$.

4   Applications to the Jacobian

4.1   General Framework on Indifferentiable Hashing into the Jacobian

Well-Distributed Encodings

Consider an encoding function F into a curve C, and $\mathbb{J}$ the Jacobian of C. Assume that C has an $\mathbb{F}_q$-rational point $\mathcal{O}$, so that one can fix an embedding $C \rightarrow \mathbb{J}$, sending a point P to the degree 0 divisor $(P) - (\mathcal{O})$. The regularity properties of function $F^{\otimes s}$ of the form below can be derived formally from the behavior of F. $F^{\otimes s} : (\mathbb{F}_q)^s \rightarrow \mathbb{J}(\mathbb{F}_q), (t_1, \ldots, t_s) \mapsto F(t_1) + \ldots F(t_s)$.

Then Farashahi et *al.* showed with their technique that hash function constructions of the general form:

$$H(m) = F(h_1(m)) + \ldots F(h_s(m))$$

are well-behaved as soon as s is greater than the genus of the target curve (that is $s \geq 3$ for genus 2).

Let χ_q be any character χ_q of the abelian group $\mathbb{J}(\mathbb{F}_q)$. Let us introduce the character sums

$$S(\chi_q) = \sum_{t \in \mathbb{F}_q} \chi_q(F(t)). \tag{1}$$

Definition 4 (well-distributed, [8]). *Let C be a smooth projective curve over a finite field $\mathbb{F}_q$, $\mathbb{J}$ its Jacobian, F a function $\mathbb{F}_q \rightarrow C(\mathbb{F}_q)$ and B a positive constant. F is B-well-distributed if for any nontrivial character χ_q of $\mathbb{J}(\mathbb{F}_q)$, the following holds:*

$$|S(\chi_q)| \leq B\sqrt{q}.$$

F is well-distributed if it is B-well-distributed for some B bounded independently of the security parameter.

Lemma 4 (statistical distance). *If $F : \mathbb{F}_q \rightarrow C(\mathbb{F}_q)$ is a B-well-distributed encoding into a curve C, then for all $D \in \mathbb{J}(\mathbb{F}_q)$, the statistical distance between the distribution defined by $F^{\otimes s}$ and the uniform distribution on $\mathbb{J}(\mathbb{F}_q)$ is bounded as:*

$$\sum_{D \in \mathbb{J}(\mathbb{F}_q)} \left| \frac{N_s(D)}{q^s} - \frac{1}{\#\mathbb{J}(\mathbb{F}_q)} \right| \leq \frac{B^s}{q^{s/2}} \sqrt{\#\mathbb{J}(\mathbb{F}_q)}.$$

where $N_s(D)$ denote the number of preimages of D under $F^{\otimes s}$: $N_s(D) = \#\{(t_1, \ldots, t_s) \in (\mathbb{F}_q)^s | D = F(t_1) + \ldots F(t_s)\}$.

Proof. See [8].   □

Character Sums on Curves

Let χ_q denote an Artin character on C. Let us introduce the character sums $S_C(\chi_q) = \sum_{P \in C(\mathbb{F}_q)} \chi_q(P)$.

Lemma 5. *If χ_q is any nontrivial character on C of genus g, then*

$$|S_C(\chi_q)| \leq (2g - 2 + \deg(f(\chi_q)))\sqrt{q}.$$

Lemma 6. *Let $h : \overline{C} \to C$ be a non constant morphism of curves, and χ_q be any nontrivial character of $\mathbb{J}(\mathbb{F}_q)$, where $\mathbb{J}$ is the Jacobian of C. Assume that h does not factor through a nontrivial unramified morphism $Z \to C$. Then:*

$$\left| \sum_{P \in \overline{C}(\mathbb{F}_q)} \chi_q(h(P)) \right| \leq (2\overline{g} - 2)\sqrt{q}$$

where $\overline{g}$ is the genus of $\overline{C}$. Furthermore, if q is odd and φ is a non constant rational function on $\overline{C}$:

$$\left| \sum_{P \in \overline{C}(\mathbb{F}_q)} \chi_q(h(P)) \left(\frac{\varphi(P)}{q} \right) \right| \leq (2\overline{g} - 2 + 2\deg\varphi)\sqrt{q}.$$

Proof. See [8].   □

Now let us use ϕ_i to construct well-behaved hash functions to $\mathbb{J}_i(\mathbb{F}_q)$ the jacobian of $\mathbb{H}_i$. We will follow the technique of Farashahi *et al.* [8].

4.2   Indifferentiable Hashing into the Jacobian of $\mathbb{H}_i$, $1 \leq i \leq 3$

We focus first on the case of the encoding ϕ_1 to hyperelliptic curve $\mathbb{H}^1 : y^2 = F_1(x) = x^5 + ax^4 + cx^2 + dx$ with $8ad^2 + c^3 = 0$ where $a, c, d \in \mathbb{F}_q^*$ over $q \equiv 7 \mod 8$, as defined in the Subsect. 3.1.

From Theorem 1, we have the following:

$$y \notin \sqrt{\mathbb{F}_q^{*2}} \iff r^2 := -\frac{2d}{cu}\left(x + \frac{2d}{c}\right) \tag{1*}$$

$$y \in \sqrt{\mathbb{F}_q^{*2}} \iff r^2 := -\frac{8d^3}{c^3 u}\frac{1}{\left(x + \frac{2d}{c}\right)} \tag{2*}$$

By Algorithm 3, the encoding function for the hyperelliptic curve $\mathbb{H}^1$ is the function $\phi_1 : \mathcal{R}_1 \to \mathbb{H}^1 : r \mapsto \phi_1(r) = (x, y)$, where

$$\mathcal{R}_1 = \left\{ r \in \mathbb{F}_q^*, \; ur^2c^2 + 4d^2 \neq 0; \right.$$

$$\left. -d - \frac{c^2}{2d}ur^2 + d\left(\frac{ur^2c^2}{4d^2} + 1\right)^3 \neq -\frac{16d^4\left(\frac{ur^2c^2}{4d^2} + 1\right)^4}{c^4} \right\}.$$

Theorem 7. *Assume that* $q = 3 \mod 4$. *Let* ϕ_1 *be the encoding function described above. For any nontrivial character* χ *of* $\mathbb{J}_1(\mathbb{F}_q)$, *the character sum* $S(\chi)$ *given by (1) satisfies*

$$|S_{\phi_1}(\chi)| \leq 32\sqrt{q} + 119.$$

Proof. Put $\mathcal{S}_1 = \mathbb{F}_q \setminus \mathcal{R}_1$. Since $q \equiv 7 (\mod 8)$ then $y \in \sqrt{\mathbb{F}_q^{*2}} \Leftrightarrow \chi_q(y) = 1$ and $y \notin \sqrt{\mathbb{F}_q^{*2}}, y \neq 0 \Leftrightarrow \chi_q(y) = -1$. By Eqs. (1*) and (2*) we define the following coverings $h_{1,j} : C_{1,j} \to \mathbb{H}_1$, $j = 0, 1$ of the smooth projective curves. As in [8], the rational function r on the $C_{1,j}$ allows to have a morphism $g_{1,j} : C_{1,j} \to \mathbb{P}^1$ such that any point in $\mathcal{R}_1$ has exactly two preimages (which are conjugate under $y \to -y$) in one of the two curves $C_{1,j}, j = 0, 1$.

Since $q \equiv 3 (\mod 4)$, then y or $-y$ is a square therefore only one of the previous preimages verify $\chi(y) = (-1)^j$ over $C_{1,j}$. Let $P \in C_j(\mathbb{F}_q)$ be that preimage.

Then, $\phi_1(r) = h_{1,j}(P)$. Let us show that ϕ_1 is a well-distributed encoding. Denote by $\mathbb{J}_1$ the Jacobian of $\mathbb{H}_1$, and fix an embedding $\mathbb{H}_1 \to \mathbb{J}_1$. We can assume that S_1 is a subset of $\mathbb{P}^1$ and put $S_{1,j} = g_{1,j}^{-1}(S_1 \cup \infty)$.

For all $r \in \mathcal{R}_1$, the character sum $S_{\phi_1}(\chi)$ can be written as:

$$\sum_{r \in \mathcal{R}_1} \chi(\phi_1(r)) = \sum_{\substack{P \in C_{10}(\mathbb{F}_q) \setminus S_{10} \\ \chi_q(y)=+1}} \chi(h_{10}(P)) + \sum_{\substack{P \in C_{11}(\mathbb{F}_q) \setminus S_{11} \\ \chi_q(y)=-1}} \chi(h_{11}(P))$$

Observe that:

$$\sum_{P \in C_{1,j}(\mathbb{F}_q)} \chi(h_{1,j}(P)) \cdot \left(\frac{1+(-1)^j \chi_q(y)}{2} \right) = \sum_{\substack{P \in C_{1,j}(\mathbb{F}_q) \\ \chi_q(y)=(-1)^j}} \chi(h_{1,j}(P)) + \frac{1}{2} \sum_{\substack{P \in C_{1,j}(\mathbb{F}_q) \\ \chi_q(y)=0}} \chi(h_{1,j}(P))$$

The expression $\frac{1}{2} \sum_{P \in C_{1,j}(\mathbb{F}_q), \chi_q(y)=0} \chi(h_{1,j}(P))$ contains at most $2 \cdot 5 = 10$ elements. Let us compute:

$$\sum_{\substack{P \in C_{1,j}(\mathbb{F}_q) \\ \chi_q(y)=(-1)^j}} \chi(h_{1,j}(P)) = \sum_{P \in C_{1,j}(\mathbb{F}_q)} \chi(h_{1,j}(P)) \cdot \left(\frac{1+(-1)^j \chi_q(y)}{2} \right) - \frac{1}{2} \sum_{\substack{P \in C_{1,j}(\mathbb{F}_q) \\ \chi_q(y)=0}} \chi(h_{1,j}(P)).$$

By the Eisenstein criterion h_0 and h_1 are totally ramified over points in $\mathbb{H}_1$ such that $x = -\frac{2d}{c}$, so they cannot factor through any unramified covering of $\mathbb{H}_1$. Therefore by Lemma 6.

$$\left| \sum_{P \in C_j(\mathbb{F}_q)} \chi(h_j(P)) \cdot \left(\frac{1+(-1)^j \chi(y)}{2} \right) \right| \leq (2g_{C_j} - 2 + \deg(y))\sqrt{q}.$$

Let us compute g_{C_j} the genus of C_j.

$$\deg(y) = [\mathbb{F}_q(x,y,r) : \mathbb{F}_q(x,y)] \cdot [\mathbb{F}_q(x,y) : \mathbb{F}_q(y)] = 2 \cdot 5 = 10.$$

Now $h_{1,j} : C_j \to \mathbb{H}_1$ is only ramified at points with $x = -\frac{2d}{c}$. Therefore by the Riemann-Hurwitz formula, we get $2g_{C_j} - 2 = 2(2(2) - 2) + 2.(2 - 1) = 6$: thus the curves C_j are of genus 4.

Instead of Riemann-Hurwitz formula, we can compute g_{C_j} by birational equivalence. For example with C_0 we have $x = \frac{N}{r^2} - \frac{2d}{c}$ where $N = -\frac{8d^3}{uc^3}$, then from $y^2 = x^5 + ax^4 + cx^2 + dx$ we can have the following relation $\lambda^2 = r^{10}y^2$ and

$\cdot$ $\lambda^2 = g(r) = (N - \frac{2d}{c}r^2)^5 + ar^2(N - \frac{2d}{c}r^2)^4 + cr^6(N - \frac{2d}{c}r^2)^2 + dr^8(N - \frac{2d}{c}r^2)$

where g is a polynomial of degree 10, hence C_0 is hyperelliptic of genus 4 as desired [note that the leading coefficient of g is $F_1(\frac{-2d}{c})$ and for the extension of the encoding ϕ_1 at $\pm\sqrt{\frac{-4d^2}{uc^2}}$, we have chosen d or c such that $F_1(\frac{-2d}{c})$ is a nonzero square and put $\phi_1(0) = (-w, \sqrt{F_1(\frac{-2d}{c})})$].

Now, we need to compute $\deg(y)$ the degree of y as a rational function on C_j. We have: $\deg(y) = [\mathbb{F}_q(x, y, r) : \mathbb{F}_q(x, y)] \cdot [\mathbb{F}_q(x, y) : \mathbb{F}_q(y)] = 2 \cdot 5 = 10$.
Thus:

$$\left| \sum_{P \in C_j(\mathbb{F}_q)} \chi(h_j(P)) \cdot \left(\frac{1 + (-1)^j \chi(y)}{2} \right) \right| \leq 16\sqrt{q}.$$

Therefore using the previous formulas, we have

$$\left| \sum_{r \in \mathcal{R}_2} \chi(\phi_2(r)) \right| \leq (16\sqrt{q} + \#S_{10} + 5) + (16\sqrt{q} + \#S_{11} + 5)$$

$$= (32\sqrt{q} + 10 + \#S_{10} + \#S_{11}).$$

Thus $|S_{\phi_2}(\chi)| \leq 32\sqrt{q} + 10 + \#S_{10} + \#S_{11} + \#S_1$. We have $\#S_1 = 1 + 2 \cdot 10 = 21$. Since $g_{1,j}$ is a map of degree 2, $\#S_{1,j} \leq 2(\#S + 1) \leq 44$.
Then

$$|S_{\phi_2}(\chi)| \leq 32\sqrt{q} + 10 + 44 + 44 + 21 = 32\sqrt{q} + 119$$

as desired.

In other words, ϕ_1 is a $(32 + 119q^{-1/2})$-well-distributed encoding.   □

Theorem 8. *Assume that $q = 3 \mod 4$. Let $\mathbb{J}_1(\mathbb{F}_q)$ be the jacobian of H_1. For any $D \in \mathbb{J}_1(\mathbb{F}_q)$, let $N_3(D)$ denote the number of preimage of D under $\phi_2^{\otimes 3}$.*

The statistical distance between the distribution defined by $\phi_2^{\otimes 3}$ and the uniform distribution in $\mathbb{J}_1(\mathbb{F}_q)$ is bounded by:

$$\sum_{D \in \mathbb{J}_1(\mathbb{F}_q)} \left| \frac{N_3(D)}{q^3} - \frac{1}{\#\mathbb{J}_1(\mathbb{F}_q)} \right| \leq \frac{(32 + 119q^{-1/2})^3}{q^{3/2}} \sqrt{\#\mathbb{J}_1(\mathbb{F}_q)}.$$

Proof. Follows from the above Theorem 7 and the proof in [8].   □

Therefore, the distribution defined by $\phi_2^{\otimes 3}$ on $\mathbb{J}_1(\mathbb{F}_q)$ is statistically indistinguishable from the uniform distribution. In particular, the following construction:

$$m \mapsto H(m) = \phi_2(h_1(m)) + \phi_2(h_2(m)) + \phi_2(h_3(m)) \in \mathbb{J}_1(\mathbb{F}_q)$$

is indifferentiable from a random oracle when h_1, h_2, h_3 are seen as random oracles to $\mathbb{F}_q$.

Indifferentiable hashing into the Jacobian of $\mathbb{H}_i$ for $2 \leq i \leq 3$

As above we have the following theorem:

Theorem 9. *Assume that $q = 7 \mod 8$. Let $\mathbb{J}_i(\mathbb{F}_q)$ be the jacobian of H_i, for $2 \leq i \leq 3$. Let ψ_i for $2 \leq i \leq 3$ be the encoding functions described above. For any nontrivial character χ of $\mathbb{J}_i(\mathbb{F}_q)$, the character sum $S_{\psi_i}(\chi)$ given by (1) satisfies $|S_{\psi_i}(\chi)| \leq 32\sqrt{q} + 119$, for $2 \leq i \leq 3$.*

5 Conclusion

In this paper, we have designed new encodings into the Jacobian of three family of hyperelliptic curves. These encodings, which are not covered by the previous works in this domain, have some interesting properties, such as *almost-injectivity*. This can be used to design efficient and indifferentiable hash functions into the Jacobian of the underlying curves. Nevertheless, even if one can use our encodings in order to encode a message into the Jacobian, it would be interesting to decide whether or not these special families of hyperelliptic curves are suitable for cryptographic concerns.

References

1. Daniel, J., Bernstein, M., Hamburg, A., Krasnova, T.L.: Elligator: elliptic-curve points indistinguishable from uniform random strings. In: Gligor, V., Yung, M. (eds.) CCS. ACM (2013)
2. Boneh, D., Franklin, M.: Identity-based encryption from the weil pairing. In: Kilian, J. (ed.) CRYPTO 2001. LNCS, vol. 2139, pp. 213–229. Springer, Heidelberg (2001). doi:10.1007/3-540-44647-8_13
3. Brier, E., Coron, J.-S., Icart, T., Madore, D., Randriam, H., Tibouchi, M.: Efficient indifferentiable hashing into ordinary elliptic curves. In: Rabin, T. (ed.) CRYPTO 2010. LNCS, vol. 6223, pp. 237–254. Springer, Heidelberg (2010). doi:10.1007/978-3-642-14623-7_13
4. Mac Kenzie, P.: An efficient two-party public key cryptosystem secure against adaptive chosen ciphertext attack. In: Desmedt, Y.G. (ed.) PKC 2003. LNCS, vol. 2567, pp. 47–61. Springer, Heidelberg (2003). doi:10.1007/3-540-36288-6_4
5. Fouque, P.-A., Joux, A., Tibouchi, M.: Injective encodings to elliptic curves. In: Boyd, C., Simpson, L. (eds.) ACISP 2013. LNCS, vol. 7959, pp. 203–218. Springer, Heidelberg (2013). doi:10.1007/978-3-642-39059-3_14
6. Fouque, P.-A., Tibouchi, M.: Deterministic encoding and hashing to odd hyperelliptic curves. In: Joye, M., Miyaji, A., Otsuka, A. (eds.) Pairing 2010. LNCS, vol. 6487, pp. 265–277. Springer, Heidelberg (2010). doi:10.1007/978-3-642-17455-1_17
7. Farashahi, R.R.: Hashing into Hessian curves. Int. J. Appl. Crypt. 3(2), 139–147 (2014)
8. Farashahi, R.R., Fouque, P.-A., Shparlinski, I.E., Tibouchi, M., Voloch, J.F.: Indifferentiable deterministic hashing to elliptic and hyperelliptic curves. Math. Comput. 82(281), 491–512 (2013)

9. Hamburg, M.: Decaf: eliminating cofactors through point compression. In: Proceedings of the 35th Annual Cryptology Conference, Santa Barbara, CA, USA, 16–20 August 2015
10. Haneda, M., Kawazoe, M., Takahashi, T.: Suitable curves for genus-4 HCC over prime fields: point counting formulae for hyperelliptic curves of type $y^2 = x^{2k+1} + ax$. In: Caires, L., Italiano, G.F., Monteiro, L., Palamidessi, C., Yung, M. (eds.) ICALP 2005. LNCS, vol. 3580, pp. 539–550. Springer, Heidelberg (2005). doi:10.1007/11523468_44
11. Horwitz, J., Lynn, B.: Toward hierarchical identity-based encryption. In: Knudsen, L.R. (ed.) EUROCRYPT 2002. LNCS, vol. 2332, pp. 466–481. Springer, Heidelberg (2002). doi:10.1007/3-540-46035-7_31
12. Icart, T.: How to hash into elliptic curves. In: Halevi, S. (ed.) CRYPTO 2009. LNCS, vol. 5677, pp. 303–316. Springer, Heidelberg (2009). doi:10.1007/978-3-642-03356-8_18
13. Kammerer, J.-G., Lercier, R., Renault, G.: Encoding points on hyperelliptic curves over finite fields in deterministic polynomial time. In: Joye, M., Miyaji, A., Otsuka, A. (eds.) Pairing 2010. LNCS, vol. 6487, pp. 278–297. Springer, Heidelberg (2010). doi:10.1007/978-3-642-17455-1_18
14. Menezes, A.J., Wu, Y.-H., Zuccherato, R.J.: An elementary introduction to hyperelliptic curves. In: Koblitz, N. (ed.) Algebraic Aspects of Cryptography. Algorithms and Computation in Mathematics, vol. 3, pp. 155–178. Springer, Heidelberg (1998)
15. Möller, B.: A public-key encryption scheme with pseudo-random ciphertexts. In: Samarati, P., Ryan, P., Gollmann, D., Molva, R. (eds.) ESORICS 2004. LNCS, vol. 3193, pp. 335–351. Springer, Heidelberg (2004). doi:10.1007/978-3-540-30108-0_21
16. Satoh, T.: Generating genus two hyperelliptic curves over large characteristic finite fields. In: Joux, A. (ed.) EUROCRYPT 2009. LNCS, vol. 5479, pp. 536–553. Springer, Heidelberg (2009). doi:10.1007/978-3-642-01001-9_31
17. Shallue, A., Woestijne, C.E.: Construction of rational points on elliptic curves over finite fields. In: Hess, F., Pauli, S., Pohst, M. (eds.) ANTS 2006. LNCS, vol. 4076, pp. 510–524. Springer, Heidelberg (2006). doi:10.1007/11792086_36
18. Tibouchi, M.: Hachage vers les courbes elliptiques et cryptanalyse de schémas RSA. Thèse de doctorat de l'Université Paris-Diderot-Luxembourg, Septembre 2011
19. Ulas, M.: Rational points on certain hyperelliptic curves over finite fields. Bull. Pol. Acad. Sci. Math. **55**(2), 97–104 (2007)

Cryptanalysis of Some Protocols
Using Matrices over Group Rings

Mohammad Eftekhari$^{(\boxtimes)}$

LAMFA, CNRS UMR 7352, Université de Picardie – Jules Verne,
33 rue Saint-Leu, 80039 Amiens, France
mohamed.eftekhari@u-picardie.fr

Abstract. We address a cryptanalysis of two protocols based on the
supposed difficulty of discrete logarithm problem on (semi) groups of
matrices over a group ring. We can find the secret key and break entirely
the protocols.

Keywords: Key exchange · Symmetric groups · Representation of
algebras

1   Introduction

The Diffie-Hellman key agreement protocol is the first published practical solution to the key distribution problem, allowing two parties that have never met to
exchange a secret key over an open channel. It uses the cyclic group $\mathbb{F}_q^*$, where $\mathbb{F}_q$
is the finite field of q elements. The security of this protocol is based on the difficulty of computing discrete logarithms (DL) in the group $\mathbb{F}_q^*$. There are several
algorithms for computing discrete logarithms, some of them are subexponential
when applied to $\mathbb{F}_q^*$.

It is important to search for easily implementable groups, for which the DL
problem is hard and there is no known subexponential time algorithm for computing DL. The group of points over $\mathbb{F}_q$ of an elliptic curve is such a group. In [8],
the group of invertible matrices with coefficients in a finite field was considered
for such a key exchange. In [6], using the Jordan form it was shown that the discrete logarithm problem on such matrices can be reduced to the same problem
over some small extensions of the finite base field.

In [4], the authors consider the semigroup of matrices (3-by-3 matrices) over
the group ring $\mathbb{F}_7[S_5]$, where S_5 is the group of permutation of $\{1, 2, 3, 4, 5\}$.
The security of this protocol is based on the supposed difficulty of the discrete
logarithm problem in the (semi) group of matrices with coefficients in $\mathbb{F}_7[S_5]$.

Moreover in [5], the authors propose the same semigroup as a platform for
the Cramer-Shoup cryptosystem which is a generalization of ElGamal's protocol.
Here the security is based on the supposed difficulty of the discrete logarithm
problem in the group of invertible 3-by-3 matrices with coefficients in $\mathbb{F}_7[S_5]$.

In [1,2,7] a cryptanalysis of [4] is proposed. Their methods are somehow
different. In [1], the problem of discrete logarithm in a semigroup is reduced

© Springer International Publishing AG 2017
M. Joye and A. Nitaj (Eds.): AFRICACRYPT 2017, LNCS 10239, pp. 223–229, 2017.
DOI: 10.1007/978-3-319-57339-7_13

to the same problem in a subgroup of the same semigroup. In [2] one uses a slight modification of Shor's quantum algorithm to find the period of a singular matrix (there is no notion of order for such a matrix) and thereby solving the discrete logarithm problem in semigroups. In [7], $\text{Mat}_3(\mathbb{F}_7[S_5])$ is embedded in $\text{Mat}_{360}(\mathbb{F}_7)$ and then one uses the same procedure as in [6] (adapted to singular matrices). The conclusion of all three papers above is that using a quantum computer one can break the key exchange protocol of [4].

In contrast to the above analysis we use the irreducible representations of the group S_5; then using the fact that the algebra $\mathbb{F}_7[S_5]$ is semi-simple, we give an isomorphism between this algebra and an algebra of block matrices with coefficients in $\mathbb{F}_7$. Then we use this isomorphism to give an isomorphism between $\text{Mat}_3(\mathbb{F}_7[S_5])$, and still another algebra of block matrices over $\mathbb{F}_7$. To do so, we combine the same blocks of the first isomorphism.

This way we reduce the discrete logarithm problem over $\text{Mat}_3(\mathbb{F}_7[S_5])$, to the same problem over block matrices with coefficients in $\mathbb{F}_7$. The maximum size of a block is 18, reducing dramatically the computations. Now we can apply the same procedure (eventually modified for singular matrices) as in [4], to each block and resolve the problem of discrete logarithm entirely (using actual computers) and find the secret key. So the conclusion is that the platform proposed in [4] and [5] are simply insecure.

The rest of this paper is organized as follows. Section 2, will be devoted to the irreducible representations of S_5. In Sect. 3, we explain the isomorphism between matrices with coefficients in $\mathbb{F}_7[S_5]$, and block matrices with coefficients in $\mathbb{F}_7$, and show that the protocols proposed in [4,5] can be broken. In Sect. 4, we give an example to illustrate our analysis. Finally we conclude with some remarks in Sect. 5.

2   Irreducible Representations of S_5

For our purpose, it will be easier to use the following presentation of S_5. We note $W := (12)$ and $Z := (12345)$. The group S_5 is defined by generators W, Z and relations T, where T is the following set of relations:

$$W^2 = \text{id}$$
$$Z^5 = \text{id}$$
$$(ZW)^4 = \text{id}$$
$$WZ^{-1}WZW = Z^{-1}WZWZ^{-1}WZ$$
$$[W, Z^{-2}WZ^2] = \text{id}$$
$$[W, Z^{-3}WZ^3] = \text{id}$$

The group S_5 has two distinct representations of dimension one (namely the trivial one and the signature), two non isomorphic irreducible representations of dimension four, two non isomorphic irreducible representations of dimension five, and one irreducible representation of dimension six. We give the images of the generators Z and W by these representations, and one can verify the relations T, for the images, thereby proving that one defines morphisms from S_5 to matrix groups. One can compare the trace of these morphisms with the character table of S_5 to be sure we obtain all the irreducible representations of S_5.

To construct these representations one can follow the general description of [3], using Young polytabloids, to construct the Specht modules which give the irreducible representation of S_5.

$$W = (12) \longmapsto A_1 \oplus A_1' \oplus A_4 \oplus A_4' \oplus A_5 \oplus A_5' \oplus A_6$$

where

$$A_1 = 1; \quad A_1' = -1; \quad A_4 = \begin{pmatrix} -1 & 0 & 0 & -1 \\ 0 & -1 & 0 & 1 \\ 0 & 0 & -1 & -1 \\ 0 & 0 & 0 & 1 \end{pmatrix}; \quad A_4' = \begin{pmatrix} 1 & 0 & 0 & 1 \\ 0 & 1 & 0 & -1 \\ 0 & 0 & 1 & 1 \\ 0 & 0 & 0 & -1 \end{pmatrix}$$

$$A_5 = \begin{pmatrix} -1 & 0 & 1 & 0 & -1 \\ 0 & -1 & -1 & 0 & 0 \\ 0 & 0 & 1 & 0 & 0 \\ 0 & 0 & 0 & -1 & -1 \\ 0 & 0 & 0 & 0 & 1 \end{pmatrix}; \quad A_5' = \begin{pmatrix} 1 & 0 & -1 & 0 & 1 \\ 0 & 1 & 1 & 0 & 0 \\ 0 & 0 & -1 & 0 & 0 \\ 0 & 0 & 0 & 1 & 1 \\ 0 & 0 & 0 & 0 & -1 \end{pmatrix}$$

$$A_6 = \begin{pmatrix} -1 & 0 & 1 & 0 & 1 & 0 \\ 0 & -1 & -1 & 0 & 0 & 1 \\ 0 & 0 & 1 & 0 & 0 & 0 \\ 0 & 0 & 0 & -1 & -1 & -1 \\ 0 & 0 & 0 & 0 & 1 & 0 \\ 0 & 0 & 0 & 0 & 0 & 1 \end{pmatrix}$$

$$Z = (12345) \longmapsto B_1 \oplus B_1' \oplus B_4 \oplus B_4' \oplus B_5 \oplus B_5' \oplus B_6$$

where

$$B_1 = 1; \quad B_1' = 1; \quad B_4 = \begin{pmatrix} 0 & 0 & 0 & 1 \\ -1 & 0 & 0 & -1 \\ 0 & -1 & 0 & 1 \\ 0 & 0 & -1 & -1 \end{pmatrix}; \quad B_4' = \begin{pmatrix} 0 & 0 & 0 & 1 \\ -1 & 0 & 0 & -1 \\ 0 & -1 & 0 & 1 \\ 0 & 0 & -1 & -1 \end{pmatrix}$$

$$B_5 = \begin{pmatrix} 0 & 0 & -1 & -1 & -1 \\ 0 & 0 & 0 & 1 & 0 \\ 0 & 0 & 0 & -1 & -1 \\ 1 & 0 & -1 & -1 & 0 \\ 0 & 1 & 1 & 1 & 1 \end{pmatrix}; \quad B_5' = \begin{pmatrix} 0 & 0 & -1 & -1 & -1 \\ 0 & 0 & 0 & 1 & 0 \\ 0 & 0 & 0 & -1 & -1 \\ 1 & 0 & -1 & -1 & 0 \\ 0 & 1 & 1 & 1 & 1 \end{pmatrix}; \quad B_6 = \begin{pmatrix} 0 & 0 & 1 & 0 & 0 & 0 \\ 0 & 0 & 0 & 0 & 1 & 0 \\ 0 & 0 & 0 & 0 & 0 & 1 \\ 1 & 0 & -1 & 0 & -1 & 0 \\ 0 & 1 & 1 & 0 & 0 & -1 \\ 0 & 0 & 0 & 1 & 1 & 1 \end{pmatrix}$$

3 Cryptanalysis of Protocols

In [4] the authors propose the Diffie-Hellman key exchange using 3-by-3 matrices over $\mathbb{F}_7[S_5]$. So Alice and Bob, take a public matrix $M \in \mathrm{Mat}_3(\mathbb{F}_7[S_5])$ which may be non-invertible. Alice chooses a secret integer n, computes M^n and sends it to Bob. Bob chooses a secret integer n', computes $M^{n'}$ and sends it to Alice. Every party can now compute the common key $M^{nn'}$.

In [5], they use the same platform for the Cramer-Shoup cryptosystem which we do not recall. We underline only that there is a public key M as above, and during the protocol among other data sent, there is M^n where n is the secret key. So if we are able to give a solution for the discrete logarithm problem in the case of $M \in \mathrm{Mat}_3(\mathbb{F}_7[S_5])$, in both cases the platform proposed is not secure. That is what we are going to explain.

As 7 does not divide $|S_5| = 120$, the algebra $\mathbb{F}_7[S_5]$ is semi-simple and Maschke's theorem asserts that this algebra is isomorphic to a direct sum of matrix algebras (over $\mathbb{F}_7$), in other words it is isomorphic to an algebra of block matrices over $\mathbb{F}_7$. Let us denote by f this isomorphism. To be of any use for our purpose, we have to make precise this isomorphism explicitly. The $\mathbb{F}_7$-linear extension (to $\mathbb{F}_7[S_5]$) of the morphism of S_5 using the irreducible representations of S_5 given on generators $W = (12), Z = (12345)$ in Sect. 2, gives the isomorphism f between $\mathbb{F}_7[S_5]$ and its image. So for any element $x = \sum_{i=1}^{120} a_i x_i \in \mathbb{F}_7[S_5]$, $a_i \in \mathbb{F}_7$ and $x_i \in S_5$ we can compute its image as a direct sum of matrices with coefficients in $\mathbb{F}_7$. Note that this decomposition is a special case of Wedderburn's theorem asserting that every semi-simple algebra can be decomposed as a direct sum of all its distinct simple submodules.

Up to now we have represented a matrix $M \in \mathrm{Mat}_3(\mathbb{F}_7[S_5])$ as a matrix with coefficients in $\mathbb{F}_7$ by replacing each coefficient M_{ij} of M by $f(M_{ij})$. For example M_{11} is replaced by

$$A = \begin{pmatrix} a_1 & & & & & & \\ & a_1' & & & & & \\ & & a_4 & & & & \\ & & & a_4' & & & \\ & & & & a_5 & & \\ & & & & & a_5' & \\ & & & & & & a_6 \end{pmatrix}$$

where a_i, a_i' are block matrices with coefficients in $\mathbb{F}_7$ and the indexes denote the size of the block.

Let us denote by $A, B, C, E, F, G, H, I, J$ the block matrices corresponding to $M_{11}, M_{12}, M_{13}, M_{21}, M_{22}, M_{23}, M_{31}, M_{32}, M_{33}$. Then B is a block matrix which we represent the same way as A by denoting $b_1, b_1', b_4, b_4', \ldots$ its blocks. We use the same notations for $C, D, \ldots$. It is an easy computation to prove that there is a natural isomorphism between matrices

$$\begin{pmatrix} A & B & C \\ D & E & F \\ H & I & J \end{pmatrix}$$

and the block matrix whose first block is obtained by composing (side by side) the first blocks of $A, B, C, D, \ldots$, namely

$$
\begin{pmatrix}
a_1 & b_1 & c_1 \\
d_1 & e_1 & f_1 \\
h_1 & i_1 & j_1
\end{pmatrix},
$$

which gives a 3×3 matrix over $\mathbb{F}_7$.

The second block is obtained by composing the second blocks of $A, B, C, D, \ldots$, namely

$$
\begin{pmatrix}
a'_1 & b'_1 & c'_1 \\
d'_1 & e'_1 & f'_1 \\
h'_1 & i'_1 & j'_1
\end{pmatrix},
$$

and so on.

To sum up, we represent the matrix $M \in \mathrm{Mat}_3(\mathbb{F}_7[S_5])$ by a block matrix in $\mathbb{F}_7$ whose blocks are of size $3, 3, 12, 12, 15, 15, 18$. We represent also the matrix M^n by a block matrix with the same size $3, 3, 12, 12, 15, 15, 18$ in $\mathbb{F}_7$. Now we can apply the same techniques as in [6], namely write the Jordan form of each block in some small extension base $\mathbb{F}_{7^\alpha}$ and find the secret key n. Note that for singular blocks, we need a slight modification of the procedure of [6], as proposed in [7].

4   An Example

We use the notations of Sects. 2 and 3.

Let us denote e the identity element of S_5.

Let

$$
M =
\begin{pmatrix}
2e + W + S & 3e + WS & e + S^2 \\
5e + 2SWS & e + W + S^3 & S \\
W + S^2 & 2e + S & e + W
\end{pmatrix}
\in \mathbb{F}_7[S_5]
$$

and $N = M^n$ (n is unknown) two given matrices. Our goal is to find $l \in \mathbb{N}$ such that $M^l = N$.

We represent every coefficient of M as a block matrix as follows:

$2e + W + S = (2A_1) + (A_1 \oplus A'_1 \oplus A_4 \oplus A'_4 \oplus A_5 \oplus A'_5 \oplus A_6) + (B_1 \oplus B'_1 \oplus B_4 \oplus B'_4 \oplus B_5 \oplus B'_5 \oplus B_6) = (2A_1 + A_1 + B_1) \oplus (A'_1 + B'_1) \oplus (A_4 + B_4) \oplus A'_4 + B'_4) \oplus (A_5 + B_5) \oplus (A'_5 + B'_5) \oplus (A_6 + B_6) = (4) \oplus (0) \oplus (A_4 + B_4) \oplus (A'_4 + B'_4) \oplus (A_5 + B_5) \oplus (A'_5 + B'_5) \oplus (A_6 + B_6).$

$3e + WS = (3A_1) + (A_1 B_1 \oplus A'_1 B'_1 \oplus (A_4 B_4 \oplus A'_4 B'_4 \oplus A_5 B_5 \oplus A'_5 B'_5 \oplus A_6 B_6) = (3A_1 + A_1 B_1) \oplus (A'_1 B'_1) \oplus A_4 B_4) \oplus (A'_4 B'_4) \oplus (A_5 B_5) \oplus (A'_5 B'_5) \oplus (A_6 B_6) = (4) \oplus (-1) \oplus (A_4 B_4) \oplus (A'_4 B'_4) \oplus (A_5 B_5) \oplus (A'_5 B'_5) \oplus (A_6 B_6).$

$$e + S^2 = (2) \oplus (1) \oplus B_4^2 \oplus B'^2_4 \oplus B_5^2 \oplus B'^2_5 \oplus B_6^2.$$

$$5 + 2SWS = (0) \oplus (2) \oplus 2B_4A_4B_4 \oplus 2B'_4A'_4B'_4 \oplus 2B_5A_5B_5 \oplus 2B'_5A'_5B'_5 \oplus 2B_6A_6B_6.$$

$$e + W + S^3 = (3) \oplus (0) \oplus (A_4 + B_4^3) \oplus (A'_4 + B'^3_4) \oplus (A_5 + B_5^3) \oplus (A'_5 + B'^3_5) \oplus (A_6 + B_6^3).$$

$$S = (1) \oplus (-1) \oplus B_4 \oplus B'_4 \oplus B_5 \oplus B'_5 \oplus B_6.$$

$$W + S^2 = (2) \oplus (2) \oplus (A_4 + B_4^2) \oplus (A'_4 + B'^2_4) \oplus (A_5 + B_5^2) \oplus (A'_5 + B'^2_5) \oplus (A_6 + B_6^2).$$

$$2e + S = (3) \oplus (1) \oplus B_4 \oplus B'_4 \oplus B_5 \oplus B'_5 \oplus B_6.$$

$$e + W = (2) \oplus (1) \oplus A_4 \oplus A'_4 \oplus A_5 \oplus A'_5 \oplus A_6.$$

So far we have represented M by a matrix whose coefficients are block matrices as above with coefficients in $\mathbb{F}_7$. It is straightforward that this matrix is isomorphic to the block matrix we form as follows:

Take the first component of each coefficients to form the matrix

$$M_1 = \begin{pmatrix} 4 & 4 & 2 \\ 0 & 3 & 1 \\ 2 & 3 & 2 \end{pmatrix},$$

then take the second component of each coefficient to form the matrix

$$M'_1 = \begin{pmatrix} 2 & -1 & 1 \\ 2 & 0 & -1 \\ 2 & 1 & 1 \end{pmatrix}.$$

Take the third component of the coefficients to obtain

$$M_4 = \begin{pmatrix} A_4 + B_4 & A_4B_4 & B_4^2 \\ 2B_4A_4B_4 & A_4 + B_4^3 & B_4 \\ A_4 + B - 4^2 & B_4 & A_4 \end{pmatrix}$$

Note that this matrix is of size 12. Continuing this way we obtain another matrix of size 12 which we denote by M'_4, two matrices of size 15 denoted by M_5 and M'_5 and a last matrix of size 18 denoted by M_6. We have $M = M_1 \oplus M'_1 \oplus M_4 \oplus M'_4 \oplus M_5 \oplus M'_5 \oplus M_6$.

We do the same operation on matrix $N = M^n$ to express it as a block matrix of the same size as above. Now we can separately work on corresponding blocks of M and N of the same size, computing the characteristic polynomials, Jordan forms... as suggested in [6], and reduce the discrete logarithm problem to the one on some small extension of the field $\mathbb{F}_7$. It may happen that some block is not invertible. We can still compute the Jordan forms and with a slight modification as suggested in [7] finish the work.

Note that the size of blocks in the above decomposition of M are the product of the size of M as a matrix with coefficients in $\mathbb{F}_7[S_5]$ (namely 3) and the degrees of irreducible representations of S_5. So if we replace the group S_5 by some other finite group G and the field $\mathbb{F}_7$ by $\mathbb{F}_p$, such that $\mathbb{F}_p[G]$ is a semi-simple algebra, the same procedure works. In fact representations of finite groups are very well known (techniques for constructing the irreducible representations will be different). If $n_1, n_2, \ldots, n_k$ are the degrees of all distinct irreducible representations of G we know that $|G| = n_1^2 + n_2^2 + \ldots n_k^2$ and each n_j divides $|G|$ and even more... (see [9]), such that these degrees are small enough comparing to $|G|$, and a matrix $M \in \mathbb{F}_p[G]$ of size 3 for example, will be decomposed in block matrices with coefficients in $\mathbb{F}_p$ and sizes $3n_1, 3n_2, \ldots, 3n_k$.

5   Conclusion

We showed that using matrices with coefficients in $\mathbb{F}_7[S_5]$ as a platform for Diffie-Hellman key exchange is not secure. One may wonder if replacing $\mathbb{F}_7$ by $\mathbb{F}_2, \mathbb{F}_3$ or $\mathbb{F}_5$ give something essentially different. In fact in these cases the group algebra is not semi-simple anymore and Wedderburn's theorem cannot be applied. But these new algebras are not far from being semi simple; in fact they differ from being semi simple by a nilpotent radical, and the quotient is semi simple and then the same procedure as explained in Sect. 2 can be applied. To sum up we believe that no secure cryptographic protocol can be based upon these algebras.

Furthermore replacing the group S_5 by some other finite group G, can be cryptanalyzed the same way using the irreducible representations of G.

References

1. Banin, M., Tsaban, B.: A reduction of semigroup DLP to classic DLP. Des. Codes Crypt. **81**, 75–82 (2006)
2. Childs, A., Ivanyos, G.: Quantum computation of discrete logarithms in semigroups. J. Math. Cryptology **8**(4), 405–416 (2014)
3. James, G.D.: The Representation Theory of the Symmetric Groups, vol. 682. Springer, Heidelberg (1978). SLN
4. Kahrobaei, D., Koupparis, C., Shpilrain, W.: Public key exchange using matrices over group rings. G.C.C. **5**(1), 97–115 (2013)
5. Kahrobaei, D., Koupparis, C., Shpilrain, W.: A CCA secure cryptosystem using matrices over group rings. Amer. Math. Soc. Contemp. Math. **633**, 73–80 (2015)
6. Menezes, A.J., Wu, Y.-H.: The discrete logarithm problem in $GL_n(\mathbb{F}_q)$. ARS Combinatorica **47**, 23–32 (1997)
7. Myasnikov, A., Ushakov, A.: Quantum algorithm for discrete logarithm problem for matrices over finite group rings.
8. Odonne, R., Varadharajan, D., Sanders, P.: Public key distribution in matrix rings. Electron. Lett. **20**, 386–387 (1984)
9. Serre, J.P.: Représentations linéaires des groupes finis. Hermann, Paris (1967)

Author Index

Abdelkhalek, Ahmed 117, 135

Bellizia, Davide 79
Bos, Joppe W. 184
Boudjou, Hortense 205

Castryck, Wouter 184
Chen, Liqun 40

Diarra, Nafissatou 205
Djukanovic, Milena 79

Eftekhari, Mohammad 223

Heuser, Annelie 61

Iliashenko, Ilia 184

Jovic, Alan 61

Khlil, Ahmed Youssef Ould Cheikh 205
Koshiba, Takeshi 149
Kosuge, Haruhisa 95

Laing, Thalia M. 40
Legay, Axel 61

Martin, Keith M. 40
Mohamed, Mohamed Saied Emam 3

Nargis, Isheeta 165

Petzoldt, Albrecht 3
Picek, Stjepan 61

Saha, Tushar Kanti 149
Scotti, Giuseppe 79
Seck, Michel 205

Tanaka, Hidema 95
Tolba, Mohamed 117, 135
Trifiletti, Alessandro 79

Veeningen, Meilof 21
Vercauteren, Frederik 184

Youssef, Amr M. 117, 135

Printed in the United States
By Bookmasters